# Physical and Chemical Characteristics of Oils, Fats, and Waxes

## Second Edition

### Editor: David Firestone
U.S. FDA
Washington, D.C.

**AOCS Mission Statement**

To be a global forum to promote the exchange of ideas, information, and experience, to enhance personal excellence, and to provide high standards of quality among those with a professional interest in the science and technology of fats, oils, surfactants, and related materials.

Copyright © 2006 by AOCS Press. All rights reserved. No part of this book may be reproduced or transmitted in any form or by any means without written permission of the publisher.

ISBN   978-1-893997-99-8

      Library of Congress Cataloging-in-Publication Data

Physical and chemical characteristics of oils, fats, and waxes / editor, David Firestone. — 2nd ed.
    p. cm.
 Includes bibliographical references and index.
 ISBN-13: 978-1-893997-99-8
1. Oils and fats—Composition. 2. Oils and fats—Statistics. 3. Oils and fats—Databases. 4. Waxes—Composition. 5. Waxes—Statistics.
6. Waxes—Databases. I. Firestone, David.

TP670.P47 2006
 665—dc22
                                                    2006019782

The paper used in this book is acid-free and falls within the guidelines established to ensure permanence and durability.

Printed in the United States of America.

# Table of Contents

Introduction .................................................................. v
Characteristics of Oils and Fats of Plant Origin ................................. 1
Characteristics of Oils and Fats of Animal Origin ............................. 171
Triglyceride Molecular Species of Selected Oils & Fats ....................... 211
Index ....................................................................... 228

# Introduction

For many years, AOCS maintained a section of the *Official Methods and Recommended Practices of the AOCS* detailing the physical and chemical properties of a number of fats and oils; these were collected in *Section I* of that publication. In 1996, the section was reintroduced as an additional section of the 4th edition of the *Official Methods and Recommended Practices of the AOCS*. It contained information relating to approximately 200 plant oils and fats.

In 1997, additions to this compendium were released covering an additional 50 fats derived mainly from animal sources. In 1998, it was decided to consolidate the 1996 and 1997 versions of *Section I*. Furthermore, comments from users of this publication convinced us that it would be advantageous if a more versatile and flexible product was developed. Through the encouragement of Dr. Gerry Szajer and other members of the AOCS technical committees, the database programming was undertaken to make this large body of information more readily accessible. In the 1999 version of the *Physical and Chemical Characteristics of Oils, Fats, and Waxes*, we brought together the original and newer versions of *Section I*, and incorporated a number of additional plant and animal oils and fats. This brought the total number of entries to over 350.

In the second edition of *Physical and Chemical Characteristics of Oils, Fats, and Waxes*, we have increased the number of entries by almost 30%. Using the format previously developed for the first edition for plant-derived oils and fats and animal fats, we retained separate sections for physical properties, fatty acid composition, sterols, tocopherols, tocotrienols and triglyceride patterns. Where multiple samples or different references were found for the same oil or fat, a range of values is presented. Care has been taken to present the data in an accurate, useful, and representative fashion. In some cases, it was necessary to edit or sum the isomers of certain fatty acids in order to preserve the size and integrity of the database. Users are asked to consult the original references to obtain full details. The fully-revised database and a new associated program are contained on the accompanying CD-ROM. It contains the same data as the printed version of the second edition.

The basis for the information presented was found in the original *Section I*, and the reference resources of Dr. David Firestone, Editor-in-Chief of the *Official Methods and Recommended Practices of the AOCS*, and Dr. David Berner, former technical director of the AOCS. In compiling the references, it was clear that a large body of information was gathered from a number of valuable sources. The main sources are listed below:

- Eckey, E.W., *Vegetable Fats and Oils*, Reinhold Publishing Company, New York, 1954
- Hilditch, J.P., and Williams, P.N., *The Chemical Constitution of Natural Fats*, 4th edition, Chapman and Hall, London, 1964
- Section I, *Physical and Chemical Characteristics of Oils, Fats, and Waxes*, AOCS 1996/7
- *Analysis of Oilseeds, Fats, and Fatty Foods* (Rossell, J.B., and Pritchard, J.L.R., eds.) Elsevier Applied Science, New York, 1991

Note: These references are not repeated in the database or printed version.

A number of entries are derived from national and international trade guidelines and standards. The following list may serve as a useful guide:

- Codex Alimentarius Commission, World Health Organization, ALINORM 97/17 and later revisions
- USDA Agricultural Handbook no. 8-4 and supplements, Composition of Foods, Fats and Oils, Raw, Processed, Prepared, Human Nutrition Information Service, USDA

- Guideline Specifications, FOSFA International, 2nd edition, 1994
- Recueil de Normes Françaises des Corps Gras, Graines Oléagineuses et Produits dérivés, 2nd edition, AFNOR, Paris, 1981

The latin names used have been harmonized where possible with the terms presented in the latest version of ISO 5527:1997 Oilseeds—Nomenclature. For more details see also Fat Sci. Technol. 97: 539 (1995).

## Other resources

AOCS maintains three very useful publications, *INFORM*, *Lipids*, and *JAOCS* (*Journal of the American Oil Chemists' Society*), that regularly contain articles reporting the composition of fats and oils. Other journals referred to in this publication provide further data regularly.

## Naming fatty acids

A number of conventions exist for naming individual fatty acids. Many of the more common fatty acids have trivial names, for example palmitic acid, oleic acid, linoleic acid, etc., whereas all have systematic names based on their carbon number, such as decanoic acid, and eicosatetraenoic acid. In this product we have used the shorthand convention of number of carbons in the fatty acid chain followed by the number of double bonds; thus palmitic acid is 16:0, oleic acid is 18:1, linoleic acid is 18:2. The arrangement of double bonds within the fatty acid chain is also subject to two different naming systems. The IUPAC convention names the position of the double bond based on its position relative to the carboxyl carbon. The position may be denoted using $\Delta$; thus oleic acid is $\Delta$9-18:1 and linoleic acid is $\Delta$9,12-18:2, but it is more common to identify the configuration of the bond using *cis* or *trans*; thus oleic acid is *cis*9-18:1. Two other similar conventions are based on the position of the double bonds relative to the methyl terminal of the fatty acid chain. These are either the ω (omega) or n- ("n-minus") conventions, where ω counts the number of carbon atoms from the methyl carbon as position 1, and n refers to the total number of carbons in the fatty acid. Within the n- and ω conventions there are recognized families of naturally occuring fatty acids based on the position of the first double bond; the most common series are n-3, n-6 and n-9 (ω3, ω6, ω9). Using these conventions, oleic acid is 18:1 n-9 or 18:1 ω9 and linoleic acid is 18:2 n-6 or 18:2 ω6. Further information regarding fatty acid nomenclature may be found in any biochemistry reference text or specifically:

Christie, W.W., *Lipid Analysis*, Pergammon Press, Oxford, UK, 1982, pp. 1.

In the following sections and in the database, we have used the shorthand notation for chainlength and number of double bonds. Double bond positions and configuration are indicated according to the IUPAC convention and identified as *cis* (c), or *trans* (t), where this is stated in the reference. Acetylenic (a) and epoxy and conjugated bonds are also identified where known.

Table 1 shows the systematic name, common name, triglyceride code, and shorthand convention (carbon number) for the fatty acids regularly presented in the database. Rarer, but more remarkable, fatty acids may be found labeled "other." For the TG code, many are based on the trivial names of the fatty acids and may have different meanings in different oils so please refer to the fatty acid pattern to avoid misinterpretation.

**Molecular Structure of Triglycerides**

With the development of reversed phase high performance liquid chromatography, it is possible to fractionate triglycerides into individual molecular species. Separation is a function of the total number of carbon atoms and the total number of double bonds. Thus for each possible triglyceride structure a unique identification may be assigned. This

# Introduction

is normally termed the "expected carbon number" or ECN and is calculated as:

(total carbon number) - 2(number of double bonds)

In the tables of triglyceride composition, the different molecular species have been identified by their single letter codes. The order of the letters does not imply the position of the fatty acids, thus POL may be a mixture of POL, OPL, and OLP. Generally positional isomers were not described in the references used, however, where such information was available, individual isomers were summed to maintain the database. The individual letter codes used when identifying the constituent fatty acids in triglyceride molecular species are given in Table 1.

**Table 1**

| Systematic Name | Common Name | TG Shorthand | Structure |
|---|---|---|---|
| Tetranoic | Butyric | | 4:0 |
| Pentanoic | Valeric | | 5:0 |
| Hexanoic | Caproic | | 6:0 |
| Octanoic | Caprylic | | 8:0 |
| Decanoic | Capric | | 10:0 |
| Dodecanoic | Lauric | La | 12:0 |
| Dodecenoic | | | 12:1 |
| | Lauroleic | | 9c-12:1 |
| Trisdecanoic | | | 13:0 |
| Tetradecanoic | Myristic | M | 14:0 |
| Tetradenenoic | | | 14:1 |
| | Myristoleic | | 9c-14:1 |
| Pentadecanoic | | | 15:0 |
| Hexadecanoic | Palmitic | P | 16:0 |
| Hexadecenoic | Palmitoleic | Po | 9c-16:1 |
| Hexadecadienoic | | | 16:2 |
| Heptadecanoic | Margaric | | 17:0 |
| Octadecanoic | Stearic | S | 18:0 |
| Octadecenoic | | | 18:1 |
| | Oleic | O | 9c-18:1 |
| | Elaidic | | 9t-18:1 |
| | Petroselenic | | 6c-18:1 |
| | cis-Vaccenic | | 11c-18:1 |
| | Vaccenic | | 11t-18:1 |
| Octadecadienoic | Linoleic | L | 9c,12c-18:2 |
| Octadecatrienoic | Linolenic | Ln | 18:3 |
| | γ-Linolenic | Lng | 6c,9c,12c-18:3 |
| | α-Linolenic | | 9c,12c,15c-18:3 |
| | Pinolenic | | 5c,9c,12c-18:3 |
| | α-Eleostearic | E | 9c,11t,13t-18:3 |
| Octadecatetraenoic | Moroctic/morotic | | 4c,8c,12c,15c-18:4 |
| | Stearidonic | | 6c,9c,12c,15c-18:4 |
| Nonadecanoic | | | 19:0 |
| Eicosanoic | Arachidic | A | 20:0 |
| Eicosenoic | | | 5c-20:1 |

*(continued)*

**Table 1** *(continued)*

| Systematic Name | Common Name | TG Shorthand | Structure |
|---|---|---|---|
| | Gadoleic/gondoleic | G | 9c-20:1 |
| | Gondoic | | 11c-20:1 |
| Eicosadienoic | | | 20:2 |
| Eicosatrienoic | | | 20:3 |
| | Dihomo-γ-linolenic/ | | |
| | homo-γ-linolenic | | 8c,11c,14c-20:3 |
| | Mead's acid | | 5c,8c,11c-20:3 |
| Eicosatetraenoic | Arachidonic | | 5c,8c,11c,14c-20:4 |
| Eicosapentaenoic | | | 5c,8c,11c,14c,17c-20:5 |
| Docosanoic | Behenic | B | 22:0 |
| Docosenoic | | | 22:1 |
| | Cetolic | | 11c-22:1 |
| | Erucic | E | 13c-22:1 |
| Docosadienoic | | | 22:2 |
| | | | 5c,13c-22:2 |
| | | | 16c,19c,-22:2 |
| Docosapentaenoic | | | 7c,10c,13c,16c,19c-22:5 |
| Docosahexaenoic | | | 4c,7c,10c,13c,16c,19c-22:6 |
| Tetracosanoic | Lignoceric | | 24:0 |
| Tetracosenoic | Nervonic | | 15c-24:1 |
| Tetracosapentaenoic | Scoliodonic | | 24:5 |
| Tetracosahexaenoic | Nisinic | | 24:6 |
| Hexacosanoic | Cerinic/cerotic | | 26:0 |
| Octacosanoic | | | 28:0 |

In TG patterns: D = Dimorphecolic, R = Ricinoleic
Compiled from Christie, W.W., Lipid Analysis, Pergammon Press, Oxford, UK, 1982, Ch. 1, pp. 1; Gunstone, F.D., in *Lipid Technologies and Applications* (Gunstone, F.D., and Padley, F.B., eds.), Marcel Dekker, New York, 1997, Ch. 1, pp. 1; AOCS Analytical Division Home page (www.aocs.org) with permission from R.O.Adlof.

# Characteristics of Oils and Fats of Plant Origin

## Acacia arabica
*Acacia arabica*

Specific Gravity (SG)
   15.5/15.5°C
   25/25°C
   Other SG
Refractive Index (RI)
   25°C . . . . . . . . . . . . . . . . . . . . . . 1.4723
   40°C
   Other RI
Iodine Value . . . . . . . . . . . . . . . . . . . . . 105.6
Saponification Value . . . . . . . . . . . . . . . 194.4
Titer °C
% Unsaponifiable . . . . . . . . . . . . . . . . . . . 2.4
Melting Point °C

**Fatty Acid Composition (%)**
   16:0 . . . . . . . . . . . . . . . . . . . . . . . . 14.6
   18:0 . . . . . . . . . . . . . . . . . . . . . . . . . 6.2
   9c-18:1 . . . . . . . . . . . . . . . . . . . . . . 32.2
   9c,12c-18:2 . . . . . . . . . . . . . . . . . . . 39.2
   Undefined 18:3 . . . . . . . . . . . . . . . . . 3.1

**References** *JAOCS 67:* 433–434 (1990)


## Acacia Auriculiformis Seed Oil
*Acacia auriculiformis*

Specific Gravity (SG)
   15.5/15.5°C
   25/25°C
   Other SG
Refractive Index (RI)
   25°C
   40°C
   Other RI
Iodine Value
Saponification Value
Titer °C
% Unsaponifiable
Melting Point °C

**Fatty Acid Composition (%)**
   16:0 . . . . . . . . . . . . . . . . . . . . . . . . . 19
   18:0 . . . . . . . . . . . . . . . . . . . . . . . . . . 4
   Total 18:1 . . . . . . . . . . . . . . . . . . . . . 24
   epoxy 18:1 . . . . . . . . . . . . . . . . . . . . . 5
   9c,12c-18:2 . . . . . . . . . . . . . . . . . . . . 44
   20:0 . . . . . . . . . . . . . . . . . . . . . . . . . . 2
   Total 20:1 . . . . . . . . . . . . . . . . . . . . . . 1

**References** *J. Am. Oil Chem. Soc. 60:* 1893 (1983)


## Acacia Coriacea Seed Oil
*Acacia coriacea*

Specific Gravity (SG)
   15.5/15.5°C
   25/25°C
   Other SG
Refractive Index (RI)
   25°C
   40°C
   Other RI
Iodine Value
Saponification Value
Titer °C
% Unsaponifiable
Melting Point °C

**Fatty Acid Composition (%)**
   16:0 . . . . . . . . . . . . . . . . . . . . . . . . . 13
   18:0 . . . . . . . . . . . . . . . . . . . . . . . . . . 3
   Total 18:1 . . . . . . . . . . . . . . . . . . . . . 42
   epoxy 18:1 . . . . . . . . . . . . . . . . . . . . . 2
   9c,12c-18:2 . . . . . . . . . . . . . . . . . . . . 38
   20:0 . . . . . . . . . . . . . . . . . . . . . . . . . . 2

**References** *J. Am. Oil Chem. Soc. 60:* 1893 (1983)


## Acacia Lenticularis Seed Oil
*Acacia lenticularis*

Specific Gravity (SG)
   15.5/15.5°C
   25/25°C
   Other SG
Refractive Index (RI)
   25°C
   40°C
   Other RI . . . . . . . . . . . . . . . (30) 1.4700
Iodine Value . . . . . . . . . . . . . . . . . . . . . . 225
Saponification Value . . . . . . . . . . . . . . . 193.5
Titer °C

% Unsaponifiable
Melting Point °C

**Fatty Acid Composition (%)**
- 14:0 . . . . . . . . . . . . . . . . . . . . . . . . . 0.5
- 15:0 . . . . . . . . . . . . . . . . . . . . . . . . . 0.4
- 16:0 . . . . . . . . . . . . . . . . . . . . . . . . . . . 7
- 9c-16:1 . . . . . . . . . . . . . . . . . . . . . 0.3
- 18:0 . . . . . . . . . . . . . . . . . . . . . . . . . . . 1
- Total 18:1 . . . . . . . . . . . . . . . . . . . . . 7
- epoxy 18:1 . . . . . . . . . . . . . . . . . . . . 2
- 9c,12c-18:2 . . . . . . . . . . . . . . . . . . . . 1
- Undefined 18:3 . . . . . . . . . . . . . . . . 80
- 20:0 . . . . . . . . . . . . . . . . . . . . . . . . . . tr

**References** *J. Am. Oil Chem. Soc.* 65: 1959 (1988)

## Acacia Mellifera Seed Oil
*Acacia mellifera*

Specific Gravity (SG)
   15.5/15.5°C
   25/25°C
   Other SG
Refractive Index (RI)
   25°C
   40°C . . . . . . . . . . . . . . . . . . . . . . . 1.4650
   Other RI
Iodine Value . . . . . . . . . . . . . . . . . . . . . . 75
Saponification Value . . . . . . . . . . . . . . . 186
Titer °C
% Unsaponifiable
Melting Point °C

**Fatty Acid Composition (%)**
- 14:0 . . . . . . . . . . . . . . . . . . . . . . . . . 0.6
- 16:0 . . . . . . . . . . . . . . . . . . . . . . . . . . 18
- 9c-16:1 . . . . . . . . . . . . . . . . . . . . . . 0.5
- 18:0 . . . . . . . . . . . . . . . . . . . . . . . . . . . 8
- Total 18:1 . . . . . . . . . . . . . . . . . . . . . 24
- epoxy 18:1 . . . . . . . . . . . . . . . . . . . . 0.6
- 9c,12c-18:2 . . . . . . . . . . . . . . . . . . . . 44
- 20:0 . . . . . . . . . . . . . . . . . . . . . . . . . . . 4

**References** *J. Am. Oil Chem. Soc.* 60: 1893 (1983)

## Acacia Minhassai Seed Oil
*Acacia minhassai*

Specific Gravity (SG)
   15.5/15.5°C
   25/25°C
   Other SG
Refractive Index (RI)
   25°C
   40°C
   Other RI
Iodine Value
Saponification Value
Titer °C
% Unsaponifiable
Melting Point °C

**Fatty Acid Composition (%)**
- 14:0 . . . . . . . . . . . . . . . . . . . . . . . . . 0.6
- 16:0 . . . . . . . . . . . . . . . . . . . . . . . . . 6–7
- 18:0 . . . . . . . . . . . . . . . . . . . . . . . . . . . 1
- Total 18:1 . . . . . . . . . . . . . . . . . . . . . 18
- 9c,12c-18:2 . . . . . . . . . . . . . . . . . . . . 72
- 20:0 . . . . . . . . . . . . . . . . . . . . . . . . . . . 2

**References** *J. Am. Oil Chem. Soc.* 60: 1893 (1983)

## Acacia Mollissima Seed Oil
*Acacia mollissima*

Specific Gravity (SG)
   15.5/15.5°C
   25/25°C
   Other SG
Refractive Index (RI)
   25°C
   40°C
   Other RI . . . . . . . . . . . . . . . (30) 1.4875
Iodine Value . . . . . . . . . . . . . . . . . . . . 140
Saponification Value . . . . . . . . . . . . . . 192
Titer °C
% Unsaponifiable
Melting Point °C

**Fatty Acid Composition (%)**
- 14:0 . . . . . . . . . . . . . . . . . . . . . . . . . 0.1

16:0 . . . . . . . . . . . . . . . . . . . . . . . . 10.5
18:0 . . . . . . . . . . . . . . . . . . . . . . . . . . 1
Total 18:1 . . . . . . . . . . . . . . . . . . . . 17
epoxy 18:1 . . . . . . . . . . . . . . . . . . . . 2
9c,12c-18:2 . . . . . . . . . . . . . . . . . . . 68
Undefined 18:3 . . . . . . . . . . . . . . . . . 1
20:0 . . . . . . . . . . . . . . . . . . . . . . . . 0.6
Total 20:1 . . . . . . . . . . . . . . . . . . . . 0.1

**References** *J. Am. Oil Chem. Soc. 65:* 1959 (1988)

## Acacia Tortilis Seed Oil
*Acacia tortilis*
Specific Gravity (SG)
  15.5/15.5°C
  25/25°C
  Other SG
Refractive Index (RI)
  25°C
  40°C
  Other RI . . . . . . . . . . . . . . . (30) 1.4745
Iodine Value . . . . . . . . . . . . . . . . . . . . . 203
Saponification Value . . . . . . . . . . . . . . . 193
Titer °C
% Unsaponifiable
Melting Point °C

**Fatty Acid Composition (%)**
12:0 . . . . . . . . . . . . . . . . . . . . . . . . 2.5
13:0 . . . . . . . . . . . . . . . . . . . . . . . . 2.5
14:0 . . . . . . . . . . . . . . . . . . . . . . . . 0.1
9c-14:1 . . . . . . . . . . . . . . . . . . . . . 0.1
15:0 . . . . . . . . . . . . . . . . . . . . . . . . 0.3
16:0 . . . . . . . . . . . . . . . . . . . . . . . . 9.5
9c-16:1 . . . . . . . . . . . . . . . . . . . . . 0.1
18:0 . . . . . . . . . . . . . . . . . . . . . . . . 2.4
Total 18:1 . . . . . . . . . . . . . . . . . . . . . 6
epoxy 18:1 . . . . . . . . . . . . . . . . . . . . 2
9c,12c-18:2 . . . . . . . . . . . . . . . . . . . . 2
Undefined 18:3 . . . . . . . . . . . . . . . . . 72
20:0 . . . . . . . . . . . . . . . . . . . . . . . . . . 1
Total 20:1 . . . . . . . . . . . . . . . . . . . . 0.2

**References** *J. Am. Oil Chem. Soc. 65:* 1959 (1988)

## Achras sapota
*Achras sapota*
Specific Gravity (SG)
  15.5/15.5°C
  25/25°C
  Other SG
Refractive Index (RI)
  25°C
  40°C
  Other RI
Iodine Value
Saponification Value
Titer °C
% Unsaponifiable
Melting Point °C

**Fatty Acid Composition (%)**
14:0 . . . . . . . . . . . . . . . . . . . . . . . 0.13
16:0 . . . . . . . . . . . . . . . . . . . . . . 20.31
9c-16:1 . . . . . . . . . . . . . . . . . . . . 0.03
18:0 . . . . . . . . . . . . . . . . . . . . . . . 9.29
9c-18:1 . . . . . . . . . . . . . . . . . . . 55.08
7c-18:1 . . . . . . . . . . . . . . . . . . . . . 0.4
9c,12c-18:2 . . . . . . . . . . . . . . . . . 11.80
9c,12c,15c-18:3 . . . . . . . . . . . . . . 0.42
20:0 . . . . . . . . . . . . . . . . . . . . . . . 0.78
11c-20:1 . . . . . . . . . . . . . . . . . . . 0.67
22:0 . . . . . . . . . . . . . . . . . . . . . . . 0.24
22:2 . . . . . . . . . . . . . . . . . . . . . . . 0.02
24:0 . . . . . . . . . . . . . . . . . . . . . . . 0.39

Tocopherol Composition, mg/kg
  α-Tocopherol . . . . . . . . . . . . . . . . . . . . 57
  γ-Tocopherol . . . . . . . . . . . . . . . . . . . . 40

**References** *JAOCS 68:* 183–189 (1991)

## Acioa edulis
*Acioa edulis*
Specific Gravity (SG)
  15.5/15.5°C
  25/25°C
  Other SG
Refractive Index (RI)
  25°C . . . . . . . . . . . . . . . . . . . . . . 1.4835

40°C
Other RI
Iodine Value ..................... 115.6
Saponification Value ............... 198.8
Titer °C
% Unsaponifiable .................... 1.2
Melting Point °C

**References** *JAOCS 80:* 1013–1020 (2003)

## Acorn Oil
*Acer spp.*
Specific Gravity (SG)
  15.5/15.5°C
  25/25°C................. 0.908–0.918
  Other SG
Refractive Index (RI)
  25°C .................. 1.462–1.470
  40°C .................. 1.458–1.465
  Other RI
Iodine Value ................... 81–107
Saponification Value ........... 184–197
Titer °C
% Unsaponifiable ............... 0.8–2.3
Melting Point °C

**Fatty Acid Composition (%)**
  16:0........................... 13
  18:0............................ 2
  Total 18:1 ..................... 58
  9c,12c-18:2..................... 27

**References**

## Adonsonia Digitata Seed Oil
*Adonsonia digitata*
Specific Gravity (SG)
  15.5/15.5°C
  25/25°C
  Other SG
Refractive Index (RI)
  25°C
  40°C
  Other RI
Iodine Value
Saponification Value
Titer °C
% Unsaponifiable
Melting Point °C

**Fatty Acid Composition (%)**
  14:0........................... 0.2
  16:0........................... 15.5
  9c-16:1........................ 0.2
  16:2........................... 0.7
  18:0........................... 3.1
  9c-18:1........................ 24.7
  11c-18:1....................... 0.7
  9c,12c-18:2.................... 19.1
  6c,9c,12c-18:3 ................ 0.4
  9c,12c,15c-18:3 ............... 1.6
  20:0........................... 0.7
  11c-20:1....................... 0.2
  22:0........................... 0.4
  24:0........................... 0.3

**References** *J. Am. Oil Chem. Soc. 75:* 1031 (1998)

## Aesculus sinensis
*Aesculus sinensis*
Specific Gravity (SG)
  15.5/15.5°C
  25/25°C
  Other SG
Refractive Index (RI)
  25°C
  40°C
  Other RI
Iodine Value
Saponification Value
Titer °C
% Unsaponifiable
Melting Point °C

**Fatty Acid Composition (%)**
  14:0........................... 0.18
  16:0........................... 12.60
  9c-16:1........................ 2.66
  18:0........................... 1.58
  9c-18:1........................ 29.99
  7c-18:1........................ 4.12
  9c,12c-18:2.................... 16.17
  9c,12c,15c-18:3 ............... 23.33

20:0 . . . . . . . . . . . . . . . . . . . . . . . . . 0.11
11c-20:1 . . . . . . . . . . . . . . . . . . . . . 0.26
22:0 . . . . . . . . . . . . . . . . . . . . . . . . . 0.32
13c-22:1 . . . . . . . . . . . . . . . . . . . . . 0.35
24:0 . . . . . . . . . . . . . . . . . . . . . . . . . 0.16

Tocopherol Composition, mg/kg
α-Tocopherol . . . . . . . . . . . . . . . . . . 195
β-Tocopherol
γ-Tocopherol . . . . . . . . . . . . . . . . . . . 88
δ-Tocopherol
Total, mg/kg

Tocotrienols Composition, mg/kg
α-Tocotrienol . . . . . . . . . . . . . . . . . . . 97
β-Tocotrienol
γ-Tocotrienol . . . . . . . . . . . . . . . . . . 626
δ-Tocotrienol . . . . . . . . . . . . . . . . . . 336
Total Tocotrienols, mg/kg

**References** *JAOCS 80:* 1013–1020 (2003)

## African Mango
*Irvingia gabonensis*
Specific Gravity (SG)
15.5/15.5°C
25/25°C . . . . . . . . . . . . . . . . . . . . . . 0.9073
Other SG
Refractive Index (RI)
25°C
40°C . . . . . . . . . . . . . . . . . . . . . . . . 1.4515
Other RI
Iodine Value . . . . . . . . . . . . . . . . . . 86–103
Saponification Value . . . . . . . . . . . 243–248
Titer °C
% Unsaponifiable
Melting Point °C

**Fatty Acid Composition (%)**
12:0 . . . . . . . . . . . . . . . . . . . . . . . . . 5.7
14:0 . . . . . . . . . . . . . . . . . . . . . . . . 21.2
16:0 . . . . . . . . . . . . . . . . . . . . . . . . 33.9
18:0 . . . . . . . . . . . . . . . . . . . . . . . . . 5.9
Total 18:1 . . . . . . . . . . . . . . . . . . . 24.5
9c,12c-18:2 . . . . . . . . . . . . . . . . . . . 6.0
Undefined 18:3 . . . . . . . . . . . . . . . . 0.7
20:0 . . . . . . . . . . . . . . . . . . . . . . . . . 0.5
Total 20:1 . . . . . . . . . . . . . . . . . . . . 0.5

**References** *Riv. Ital. Sost. Grasse 61:* 569 (1984)

## Afzelia Bella Seed Oil
*Afzelia bella*
Specific Gravity (SG)
15.5/15.5°C
25/25°C
Other SG
Refractive Index (RI)
25°C
40°C
Other RI
Iodine Value
Saponification Value
Titer °C
% Unsaponifiable
Melting Point °C

**Fatty Acid Composition (%)**
16:0 . . . . . . . . . . . . . . . . . . . . . . . . . 2.2
16:2 . . . . . . . . . . . . . . . . . . . . . . . . . 1.4
18:0 . . . . . . . . . . . . . . . . . . . . . . . . . 2.3
9c-18:1 . . . . . . . . . . . . . . . . . . . . . . 8.1
11c-18:1 . . . . . . . . . . . . . . . . . . . . . 0.3
9c,12c-18:2 . . . . . . . . . . . . . . . . . . 20.4
9c,12c,15c-18:3 . . . . . . . . . . . . . . . 0.3
20:0 . . . . . . . . . . . . . . . . . . . . . . . . . 1.0
Total 20:1 . . . . . . . . . . . . . . . . . . . . 1.4
22:0 . . . . . . . . . . . . . . . . . . . . . . . . . 4.0
24:0 . . . . . . . . . . . . . . . . . . . . . . . . . 8.3
Other . . . . . . . . . . . . 3 unidentified - 48.3

**References** *J. Am. Oil Chem. Soc. 75:* 1031 (1998)

## Albizia Lebbeck Seed Oil
*Albizia lebbeck*
Specific Gravity (SG)
15.5/15.5°C
25/25°C
Other SG
Refractive Index (RI)
25°C
40°C
Other RI

Iodine Value
Saponification Value
Titer °C
% Unsaponifiable
Melting Point °C

**Fatty Acid Composition (%)**
16:0 . . . . . . . . . . . . . . . . . . . . . . . . . . 5.6
9c-16:1 . . . . . . . . . . . . . . . . . . . . . . . 0.3
16:2 . . . . . . . . . . . . . . . . . . . . . . . . . . 1.4
18:0 . . . . . . . . . . . . . . . . . . . . . . . . . . 2.7
9c-18:1 . . . . . . . . . . . . . . . . . . . . . . 11.2
11c-18:1 . . . . . . . . . . . . . . . . . . . . . . 0.7
9c,12c-18:2 . . . . . . . . . . . . . . . . . . 15.8
6c,9c,12c-18:3 . . . . . . . . . . . . . . . . 0.2
9c,12c,15c-18:3 . . . . . . . . . . . . . . . 6.4
20:0 . . . . . . . . . . . . . . . . . . . . . . . . . . 1.5
Total 20:1 . . . . . . . . . . . . . . . . . . . . 0.2
22:0 . . . . . . . . . . . . . . . . . . . . . . . . . . 5.2
24:0 . . . . . . . . . . . . . . . . . . . . . . . . . . 1.3

**References** *J. Am. Oil Chem. Soc. 75:* 1031 (1998)

## Albizia Zygia Seed Oil
*Albizia zygia*
Specific Gravity (SG)
  15.5/15.5°C
  25/25°C
  Other SG
Refractive Index (RI)
  25°C
  40°C
  Other RI
Iodine Value
Saponification Value
Titer °C
% Unsaponifiable
Melting Point °C

**Fatty Acid Composition (%)**
16:0 . . . . . . . . . . . . . . . . . . . . . . . . 13.8
9c-16:1 . . . . . . . . . . . . . . . . . . . . . . 7.7
16:2 . . . . . . . . . . . . . . . . . . . . . . . . . 1.5
18:0 . . . . . . . . . . . . . . . . . . . . . . . . . 2.7
9c-18:1 . . . . . . . . . . . . . . . . . . . . . 13.4
11c-18:1 . . . . . . . . . . . . . . . . . . . . . 4.9
9c,12c-18:2 . . . . . . . . . . . . . . . . . . 13.6

9c,12c,15c-18:3 . . . . . . . . . . . . . . . 2.2
20:0 . . . . . . . . . . . . . . . . . . . . . . . . . . 0.9
Total 20:1 . . . . . . . . . . . . . . . . . . . . 0.3
22:0 . . . . . . . . . . . . . . . . . . . . . . . . . . 1.6
24:0 . . . . . . . . . . . . . . . . . . . . . . . . . . 0.6

**References** *J. Am. Oil Chem. Soc. 75:* 1031 (1998)

## Aleurites Montana Seed Oil
*Aleurites montana*
Specific Gravity (SG)
  15.5/15.5°C
  25/25°C
  Other SG
Refractive Index (RI)
  25°C
  40°C
  Other RI
Iodine Value
Saponification Value
Titer °C
% Unsaponifiable
Melting Point °C

**Fatty Acid Composition (%)**
14:0 . . . . . . . . . . . . . . . . . . . . . . . . . 0.03
16:0 . . . . . . . . . . . . . . . . . . . . . . . . . 2.54
9c-16:1 . . . . . . . . . . . . . . . . . . . . . . 0.02
18:0 . . . . . . . . . . . . . . . . . . . . . . . . . 2.41
9c-18:1 . . . . . . . . . . . . . . . . . . . . . . 8.02
7c-18:1 . . . . . . . . . . . . . . . . . . . . . . 0.36
9c,12c-18:2 . . . . . . . . . . . . . . . . . 10.25
Undefined 18:3 . . . . . . . . . . . . . . . . . 73
9c,12c,15c-18:3 . . . . . . . . . . . . . . 0.03
20:0 . . . . . . . . . . . . . . . . . . . . . . . . . 0.16
11c-20:1 . . . . . . . . . . . . . . . . . . . . . 1.01
22:0 . . . . . . . . . . . . . . . . . . . . . . . . . 0.09
13c-22:1 . . . . . . . . . . . . . . . . . . . . . 0.04
22:2 . . . . . . . . . . . . . . . . . . . . . . . . . 0.06
24:0 . . . . . . . . . . . . . . . . . . . . . . . . . 0.05

Tocopherol Composition, mg/kg
  α-Tocopherol . . . . . . . . . . . . . . . . . . . 255
  β-Tocopherol
  γ-Tocopherol . . . . . . . . . . . . . . . . . . 1206
  δ-Tocopherol . . . . . . . . . . . . . . . . . . . . 44
Total, mg/kg

Tocotrienols Composition, mg/kg
α-Tocotrienol
β-Tocotrienol
γ-Tocotrienol..................... 34
δ-Tocotrienol
Total Tocotrienols, mg/kg

**References** *JAOCS 80:* 1013–1020 (2003)

# Alfalfa Oil (Utah)
*Medicago sativa*
Specific Gravity (SG)
  15.5/15.5°C
  25/25°C....................... 0.925
  Other SG
Refractive Index (RI)
  25°C ......................... 1.4797
  40°C
  Other RI
Iodine Value .................. 161–168
Saponification Value ............ 185–188
Titer °C
% Unsaponifiable
Melting Point °C

**Fatty Acid Composition (%)**
  16:0........................... 10
  18:0............................ 5
  Total 18:1 ..................... 7–11
  9c,12c-18:2................... 43–71
  Undefined 18:3 ................ 11–32

**References**

# Alyogine Hakeifolia
*Alyogine hakeifolia*
Specific Gravity (SG)
  15.5/15.5°C
  25/25°C
  Other SG
Refractive Index (RI)
  25°C
  40°C
  Other RI
Iodine Value
Saponification Value
Titer °C
% Unsaponifiable
Melting Point °C

**Fatty Acid Composition (%)**
  14:0.......................... 0.1
  16:0.......................... 11.4
  16:1.......................... 0.1
  18:0............................ 3
  Total 18:1 .................... 12.6
  Undefined 18:2 ................ 62.1
  Undefined 18:3 ................. 2.4
  20:0.......................... 0.3

**References** *JAOCS 68:* 518–519 (1991)

# Alyogine Huegelii
*Alyogine huegelii*
Specific Gravity (SG)
  15.5/15.5°C
  25/25°C
  Other SG
Refractive Index (RI)
  25°C
  40°C
  Other RI
Iodine Value
Saponification Value
Titer °C
% Unsaponifiable
Melting Point °C

**Fatty Acid Composition (%)**
  14:0............................ 0
  16:0.......................... 12.9
  16:1.......................... 0.3
  18:0.......................... 2.4
  Total 18:1 .................... 15.5
  Undefined 18:2 ................. 60
  Undefined 18:3 ................ 1.6
  20:0.......................... 0.6

**References** *JAOCS 68:* 518–519 (1991)

## Almond Kernel Oil
*Prunus dulcis*
Specific Gravity (SG)
  15.5/15.5°C
  25/25°C . . . . . . . . . . . . . . . . . 0.910–0.916
  Other SG . . . . . . . . . (26/26) 0.915–0.920
Refractive Index (RI)
  25°C
  40°C . . . . . . . . . . . . . . . . . . 1.462–1.466
  Other RI . . . . . . . . . . . . .(26) 1.464–1.470
Iodine Value . . . . . . . . . . . . . . . . . . . . 89–101
Saponification Value . . . . . . . . . . . . 188–200
Titer °C
% Unsaponifiable . . . . . . . . . . . . . . . . . . 1–2
Melting Point °C

**Fatty Acid Composition (%)**
  16:0 . . . . . . . . . . . . . . . . . . . . . . . . . . . 4–13
  9c-16:1 . . . . . . . . . . . . . . . . . . . . . . 0.2–0.8
  18:0 . . . . . . . . . . . . . . . . . . . . . . . . . . . 1–10
  Total 18:1 . . . . . . . . . . . . . . . . . . . . . 43–70
  9c,12c-18:2 . . . . . . . . . . . . . . . . . . . 20–34
  20:0 . . . . . . . . . . . . . . . . . . . . . . . . . 0.1–0.5
  Total 20:1 . . . . . . . . . . . . . . . . . . . . . . 0–0.3

Sterol Composition, %
  Cholesterol
  Brassicasterol
  Campesterol . . . . . . . . . . . . . . . . . . . . . 2–4
  Stigmastero . . . . . . . . . . . . . . . . . . . . . 1–2
  Stigmasta-8,22-dien-3β-ol
  5α-Stigmasta-7,22-dien-3β-ol
  Δ7,25-Stigmastadienol
  β-Sitosterol . . . . . . . . . . . . . . . . . . . . . . . 80
  Δ5-Avenasterol . . . . . . . . . . . . . . . . 10–12
  Δ7-Stigmasterol . . . . . . . . . . . . . . . . . . 1–2
  Δ7-Avenasterol . . . . . . . . . . . . . . . . . . 1–2
  Δ7-Campesterol
  Δ7-Ergosterol
  Δ7,25-Stigmasterol
  Sitostanol
  Spinasterol
  Squalene
  24-Methylene Cholesterol
  Other
  % sterols in oil
  Total Sterols, mg/kg . . . . . . . . . . . . . 2660

Tocopherol Composition, mg/kg
  α-Tocopherol . . . . . . . . . . . . . . . . . . . 228
  β-Tocopherol
  γ-Tocopherol . . . . . . . . . . . . . . . . . . . . . 8
  δ-Tocopherol
  Total, mg/kg . . . . . . . . . . . . . . . . . . . 236

**References** *J. Am. Dietetic Assn. 73:* 39 (1978)
*Fat Sci. Technol. 91:* 23 (1989)
*Rev. Franc. Corps Gras 33:* 115 (1986)

## Alpine Current Seed Oil
*Ribes nigrum*
Specific Gravity (SG)
  15.5/15.5°C
  25/25°C
  Other SG
Refractive Index (RI)
  25°C
  40°C
  Other RI
Iodine Value
Saponification Value
Titer °C
% Unsaponifiable
Melting Point °C

**Fatty Acid Composition (%)**
  16:0 . . . . . . . . . . . . . . . . . . . . . . . . . . . . 5–6
  18:0 . . . . . . . . . . . . . . . . . . . . . . . . . . . . 1–2
  Total 18:1 . . . . . . . . . . . . . . . . . . . . . . . . 19
  9c,12c-18:2 . . . . . . . . . . . . . . . . . . . . . . 41
  Undefined 18:3 . . . . . . . . . . . . . . . . . . . 18

**References** *Lipids 13:* 1311 (1996)

## Amaranth Seed Oil (Various)
*Amaranthus hypochondriacus, cruentus, edulis*
Specific Gravity (SG)
  15.5/15.5°C
  25/25°C
  Other SG
Refractive Index (RI)
  25°C
  40°C

Other RI
Iodine Value
Saponification Value
Titer °C
% Unsaponifiable . . . . . . . . . . . . . . . . 4.5–7
Melting Point °C

**Fatty Acid Composition (%)**
14:0 . . . . . . . . . . . . . . . . . . . . . . 0.21–0.29
16:0 . . . . . . . . . . . . . . . . . . . . . . . 13–23.8
16:1 . . . . . . . . . . . . . . . . . . . . . . . 0.1–0.19
17:0 . . . . . . . . . . . . . . . . . . . . . . . . 0.6–1.3
18:0 . . . . . . . . . . . . . . . . . . . . . . . . . 2.5–4
Total 18:1 . . . . . . . . . . . . . . . . . . . . 19–34
Undefined 18:2 . . . . . . . . . . . . . 39.4–49.1
9c,12c-18:2 . . . . . . . . . . . . . . . . . . . 47–62
Undefined 18:3 . . . . . . . . . . . . . . 0.65–1.3
20:0 . . . . . . . . . . . . . . . . . . . . . . . 0.56–1.4
Total 20:1 . . . . . . . . . . . . . . . . . 0.18–0.24
22:0 . . . . . . . . . . . . . . . . . . . . . . 0.14–0.32

Sterol Composition, %
   Cholesterol
   Brassicasterol
   Campesterol
   Stigmasterol . . . . . . . . . . . . . . . . . . . . . 8
   Stigmasta-8,22-dien-3β-ol
   5α-Stigmasta-7,22-dien-3β-ol
   Δ7,25-Stigmastadienol
   β-Sitosterol
   Δ5-Avenasterol
   Δ7-Stigmasterol . . . . . . . . . . . . . . . . . 16
   Δ7-Avenasterol
   Δ7-Campesterol
   Δ7-Ergosterol . . . . . . . . . . . . . . . . . . . 14
   Δ7,25-Stigmasterol
   Sitostanol
   Spinasterol . . . . . . . . . . . . . . . . . . . . . 54
   Squalene . . . . . . . . . . . . . . . . . . . . . . 5–8
   24-Methylene Cholesterol
   Other
   % sterols in oil
   Total Sterols, mg/kg

Tocopherol Composition, mg/kg
   α-Tocopherol
   β-Tocopherol
   γ-Tocopherol
   δ-Tocopherol
   Total, mg/kg . . . . . . . . . . . . . . . 370–390

**References** *J. Food Sci. 46:* 1175 (1981)
*Cereal Foods World 34:* 950 (1989)
*J. Am. Oil Chem. Soc. 64:* 233 (1987)
*JAOCS 77:* 847–852 (2000)

## Amaranth Seed Oil
*Amaranthus caudatus*

Specific Gravity (SG)
   15.5/15.5°C
   25/25°C
   Other SG
Refractive Index (RI)
   25°C
   40°C
   Other RI
Iodine Value
Saponification Value
Titer °C
% Unsaponifiable
Melting Point °C

**Fatty Acid Composition (%)**
16:0 . . . . . . . . . . . . . . . . . . . . . . . . . 20–25
18:0 . . . . . . . . . . . . . . . . . . . . . . . . . 0.5–4
Total 18:1 . . . . . . . . . . . . . . . . . . . . 28–37
Undefined 18:2 . . . . . . . . . . . . . . . . 37–46
Undefined 18:3 . . . . . . . . . . . . . . . . . . 0–2

**References**

## Amaranth Seed Oil
*Amaranthus cruentus*

Specific Gravity (SG)
   15.5/15.5°C
   25/25°C
   Other SG
Refractive Index (RI)
   25°C
   40°C
   Other RI
Iodine Value
Saponification Value
Titer °C
% Unsaponifiable . . . . . . . . . . . . . . . . 8.5
% Squalene in crude oil . . . . . . . . . . . . . 4.6
Melting Point °C

Fatty Acid Composition (%)
- 14:0 .......................... 0–0.2
- 16:0 .......................... 17–22
- 16:1 .......................... 0–0.2
- 18:0 .......................... 3–4
- Total 18:1 ................... 19–36
- Undefined 18:2 ............... 37–56
- Undefined 18:3 ............... 0.6–2
- 20:0 .......................... 1
- Total 20:1 .................. 0.2–0.4
- 22:0 ........................ 0.2–0.4
- 24:0 ........................ 0–0.4

References  *J. Food Sci. 46:* 1175 (1981)
*JAOCS 65:* 942 (1988)
*Riv. Ital. Sost. Grasse 75:* 505 (1998)

## Amaranthus mangostanus
*Amaranthus mangostanus*

Specific Gravity (SG)
- 15.5/15.5°C
- 25/25°C
- Other SG

Refractive Index (RI)
- 25°C
- 40°C
- Other RI

Iodine Value
Saponification Value
Titer °C
% Unsaponifiable
Melting Point °C

Fatty Acid Composition (%)
- 14:0 .......................... 0.29
- 16:0 .......................... 19.08
- 9c-16:1 ....................... 0.19
- 18:0 .......................... 3.19
- 9c-18:1 ....................... 18.82
- 7c-18:1 ....................... 1.26
- 9c,12c-18:2 ................... 44.68
- 9c,12c,15c-18:3 ............... 0.14
- 20:0 .......................... 0.92
- 11c-20:1 ...................... 0.24
- 22:0 .......................... 0.02
- 13c-22:1 ...................... 0.18
- 22:2 ........................ 0.09–0.11
- 24:0 .......................... 0.31

Tocopherol Composition, mg/kg
- α-Tocopherol .................... 94
- β-Tocopherol
- γ-Tocopherol ................... 580
- δ-Tocopherol
- Total, mg/kg

References  *JAOCS 80:* 1013–1020 (2003)

## Ambrette Seed Oil: raw seed oil
*Hibiscus abelmoschus*

Specific Gravity (SG)
- 15.5/15.5°C
- 25/25°C
- Other SG

Refractive Index (RI)
- 25°C
- 40°C
- Other RI ................. (30) 1.4750

Iodine Value ...................... 92.5
Saponification Value .............. 193.7
Titer °C
% Unsaponifiable .................. 1.8
Melting Point °C

Fatty Acid Composition (%)
- 16:0 .......................... 20.7
- 18:0 .......................... 5.5
- Total 18:1 .................... 26.1
- Undefined 18:2 ................. 39.4
- 19:0 .......................... 1.5
- 20:0 .......................... 0.3
- Unidentified 22:1 .............. 0.2

References  *JAOCS 80:* 209–211 (2003)

## Amoora Rohituka Seed Oil
*Amoora rohituka*

Specific Gravity (SG)
- 15.5/15.5°C
- 25/25°C
- Other SG

Refractive Index (RI)
- 25°C

40°C
Other RI
Iodine Value .................... 103
Saponification Value
Titer °C ........................ 291
% Unsaponifiable ................ 5–6
Melting Point °C

**Fatty Acid Composition (%)**
    16:0 ....................... 23–25
    18:0 ....................... 12–13
    Total 18:1 .................. 21–22
    9c,12c-18:2 .................... 29
    Undefined 18:3 .............. 13–14

**References** *J. Am. Oil Chem. Soc. 53:* 478 (1976)

# Andenopus Breviflorus Seed Oil
*Andeopus breviflorus*

Specific Gravity (SG)
  15.5/15.5°C
  25/25°C .................... 0.8995
  Other SG
Refractive Index (RI)
  25°C ....................... 1.4615
  40°C
  Other RI
Iodine Value .................... 100
Saponification Value ............ 193
Titer °C
% Unsaponifiable .................. 1
Melting Point °C

**Fatty Acid Composition (%)**
    16:0 .......................... 10
    18:0 .......................... 17
    Total 18:1 ................... 0.6
    9c,12c-18:2 ................... 14
    Undefined 18:3 ................ 56
    20:0 ........................... 1

**References** *Riv. Ital. Sost. Grasse 75:* 191 (1998)

# Anethum graveolens
*Anethum graveolens*

Specific Gravity (SG)
  15.5/15.5°C
  25/25°C
  Other SG
Refractive Index (RI)
  25°C
  40°C
  Other RI
Iodine Value
Saponification Value
Titer °C
% Unsaponifiable
Melting Point °C

**Fatty Acid Composition (%)**
    12:0 ......................... 6.05
    14:0 ......................... 0.07
    16:0 ......................... 3.64
    9c-16:1 ...................... 0.21
    18:0 ......................... 0.87
    9c-18:1 ...................... 7.79
    9c,12c-18:2 .................. 5.51
    9c,12c,15c-18:3 .............. 0.34
    20:0 ......................... 0.12
    11c-20:1 ..................... 0.03
    22:0 ......................... 0.01
    13c-22:1 ..................... 0.06
    24:0 ......................... 0.04

Tocopherol Composition, mg/kg
  α-Tocopherol .................... 96
  β-Tocopherol
  γ-Tocopherol .................... 29
  δ-Tocopherol .................... 30
  Total, mg/kg

Tocotrienols Composition, mg/kg
  α-Tocotrienol .................. 102
  β-Tocotrienol
  γ-Tocotrienol ................... 69
  δ-Tocotrienol ................... 29
  Total Tocotrienols, mg/kg

**References** *JAOCS 80:* 1013–1020 (2003)

## Anise Seed Oil
*Pimpinella anisum*

Specific Gravity (SG)
  15.5/15.5°C
  25/25°C
  Other SG
Refractive Index (RI)
  25°C
  40°C
  Other RI
Iodine Value . . . . . . . . . . . . . . . . . . . . . . . 97
Saponification Value . . . . . . . . . . . . . . . 184
Titer °C
% Unsaponifiable . . . . . . . . . . . . . . . . . . . 6.7
Melting Point °C

**Fatty Acid Composition (%)**
  14:0 . . . . . . . . . . . . . . . . . . . . . . . . . . . 2.5
  16:0 . . . . . . . . . . . . . . . . . . . . . . . . . . . . 6
  9c-16:1 . . . . . . . . . . . . . . . . . . . . . . . . . . 1
  18:0 . . . . . . . . . . . . . . . . . . . . . . . . . . 1–2
  9c-18:1 . . . . . . . . . . . . . . . . . . . . . . . . . 62
  6c-18:1 . . . . . . . . . . . . . . . . . . . . . . . . . 25

**References** *Fette Seifen Anstrichm. 85:* 23 (1983)

## Apple Seed Oil
*Malus deomestica L.*

Specific Gravity (SG)
  15.5/15.5°C
  25/25°C
  Other SG . . . . . . . . . (20/20) 0.902–0.923
Refractive Index (RI)
  25°C
  40°C . . . . . . . . . . . . . . . . . . 1.466–1.468
  Other RI
Iodine Value . . . . . . . . . . . . . . . . . 104–123
Saponification Value . . . . . . . . . . . 186–197
Titer °C
% Unsaponifiable . . . . . . . . . . . . . . 0.8–1.8
Melting Point °C

**Fatty Acid Composition (%)**
  14:0 . . . . . . . . . . . . . . . . . . . . . . . . . . . 0.5
  16:0 . . . . . . . . . . . . . . . . . . . . . . . . . . . 7.2

  16:1 . . . . . . . . . . . . . . . . . . . . . . . . . . . 1.4
  18:0 . . . . . . . . . . . . . . . . . . . . . . . . . . . 1.1
  Total 18:1 . . . . . . . . . . . . . . . . . . . . . . 29
  Undefined 18:2 . . . . . . . . . . . . . . . . . . 61
  Undefined 18:3 . . . . . . . . . . . . . . . . . 0.3

Sterol Composition, %
  Cholesterol . . . . . . . . . . . . . . . . . . . . . 0.3
  Brassicasterol . . . . . . . . . . . . . . . . . . . 0.5
  Campesterol . . . . . . . . . . . . . . . . . . . . 2.6
  Stigmasterol . . . . . . . . . . . . . . . . . . . . 6.6
  Stigmasta-8,22-dien-3β-ol
  5α-Stigmasta-7,22-dien-3β-ol
  Δ7,25-Stigmastadienol
  β-Sitosterol . . . . . . . . . . . . . . . . . . . . 86.6
  Δ5-Avenasterol . . . . . . . . . . . . . . . . . . 1.4
  Δ7-Stigmasterol . . . . . . . . . . . . . . . . . 0.5
  Δ7-Avenasterol . . . . . . . . . . . . . . . . . . 0.3
  Δ7-Campesterol . . . . . . . . . . . . . . . . . 1.0
  Δ7-Ergosterol
  Δ7,25-Stigmasterol
  Sitostanol
  Spinasterol
  Squalene
  24-Methylene Cholesterol
  Other
  % sterols in oil . . . . . . . . . . . . . . . . . . 0.3
  Total Sterols, mg/kg

**References** *Riv. Ital. Sost. Grasse 75:* 405 (1998)

## Apricot Kernel Oil
*Prunus armeniaca*

Specific Gravity (SG)
  15.5/15.5°C . . . . . . . . . . . . . . 0.915–0.921
  25/25°C . . . . . . . . . . . . . . . . . 0.910–0.916
  Other SG
Refractive Index (RI)
  25°C . . . . . . . . . . . . . . . . . . . 1.467–1.470
  40°C . . . . . . . . . . . . . . . . . . . 1.462–1.466
  Other RI
Iodine Value . . . . . . . . . . . . . . . . . . 97–110
Saponification Value . . . . . . . . . . . 185–199
Titer °C . . . . . . . . . . . . . . . . . . . . . . . . 0–6
% Unsaponifiable . . . . . . . . . . . . . . 0.4–1.4
Melting Point °C

**Fatty Acid Composition (%)**
  16:0 . . . . . . . . . . . . . . . . . . . . . . . . 4.6–6
  9c-16:1 . . . . . . . . . . . . . . . . . . . . . . . 1–2
  18:0 . . . . . . . . . . . . . . . . . . . . . . . 0.5–1.2
  Total 18:1 . . . . . . . . . . . . . . . . . . 58–65.7
  Undefined 18:2 . . . . . . . . . . . . . . . . 28.5
  9c,12c-18:2 . . . . . . . . . . . . . . . . . . 29–33
  Undefined 18:3 . . . . . . . . . . . . . . 0.5–1.0
  20:0 . . . . . . . . . . . . . . . . . . . . . . . . . . 0.2

Sterol Composition, %
  Cholesterol . . . . . . . . . . . . . . . . . . 0.6–0.8
  Brassicasterol
  Campesterol . . . . . . . . . . . . . . . . . . . . . 6
  Stigmasterol . . . . . . . . . . . . . . . . . . . . 2–5
  Stigmasta-8,22-dien-3β-ol
  5α-Stigmasta-7,22-dien-3β-ol
  Δ7,25-Stigmastadienol
  β-Sitosterol . . . . . . . . . . . . . . . . . . . 60–88
  Δ5-Avenasterol . . . . . . . . . . . . . . . . . . 3–5
  Δ7-Stigmasterol . . . . . . . . . . . . . . . . . . 13
  Δ7-Avenasterol . . . . . . . . . . . . . . . . . . . . 5
  Δ7-Campesterol . . . . . . . . . . . . . . . . . . . . 1
  Δ7-Ergosterol
  Δ7,25-Stigmasterol . . . . . . . . . . . . . . . . . . 2
  Sitostanol
  Spinasterol
  Squalene
  24-Methylene Cholesterol
  Other
  % sterols in oil
  Total Sterols, mg/kg

Tocopherol Composition, mg/kg
  α-Tocopherol . . . . . . . . . . . . . . . . . 10–22
  β-Tocopherol
  γ-Tocopherol . . . . . . . . . . . . . . . . 170–794
  δ-Tocopherol . . . . . . . . . . . . . . . . . . 20–24
  Total, mg/kg . . . . . . . . . . . . . . . . 200–840

**References** *Riv. Ital. Sost. Grasse 62:* 79 (1975)
  *J. Am. Oil Chem. Soc. 53:* 713 (1976)
  *Lebensmittelchem. Gerichtl. Chem. 36:* 53 (1982)
  *Rev. Franc. Corps Gras 33:* 115 (1986)
  *Riv. Ital. Sost. Grasse 75:* 405 (1998)
  *JAOCS 69:* 492–494 (1992)

# Arabidopsis Thaliana Seed Oil
*Brassicaceae columbia*

Specific Gravity (SG)
  15.5/15.5°C
  25/25°C
  Other SG
Refractive Index (RI)
  25°C
  40°C
  Other RI
Iodine Value
Saponification Value
Titer °C
% Unsaponifiable
Melting Point °C

**Fatty Acid Composition (%)**
  16:0 . . . . . . . . . . . . . . . . . . . . . . . . . . . . 8
  18:0 . . . . . . . . . . . . . . . . . . . . . . . . . . . . 3
  Total 18:1 . . . . . . . . . . . . . . . . . . . . . . 14
  11c-18:1 . . . . . . . . . . . . . . . . . . . . . . . . 1
  Undefined 18:2 . . . . . . . . . . . . . . . . . 28
  Undefined 18:3 . . . . . . . . . . . . . . . . . 18
  20:0 . . . . . . . . . . . . . . . . . . . . . . . . . . . . 2
  Total 20:1 . . . . . . . . . . . . . . . . . . . . . 20
  20:2 . . . . . . . . . . . . . . . . . . . . . . . . . . . . 2
  22:0 . . . . . . . . . . . . . . . . . . . . . . . . . . 0.4
  Unidentified 22:1 . . . . . . . . . . . . . . . . 2

**References**

# Argan Seed Oil
*Argania spinosa*

Specific Gravity (SG)
  15.5/15.5°C
  25/25°C
  Other SG . . . . . . . . . . (20/20) 0.906–0.919
Refractive Index (RI)
  25°C
  40°C
  Other RI . . . . . . . . . . . . (20) 1.463–1.4708
Iodine Value . . . . . . . . . . . . . . . . . . 92–102
Saponification Value . . . . . . . . . . . 189–195
Titer °C
% Unsaponifiable . . . . . . . . . . . . . . . 0.3–1.1
Melting Point °C

**Fatty Acid Composition (%)**
  14:0 . . . . . . . . . . . . . . . . . . . . . . . 0.1–0.3
  16:0 . . . . . . . . . . . . . . . . . . . . . . . . 12–16
  16:1 . . . . . . . . . . . . . . . . . . . . . . . . . 0.1–1
  9c-16:1 . . . . . . . . . . . . . . . . . . . . . . . 0.1
  18:0 . . . . . . . . . . . . . . . . . . . . . . . . . . 2–7
  Total 18:1 . . . . . . . . . . . . . . . . . . . 42–55
  Undefined 18:2 . . . . . . . . . . . . . . 28–37.4
  9c,12c-18:2 . . . . . . . . . . . . . . . . . . 30–34
  Undefined 18:3 . . . . . . . . . . . . . . . . . 0–1
  20:0 . . . . . . . . . . . . . . . . . . . . . . . . . . 0–1
  Total 20:1 . . . . . . . . . . . . . . . . . . . . . 0.1

Sterol Composition, %
  Cholesterol
  Brassicasterol
  Campesterol
  Stigmasterol
  Stigmasta-8,22-dien-3β-ol . . . . . . . . . . . 4
  5α-Stigmasta-7,22-dien-3β-ol
  Δ7,25-Stigmastadienol
  β-Sitosterol . . . . . . . . . . . . . . . . . . . . . . 44
  Δ5-Avenasterol
  Δ7-Stigmasterol . . . . . . . . . . . . . . . . . . 48
  Δ7-Avenasterol . . . . . . . . . . . . . . . . . . . 4
  Δ7-Campesterol
  Δ7-Ergosterol
  Δ7,25-Stigmasterol
  Sitostanol
  Spinasterol . . . . . . . . . . . . . . . . . . . . . 44
  Squalene
  24-Methylene Cholesterol
  Other
  % sterols in oil
  Total Sterols, mg/kg

Tocopherol Composition, mg/kg
  α-Tocopherol
  β-Tocopherol
  γ-Tocopherol
  δ-Tocopherol
  Total, mg/kg . . . . . . . . . . . . . . . 167–635

**References** *Rev. Franc. Corps Gras 39:* 139 (1992)
  *JAOCS 69:* 141 (1992)
  *JAOCS 76:* 15–18 (1999)

# Argemone Oil
*Argemone mexicana L*
Specific Gravity (SG)
  15.5/15.5°C . . . . . . . . . . . . 0.9220–0.9247
  25/25°C
  Other SG
Refractive Index (RI)
  25°C
  40°C . . . . . . . . . . . . . . . . . 1.4660–1.4685
  Other RI
Iodine Value . . . . . . . . . . . . . . . . . 119–128
Saponification Value . . . . . . . . . . . 190–193
Titer °C
% Unsaponifiable . . . . . . . . . . . . . . . 1.1–1.6
Melting Point °C

**Fatty Acid Composition (%)**
  16:0 . . . . . . . . . . . . . . . . . . . . . . . . . . 8–15
  9c-16:1 . . . . . . . . . . . . . . . . . . . . . . . . 1–2
  18:0 . . . . . . . . . . . . . . . . . . . . . . . . . . . 2–5
  Total 18:1 . . . . . . . . . . . . . . . . . . . . 22–33
  9c,12c-18:2 . . . . . . . . . . . . . . . . . . . 48–62

**References** *J. Am. Oil Chem. Soc. 52:* 171 (1975)

# Avocado (pulp) Oil
*Persea americana*
Specific Gravity (SG)
  15.5/15.5°C
  25/25°C . . . . . . . . . . . . . . . . . 0.910–0.920
  Other SG
Refractive Index (RI)
  25°C . . . . . . . . . . . . . . . . . . . 1.462–1.470
  40°C . . . . . . . . . . . . . . . . . . . 1.458–1.465
  Other RI
Iodine Value . . . . . . . . . . . . . . . . . . . 85–90
Saponification Value . . . . . . . . . . . 177–198
Titer °C
% Unsaponifiable . . . . . . . . . . . . . . . . 1–12
Melting Point °C

**Fatty Acid Composition (%)**
  16:0 . . . . . . . . . . . . . . . . . . . . . . . . . . 9–18
  9c-16:1 . . . . . . . . . . . . . . . . . . . . . . . . 3–9

18:0 . . . . . . . . . . . . . . . . . . . . . . . . 0.4–1
Total 18:1 . . . . . . . . . . . . . . . . . . . . 56–74
9c,12c-18:2 . . . . . . . . . . . . . . . . . . . 10–17
Undefined 18:3 . . . . . . . . . . . . . . . . . 0–2

Sterol Composition, %
  Cholesterol . . . . . . . . . . . . . . . . . . . 0–0.2
  Brassicasterol . . . . . . . . . . . . . . . . . . . . 2
  Campesterol . . . . . . . . . . . . . . . . . . . . 6–8
  Stigmasterol . . . . . . . . . . . . . . . . . . . . . 0–2
  Stigmasta-8,22-dien-3β-ol
  5α-Stigmasta-7,22-dien-3β-ol
  Δ7,25-Stigmastadienol
  β-Sitosterol . . . . . . . . . . . . . . . . . . . . 89–92
  Δ5-Avenasterol . . . . . . . . . . . . . . . . . . . . 3
  Δ7-Stigmasterol
  Δ7-Avenasterol . . . . . . . . . . . . . . . . . . 0.2
  Δ7-Campesterol
  Δ7-Ergosterol
  Δ7,25-Stigmasterol
  Sitostanol
  Spinasterol
  Squalene
  24-Methylene Cholesterol
  Other
  % sterols in oil
  Total Sterols, mg/kg . . . . . . . . . . . . 4040

Tocopherol Composition, mg/kg
  α-Tocopherol . . . . . . . . . . . . . . . . 64–100
  β-Tocopherol
  γ-Tocopherol . . . . . . . . . . . . . . . . . . . 0–19
  δ-Tocopherol
  Total, mg/kg . . . . . . . . . . . . . . . . 83–100

**References** *J. Am. Oil Chem. Soc. 65:* 1704 (1988)
*Riv. Ital. Sost. Grasse 52:* 79 (1975)
*J. Am. Dietetic Assn. 73:* 39 (1978)
*J. Am. Oil Chem. Soc. 53:* 732 (1976)
*Lipids 9:* 658 (1974)

## Azima Tetracantha
*Azima tetracantha*
Specific Gravity (SG)
  15.5/15.5°C
  25/25°C
  Other SG

Refractive Index (RI)
  25°C
  40°C
  Other RI
Iodine Value . . . . . . . . . . . . . . . . . . . . . . 141
Saponification Value . . . . . . . . . . . . . . . 201.5
Titer °C
% Unsaponifiable . . . . . . . . . . . . . . . . . . . 2.3
Melting Point °C

**Fatty Acid Composition (%)**
  12:0 . . . . . . . . . . . . . . . . . . . . . . . . . . 3.5
  14:0 . . . . . . . . . . . . . . . . . . . . . . . . . . 4.2
  16:0 . . . . . . . . . . . . . . . . . . . . . . . . . . 5.2
  18:0 . . . . . . . . . . . . . . . . . . . . . . . . . . 1.6
  9c-18:1 . . . . . . . . . . . . . . . . . . . . . . . 15.3
  9c,12c-18:2 . . . . . . . . . . . . . . . . . . . . 28.8
  Undefined 18:3 . . . . . . . . . . . . . . . . . . . 22
  Other: ricinoleic, 9.8; malvalic, 4.0;
     sterculic, 5.6

**References** *JAOCS 68:* 978–979 (1991)

## Babassu Palm Oil (Brazil)
*Attalea speciosa Martius syn Orbignya spp.*
Specific Gravity (SG)
  15.5/15.5°C
  25/25°C
  Other SG . . . . . . . . . (25/20) 0.914–0.917
Refractive Index (RI)
  25°C
  40°C . . . . . . . . . . . . . . . . . . . 1.448–1.451
  Other RI
Iodine Value . . . . . . . . . . . . . . . . . . . . 10–18
Saponification Value . . . . . . . . . . . . 245–256
Titer °C
% Unsaponifiable . . . . . . . . . . . . . . . . . . . 1.2
Melting Point °C . . . . . . . . . . . . . . . . . 24–26

**Fatty Acid Composition (%)**
  8:0 . . . . . . . . . . . . . . . . . . . . . . . . . 2.6–7.3
  10:0 . . . . . . . . . . . . . . . . . . . . . . . . 1.2–7.6
  12:0 . . . . . . . . . . . . . . . . . . . . . . . . . 40–55
  14:0 . . . . . . . . . . . . . . . . . . . . . . . . . 11–27
  16:0 . . . . . . . . . . . . . . . . . . . . . . . 5.2–11.0
  18:0 . . . . . . . . . . . . . . . . . . . . . . . . 1.8–7.4

Total 18:1 .................... 9–20
9c,12c-18:2 ................. 1.4–6.6

Sterol Composition, %
 Cholesterol .................. 1.2–1.7
 Brassicasterol ................ 0–0.3
 Campesterol .............. 17.7–18.7
 Stigmasterol ................ 8.7–9.2
 Stigmasta-8,22-dien-3β-ol
 5α-Stigmasta-7,22-dien-3β-ol
 Δ7,25-Stigmastadienol
 β-Sitosterol ............... 48.2–53.9
 Δ5-Avenasterol ............ 16.9–20.4
 Δ7-Stigmasterol
 Δ7-Avenasterol .............. 0.4–1.0
 Δ7-Campesterol
 Δ7-Ergosterol
 Δ7,25-Stigmasterol
 Sitostanol
 Spinasterol
 Squalene
 24-Methylene Cholesterol
 Other
 % sterols in oil
 Total Sterols, mg/kg .......... 570–766

Tocotrienols Composition, mg/kg
 α-Tocotrienol ................ 25–46
 β-Tocotrienol
 γ-Tocotrienol ................. 32–80
 δ-Tocotrienol ................. 9–10
 Total Tocotrienols, mg/kg ...... 67–128

**References** *Codex* CX 1993/16

# Bacury Seed Fat
*Platonia insignis*

Specific Gravity (SG)
 15.5/15.5°C
 25/25°C
 Other SG
Refractive Index (RI)
 25°C
 40°C
 Other RI
Iodine Value ....................... 44
Saponification Value
Titer °C

% Unsaponifiable
Melting Point °C ................. 54–56

**Fatty Acid Composition (%)**
 14:0 ........................... 1
 16:0 ........................... 55
 9c-16:1 ........................ 3
 18:0 ........................... 6
 Total 18:1 ..................... 32
 9c,12c-18:2 .................... 2
 20:0 ........................... 0.3

**References**

# Baguacu Pulp Oil
*Pindarea fastuosa*

Specific Gravity (SG)
 15.5/15.5°C
 25/25°C
 Other SG ........... (26.4/26.4) 0.9057
Refractive Index (RI)
 25°C ......................... 1.463
 40°C
 Other RI
Iodine Value ...................... 21
Saponification Value
Titer °C
% Unsaponifiable ................... 1.3
Melting Point °C .................. 22.4

**Fatty Acid Composition (%)**
 14:0 ........................... 0.3
 16:0 ........................... 36
 16:1 ........................... 0.7
 18:0 ........................... 2
 Total 18:1 ..................... 47
 Undefined 18:2 ................. 1.7
 Undefined 18:3 ................. 7.2

Sterol Composition, %
 Cholesterol .................... 4.5
 Brassicasterol
 Campesterol .................... 4
 Stigmasterol ................... 77
 Stigmasta-8,22-dien-3β-ol
 5α-Stigmasta-7,22-dien-3β-ol
 Δ7,25-Stigmastadienol
 β-Sitosterol

Δ5-Avenasterol .................. 1.8
Δ7-Stigmasterol
Δ7-Avenasterol
Δ7-Campesterol
Δ7-Ergosterol
Δ7,25-Stigmasterol
Sitostanol ..................... 1.5
Spinasterol
Squalene
24-Methylene Cholesterol
Other .......................... 8.5
% sterols in oil
Total Sterols, mg/kg

**References** *Riv. Ital. Sost. Grasse 75:* 345 (1998)

## Baguacu Seed Oil
*Pindarea fastuosa*
Specific Gravity (SG)
  15.5/15.5°C
  25/25°C ...................... 0.9217
  Other SG
Refractive Index (RI)
  25°C
  40°C
  Other RI ................. (20) 1.457
Iodine Value
Saponification Value
Titer °C
% Unsaponifiable .................. 0.8
Melting Point °C

**Fatty Acid Composition (%)**
  6:0 ............................ 0.5
  8:0 .......................... 10–11
  10:0 ........................... 11
  12:0 ........................... 46
  14:0 ........................... 10
  16:0 ............................ 6
  18:0 ............................ 2
  Total 18:1 .................. 11–12
  Undefined 18:2 .................. 2

Sterol Composition, %
  Cholesterol .................... 1.5
  Brassicasterol

Campesterol ..................... 14
Stigmasterol ..................... 7
Stigmasta-8,22-dien-3β-ol
5α-Stigmasta-7,22-dien-3β-ol
Δ7,25-Stigmastadienol
β-Sitosterol .................... 69
Δ5-Avenasterol
Δ7-Stigmasterol
Δ7-Avenasterol ................. 3.8
Δ7-Campesterol
Δ7-Ergosterol
Δ7,25-Stigmasterol
Sitostanol
Spinasterol
Squalene
24-Methylene Cholesterol
Other .......................... 4.5
% sterols in oil
Total Sterols, mg/kg

**References** *Riv. Ital. Sost. Grasse 75:* 345 (1998)

## Bahera Seed Oil
*Terminalia bellirica*
Specific Gravity (SG)
  15.5/15.5°C
  25/25°C
  Other SG
Refractive Index (RI)
  25°C
  40°C
  Other RI
Iodine Value ..................... 76
Saponification Value ............ 209
Titer °C
% Unsaponifiable
Melting Point °C

**Fatty Acid Composition (%)**
  16:0 ........................... 18
  18:0 ............................ 8
  Total 18:1 ..................... 56
  9c,12c-18:2 .................... 11

**References** *J. Agric. Food Chem. 43:* 902 (1995)

## Baillonella Toxisperma Kernel Oil

*Baillonella toxisperma*

Specific Gravity (SG)
   15.5/15.5°C
   25/25°C
   Other SG
Refractive Index (RI)
   25°C
   40°C
   Other RI
Iodine Value
Saponification Value
Titer °C
% Unsaponifiable
Melting Point °C

**Fatty Acid Composition (%)**
   16:0 ............................ 19
   18:0 ............................ 22
   Total 18:1 ...................... 55
   9c,12c-18:2 ...................... 4

**References** *Rev. Franc. Corps Gras 39:* 147 (1992)

## Baobab Seed Oil

*Adansonia spp.*

Specific Gravity (SG)
   15.5/15.5°C
   25/25°C ........................ 0.937
   Other SG
Refractive Index (RI)
   25°C
   40°C ................. 1.4596–1.4633
   Other RI
Iodine Value ..................... 55–96
Saponification Value ............ 133–195
Titer °C
% Unsaponifiable ............... 2.8–3.8
Melting Point °C

**Fatty Acid Composition (%)**
   12:0 .......................... 0–0.3
   14:0 .......................... 0.3–1.5
   16:0 ........................... 25–46
   9c-16:1 ........................ 0.3–1.7
   18:0 ........................... 0–4
   Total 18:1 ..................... 21–59
   9c,12c-18:2 .................... 12–29
   Undefined 18:3 .................. 0–8
   20:0 ........................... 0.5–1.0
   Total 20:1 ..................... 0–3.6
   Other malvalic, 1–7; sterculic, 1–8; dihydrosterculic, 2–5

Sterol Composition, %
   Cholesterol ....................... 2
   Brassicasterol
   Campesterol ....................... 6
   Stigmasterol .................... 1–2
   Stigmasta-8,22-dien-3β-ol
   5α-Stigmasta-7,22-dien-3β-ol
   Δ7,25-Stigmastadienol
   β-Sitosterol ..................... 75
   Δ5-Avenasterol .................. 0.5
   Δ7-Stigmasterol ................. 0.6
   Δ7-Avenasterol ................... 12
   Δ7-Campesterol
   Δ7-Ergosterol
   Δ7,25-Stigmasterol
   Sitostanol
   Spinasterol
   Squalene
   24-Methylene Cholesterol
   Other
   % sterols in oil
   Total Sterols, mg/kg

**References** *Lipids 17:* 1 (1982)
*Riv. Ital. Sost. Grasse 60:* 747 (1983)
*Riv. Ital. Sost. Grasse 73:* 371 (1996)

## Barley Oil

*Hordeum vulgare*

Specific Gravity (SG)
   15.5/15.5°C
   25/25°C
   Other SG
Refractive Index (RI)
   25°C
   40°C
   Other RI

Iodine Value
Saponification Value
Titer °C
% Unsaponifiable .................. 5–6
Melting Point °C

**Fatty Acid Composition (%)**
14:0........................0.3–2.0
16:0..........................12–29
18:0............................1–4
Total 18:1......................4–28
9c,12c-18:2....................50–62
Undefined 18:3...................3–6

Tocopherol Composition, mg/kg
α-Tocopherol..................13–23
β-Tocopherol....................2–3
γ-Tocopherol....................1–3
δ-Tocopherol
Total, mg/kg..................16–29

Tocotrienols Composition, mg/kg
α-Tocotrienol.................44–59
β-Tocotrienol..................8–15
γ-Tocotrienol...................7–9
δ-Tocotrienol
Total Tocotrienols, mg/kg........59–83

**References** *Cereal Sci. Today 11:* 99 (1966)
*Lipids 9:* 560 (1974)
*Lipids 9:* 804 (1974)
*Anal. Biochem. 32:* 81 (1969)
*J. Agr. Food Chem. 20:* 240 (1972)

## Basil Seed Oil
*Ocimum basilicum/canum/ gratissimum/sanctum*

Specific Gravity (SG)
15.5/15.5°C
25/25°C
Other SG
Refractive Index (RI)
25°C
40°C
Other RI .............(20) 1.460–1.481
Iodine Value ................. 172–200
Saponification Value ............ 191–200

Titer °C
% Unsaponifiable
Melting Point °C

**Fatty Acid Composition (%)**
14:0.......................... 0.03
16:0.......................... 7–11
9c-16:1......................0.2–0.3
11c-16:1...................... 0.12
18:0............................2–4
Total 18:1......................9–13
9c-18:1....................... 7.43
7c-18:1....................... 0.78
9c,12c-18:2....................18–31
Undefined 18:3.................44–65
9c,12c,15c-18:3 .............. 54.58
20:0.......................... 0.16
11c-20:1...................... 0.11
22:0.......................... 0.04
24:0.......................... 0.05

Tocopherol Composition, mg/kg
α-Tocopherol.....................52
β-Tocopherol
γ-Tocopherol....................828
δ-Tocopherol.....................47
Total, mg/kg

**References** *J. Am. Oil Chem. Soc. 73:* 393 (1996)
*JAOCS 80:* 1013–1020 (2003)

## Basella Rubra Seed Oil
*Basella rubra*

Specific Gravity (SG)
15.5/15.5°C
25/25°C
Other SG
Refractive Index (RI)
25°C
40°C
Other RI
Iodine Value
Saponification Value
Titer °C
% Unsaponifiable
Melting Point °C

**Fatty Acid Composition (%)**
- 14:0 .......................... 0.11
- 16:0 .......................... 18.51
- 9c-16:1 ....................... 0.81
- 18:0 .......................... 6.43
- 9c-18:1 ....................... 47.44
- 7c-18:1 ....................... 4.27
- 9c,12c-18:2 ................... 15.82
- 9c,12c,15c-18:3 ............... 0.28
- 20:0 .......................... 1.46
- 11c-20:1 ...................... 0.27
- 22:0 .......................... 0.63
- 24:0 .......................... 3.84

Tocopherol Composition, mg/kg
- $\alpha$-Tocopherol ................... 138
- $\beta$-Tocopherol
- $\gamma$-Tocopherol .................... 290
- $\delta$-Tocopherol ..................... 29
- Total, mg/kg

**References** *JAOCS 80:* 1013–1020 (2003)

# Beechnut Kernel Oil
*Fagus orientalis Lipsky*

Specific Gravity (SG)
  15.5/15.5°C
  25/25°C
  Other SG
Refractive Index (RI)
  25°C
  40°C
  Other RI ................. (20) 1.4725
Iodine Value ..................... 106.61
Saponification Value .............. 193.50
Titer °C
% Unsaponifiable .................. 0.84
Melting Point °C

**Fatty Acid Composition (%)**
- 16:0 .......................... 8.8
- 18:0 .......................... 3.2
- Total 18:1 .................... 30.4
- Undefined 18:2 ................ 48.9
- Total 20:1 .................... 6.7

**References** *JAOCS 69:* 1274 (1992)

# Bengal Gram (Chickpea) Oil
*Cicer arietinum*

Specific Gravity (SG)
  15.5/15.5°C
  25/25°C
  Other SG
Refractive Index (RI)
  25°C
  40°C
  Other RI
Iodine Value
Saponification Value
Titer °C
% Unsaponifiable
Melting Point °C

**Fatty Acid Composition (%)**
- 14:0 .......................... 0–0.5
- 16:0 .......................... 9.5–18
- 18:0 .......................... 1.8–3
- Total 18:1 .................... 19–28
- 9c,12c-18:2 ................... 45–66
- Undefined 18:3 ................ 0–6
- 20:0 .......................... 0–0.8
- 22:0 .......................... 0–0.7

Tocopherol Composition, mg/kg
- $\alpha$-Tocopherol .................... 17
- $\beta$-Tocopherol ...................... 1
- $\gamma$-Tocopherol ..................... 92
- $\delta$-Tocopherol ...................... 4
- Total, mg/kg .................... 114

Tocotrienols Composition, mg/kg
- $\alpha$-Tocotrienol
- $\beta$-Tocotrienol
- $\gamma$-Tocotrienol
- $\delta$-Tocotrienol
- Total Tocotrienols, mg/kg ............ 2

**References** *J. Am. Oil Chem. Soc. 74:* 1603 (1997)
*Food Chem. 56:* 123 (1996)

## Bitter Almond Kernel Oil
*Prunus dulcis*

Specific Gravity (SG)
  15.5/15.5°C
  25/25°C ..................... 0.9145
  Other SG
Refractive Index (RI)
  25°C ....................... 1.4703
  40°C
  Other RI
Iodine Value ...................... 94
Saponification Value .............. 189
Titer °C ........................... 7
% Unsaponifiable .................. 1.2
Melting Point °C

**Fatty Acid Composition (%)**
  14:0 ............................ 0.4
  16:0 .............................. 7
  9c-16:1 ......................... 0.3
  18:0 .............................. 1
  Total 18:1 ....................... 74
  9c,12c-18:2 ...................... 17

**References** *Fat Sci. Technol. 89:* 305 (1987)


## Bittersweet Oil
*Celastrus scandens*

Specific Gravity (SG)
  15.5/15.5°C
  25/25°C
  Other SG ............... (20/4) 0.9772
Refractive Index (RI)
  25°C
  40°C
  Other RI ............... (20) 1.4815
Iodine Value ..................... 122
Saponification Value .............. 297
Titer °C
% Unsaponifiable .................... 3
Melting Point °C

**Fatty Acid Composition (%)**
  2:0 .............................. 19
  16:0 ............................. 10
  18:0 .............................. 2
  Undefined 18:2 ................... 44
  Undefined 18:3 ................... 24

**References**


## Bitter Vetch Seed Oil
*Lathyrus cicera*

Specific Gravity (SG)
  15.5/15.5°C
  25/25°C
  Other SG
Refractive Index (RI)
  25°C
  40°C
  Other RI
Iodine Value
Saponification Value
Titer °C
% Unsaponifiable
Melting Point °C

**Fatty Acid Composition (%)**
  14:0 ............................ 0.5
  16:0 ............................ 5–6
  9c-16:1 ......................... 0.3
  18:0 .......................... 17–20
  Total 18:1 .................... 57–58
  9c,12c-18:2 ...................... 12
  Undefined 18:3 ............... 0.6–0.8
  20:0 ......................... 1.1–1.3
  22:0 ......................... 0.8–1.0

Sterol Composition, %
  Cholesterol .................. 1.1–1.5
  Brassicasterol
  Campesterol ..................... 8–9
  Stigmasterol .................. 21–22
  Stigmasta-8,22-dien-3β-ol
  5α-Stigmasta-7,22-dien-3β-ol
  Δ7,25-Stigmastadienol
  β-Sitosterol ................... 56–59
  Δ5-Avenasterol ............... 0.3–0.4
  Δ7-Stigmasterol ................. 7–8
  Δ7-Avenasterol
  Δ7-Campesterol
  Δ7-Ergosterol
  Δ7,25-Stigmasterol

Sitostanol
Spinasterol
Squalene
24-Methylene Cholesterol
Other
% sterols in oil
Total Sterols, mg/kg

**References** *Riv. Ital. Sost. Grasse 71:* 567 (1994)

# Black Gram Oil
*Vigna mungo*
Specific Gravity (SG)
 15.5/15.5°C
 25/25°C
 Other SG
Refractive Index (RI)
 25°C
 40°C
 Other RI
Iodine Value
Saponification Value
Titer °C
% Unsaponifiable
Melting Point °C

**Fatty Acid Composition (%)**
 16:0 . . . . . . . . . . . . . . . . . . . . . . . . . . . 11
 18:0 . . . . . . . . . . . . . . . . . . . . . . . . . . . 2.6
 9c-18:1 . . . . . . . . . . . . . . . . . . . . . . . 26.1
 9c,12c-18:2 . . . . . . . . . . . . . . . . . . . . 7.2
 9c,12c,15c-18:3 . . . . . . . . . . . . . . . . 49.1
 20:0 . . . . . . . . . . . . . . . . . . . . . . . . . . 0.1
 Total 20:1 . . . . . . . . . . . . . . . . . . . . . 0.1
 22:0 . . . . . . . . . . . . . . . . . . . . . . . . . . 0.8
 11c-22:1 . . . . . . . . . . . . . . . . . . . . . . 0.1
 22:2 . . . . . . . . . . . . . . . . . . . . . . . . . . 0.2
 16c,19c,-22:3 . . . . . . . . . . . . . . . . . . 0.1
 24:0 . . . . . . . . . . . . . . . . . . . . . . . . . . 0.1

Tocopherol Composition, mg/kg
 α-Tocopherol . . . . . . . . . . . . . . . . . . . . 3
 β-Tocopherol
 γ-Tocopherol . . . . . . . . . . . . . . . . . . . 66
 δ-Tocopherol . . . . . . . . . . . . . . . . . . . . 2
 Total, mg/kg . . . . . . . . . . . . . . . . . . . 70

**References** *J. Am. Oil Chem. Soc. 74:* 1603 (1997)

# Blackcurrant Oil
*Ribes nigrum*
Specific Gravity (SG)
 15.5/15.5°C
 25/25°C
 Other SG . . . . . . . . . . (20/20) 0.921–0.928
Refractive Index (RI)
 25°C
 40°C
 Other RI . . . . . . . . . . . . .(20) 1.479–1.481
Iodine Value . . . . . . . . . . . . . . . . . . 173–182
Saponification Value . . . . . . . . . . . . 185–195
Titer °C
% Unsaponifiable . . . . . . . . . . . . . . . . . . . 1
Melting Point °C

**Fatty Acid Composition (%)**
 14:0 . . . . . . . . . . . . . . . . . . . . . . . . . . 0.1
 16:0 . . . . . . . . . . . . . . . . . . . . . . . . . . 6–8
 9c-16:1 . . . . . . . . . . . . . . . . . . . . . . 0–0.2
 18:0 . . . . . . . . . . . . . . . . . . . . . . . . . . 1–2
 Total 18:1 . . . . . . . . . . . . . . . . . . . . 9–13
 9c,12c-18:2 . . . . . . . . . . . . . . . . . . 45–50
 6c,9c,12c-18:3 . . . . . . . . . . . . . . . . 14–20
 9c,12c,15c-18:3 . . . . . . . . . . . . . . . 12–15
 6c,9c,12c,15c-18:4 . . . . . . . . . . . . . . 2–4
 20:0 . . . . . . . . . . . . . . . . . . . . . . . . . . 0.2
 Total 20:1 . . . . . . . . . . . . . . . . . . . 0.9–1.0
 22:0 . . . . . . . . . . . . . . . . . . . . . . . . . . 0.1
 24:0 . . . . . . . . . . . . . . . . . . . . . . . . . . 0.1

Sterol Composition, %
 Cholesterol . . . . . . . . . . . . . . . . . . 0.2–0.7
 Brassicasterol
 Campesterol . . . . . . . . . . . . . . . . 7.2–10.4
 Stigmasterol . . . . . . . . . . . . . . . . . 0.5–1.0
 Stigmasta-8,22-dien-3β-ol
 5α-Stigmasta-7,22-dien-3β-ol
 Δ7,25-Stigmastadienol
 β-Sitosterol . . . . . . . . . . . . . . . . . . . 70–85
 Δ5-Avenasterol . . . . . . . . . . . . . . . . . 2–3
 Δ7-Stigmasterol . . . . . . . . . . . . . . 0.4–4.5
 Δ7-Avenasterol . . . . . . . . . . . . . . . . 0.4–2
 Δ7-Campesterol

Δ7-Ergosterol
Δ7,25-Stigmasterol
Sitostanol
Spinasterol
Squalene
24-Methylene Cholesterol
Other
% sterols in oil
Total Sterols, mg/kg

Tocopherol Composition, mg/kg
 α-Tocopherol . . . . . . . . . . . . . . . . . . . 320
 β-Tocopherol. . . . . . . . . . . . . . . . . . . . . 8
 γ-Tocopherol. . . . . . . . . . . . . . . . . . . 647
 δ-Tocopherol. . . . . . . . . . . . . . . . . . . . 68
 Total, mg/kg . . . . . . . . . . . . . . . . . 1043

**References** *Rev. Franc. Corps Gras 35:* 501 (1988)
*Rev. Franc. Corps Gras 39:* 339 (1992)
*Codex* 1987/8, 1987/17
*Riv. Ital. Sost. Grasse 65:* 1 (1988)
*Lipids 31:* 131 (1996)

## Bliphia Sapida Seed Oil
*Bliphia sapida*
Specific Gravity (SG)
 15.5/15.5°C
 25/25°C. . . . . . . . . . . . . . . . . . . . . 0.942
 Other SG
Refractive Index (RI)
 25°C
 40°C
 Other RI
Iodine Value . . . . . . . . . . . . . . . . . . . . . 64
Saponification Value . . . . . . . . . . . . . . . 176
Titer °C
% Unsaponifiable
Melting Point °C

**Fatty Acid Composition (%)**
 12:0. . . . . . . . . . . . . . . . . . . . . . . . . 5–6
 14:0. . . . . . . . . . . . . . . . . . . . . . . . . . 1
 16:0. . . . . . . . . . . . . . . . . . . . . . . . . . 8
 9c-16:1 . . . . . . . . . . . . . . . . . . . . . 1–2
 18:0. . . . . . . . . . . . . . . . . . . . . . . . . . 2
 Total 18:1 . . . . . . . . . . . . . . . . . . . . 53

 9c,12c-18:2. . . . . . . . . . . . . . . . . . . . 19
 Undefined 18:3 . . . . . . . . . . . . . . . . . . 8
 20:0. . . . . . . . . . . . . . . . . . . . . . . . . . 1

**References** *Riv. Ital. Sost. Grasse 72:* 311 (1995)

## Bombax Constantum Seed Oil
*Bombax constantum*
Specific Gravity (SG)
 15.5/15.5°C
 25/25°C
 Other SG
Refractive Index (RI)
 25°C
 40°C
 Other RI
Iodine Value . . . . . . . . . . . . . . . . . . . . 103
Saponification Value . . . . . . . . . . . . . . . 285
Titer °C
% Unsaponifiable
Melting Point °C

**Fatty Acid Composition (%)**
 6:0. . . . . . . . . . . . . . . . . . . . . . . . . . . 3
 8:0. . . . . . . . . . . . . . . . . . . . . . . . . . . 7
 16:0. . . . . . . . . . . . . . . . . . . . . . . . . . 8
 18:0. . . . . . . . . . . . . . . . . . . . . . . . . . 3
 Total 18:1 . . . . . . . . . . . . . . . . . . . . 49
 9c,12c-18:2. . . . . . . . . . . . . . . . . . . . 13
 20:0. . . . . . . . . . . . . . . . . . . . . . . . . . 3
 24:0. . . . . . . . . . . . . . . . . . . . . . . . 1–2
 Other: 12,13-Epoxy-octadeca-9-enoic, 13 (vernolic)

**References** *Riv. Ital. Sost. Grasse 73:* 271 (1996)

## Bombax Munguba Seed Oil
*Bombax munguba*
Specific Gravity (SG)
 15.5/15.5°C
 25/25°C
 Other SG

Refractive Index (RI)
  25°C
  40°C
  Other RI
Iodine Value
Saponification Value
Titer °C
% Unsaponifiable
Melting Point °C

**Fatty Acid Composition (%)**
  16:0 .......................... 58.3
  18:0 .......................... 3.8
  Total 18:1 .................... 0.5
  9c-18:1 ....................... 5.3
  Undefined 18:2 ................ 5.6
  19:1 .......................... 18.4

**References**  *JAOCS 75:* 1757–1760 (1998)

# Borage Oil
*Borago officinalis*
Specific Gravity (SG)
  15.5/15.5°C
  25/25°C
  Other SG
Refractive Index (RI)
  25°C
  40°C
  Other RI
Iodine Value ................... 141–160
Saponification Value ........... 189–192
Titer °C
% Unsaponifiable ............... 1.2–1.9
Melting Point °C

**Fatty Acid Composition (%)**
  14:0 .......................... 0.1
  16:0 .......................... 9.4–11.9
  9c-16:1 ....................... 0.4
  18:0 .......................... 2.6–5.0
  9c-18:1 ....................... 14.6–21.3
  9c,12c-18:2 ................... 36.5–40.1
  6c,9c,12c-18:3 ................ 17.1–25.4
  9c,12c,15c-18:3 ............... 0.2
  6c,9c,12c,15c-18:4 ............ 0.2
  20:0 .......................... 0.2
  11c-20:1 ...................... 2.9–4.1

  13c-22:1 ...................... 1.8–2.8
  15c-24:1 ...................... 1.2–4.5
  Other ......................... 0.8

Sterol Composition, %
  Cholesterol
  Brassicasterol ................ 0–1.6
  Campesterol ................... 25–30
  Stigmasterol
  Stigmasta-8,22-dien-3β-ol
  5α-Stigmasta-7,22-dien-3β-ol
  Δ7,25-Stigmastadienol
  β-Sitosterol .................. 22–42
  Δ5-Avenasterol ................ 15–28
  Δ7-Stigmasterol
  Δ7-Avenasterol ................ 1
  Δ7-Campesterol
  Δ7-Ergosterol
  Δ7,25-Stigmasterol
  Sitostanol
  Spinasterol
  Squalene
  24-Methylene Cholesterol ...... 15–20
  Other
  % sterols in oil
  Total Sterols, mg/kg

Tocopherol Composition, mg/kg
  α-Tocopherol .................. 0–46
  β-Tocopherol
  γ-Tocopherol .................. 33–272
  δ-Tocopherol .................. 690–1013
  Total, mg/kg .................. 732–1111

**References**  *Rev. Franc. Corps Gras 39:* 135 (1992)
  *Rev. Franc. Corps Gras 36:* 279 (1989)
  *J. Am. Oil Chem. Soc. 71:* 117 (1994)
  *Rev. Franc. Corps Gras 39:* 339 (1992)
  *J. Am. Oil Chem. Soc. 65:* 979 (1988)
  *Lipids 31:* 1311 (1996)

# Borage Oil (Dwarf)
*Borago pygmea*
Specific Gravity (SG)
  15.5/15.5°C
  25/25°C
  Other SG

Refractive Index (RI)
 25°C
 40°C
 Other RI
Iodine Value .................. 148–151
Saponification Value ............ 190–196
Titer °C
% Unsaponifiable ............... 0.9–1.1
Melting Point °C

**Fatty Acid Composition (%)**
 16:0 ...................... 10.6–10.8
 9c-16:1 ........................ 0.2
 18:0 ........................ 3.8–4.2
 Total 18:1 .................. 12.9–15.3
 9c,12c-18:2 .................... 34.2
 6c,9c,12c-18:3 ............. 25.1–27.9
 9c,12c,15c-18:3 .............. 0.9–1.3
 20:0 ........................ 0.2–0.4
 Total 20:1 ................... 2.9–3.7
 22:0 .......................... 0.1
 Unidentified 22:1 ............. 0.6–3.0
 15c-24:1 .................... 1.4–2.3

**References** *Rev. Franc. Corps Gras 39:* 135 (1992)

# Borneo Tallow
*Shorea stenoptera*

Specific Gravity (SG)
 15.5/15.5°C
 25/25°C
 Other SG ......... (100/15) 0.852–0.860
Refractive Index (RI)
 25°C
 40°C ................... 1.456–1.457
 Other RI
Iodine Value .................... 29–38
Saponification Value ............ 189–200
Titer °C ........................ 51–53
% Unsaponifiable ............... 0.4–2.0
Melting Point °C ................. 37–39

**Fatty Acid Composition (%)**
 16:0 ........................ 18–21
 18:0 ........................ 39–43
 Total 18:1 .................... 34–37

 9c,12c-18:2 .................... 0.2
 20:0 .......................... 1.0

**References**

# Brachyandra Calophylla Seed Oil
*Brachyandra calophylla*

Specific Gravity (SG)
 15.5/15.5°C
 25/25°C
 Other SG
Refractive Index (RI)
 25°C
 40°C
 Other RI
Iodine Value
Saponification Value
Titer °C
% Unsaponifiable
Melting Point °C

**Fatty Acid Composition (%)**
 8:0 ........................... 0.1
 10:0 ........................... 11
 12:0 ........................... 77
 14:0 ............................ 4
 16:0 ............................ 2
 18:0 .......................... 0.1
 Total 18:1 ...................... 2
 9c,12c-18:2 ..................... 3
 Undefined 18:3 ................. 0.3

**References** *Crit. Rev. Food Sci. Nutr. 28:* 139 (1989)

# Brachystegia Nigerica
*Brachystegia nigerica*

Specific Gravity (SG)
 15.5/15.5°C
 25/25°C
 Other SG
Refractive Index (RI)
 25°C
 40°C

Other RI .................. (27) 1.4641
Iodine Value ....................... 9.79
Saponification Value .............. 145.9
Titer °C
% Unsaponifiable ................. 14.4
Melting Point °C

**Fatty Acid Composition (%)**
    12:0......................... 0.24
    14:0......................... 0.45
    16:0........................ 13.18
    18:0........................ 19.80
    9c-18:1 ..................... 20.84
    9c,12c-18:2.................. 43.65
    20:0......................... 1.14
    5c,8c,11c,14c-20:4 ........... 0.90

**References** *JAOCS 68:* 649 (1991)

22:2.......................... 0.52
24:0.......................... 0.39

Tocopherol Composition, mg/kg
  α-Tocopherol ................... 140
  β-Tocopherol
  γ-Tocopherol.................... 415
  δ-Tocopherol..................... 11
  Total, mg/kg

Tocotrienols Composition, mg/kg
  α-Tocotrienol .................... 1
  β-Tocotrienol
  γ-Tocotrienol
  δ-Tocotrienol
  Total Tocotrienols, mg/kg

**References** *JAOCS 80:* 1013–1020 (2003)

## Brassica Chinensis Seed Oil
*Brassica chinensis*

Specific Gravity (SG)
  15.5/15.5°C
  25/25°C
  Other SG
Refractive Index (RI)
  25°C
  40°C
  Other RI
Iodine Value
Saponification Value
Titer °C
% Unsaponifiable
Melting Point °C

**Fatty Acid Composition (%)**
    14:0......................... 0.03
    16:0......................... 1.99
    9c-16:1 ...................... 0.16
    18:0......................... 1.22
    9c-18:1 ..................... 18.13
    7c-18:1 ...................... 1.02
    9c,12c-18:2.................. 11.66
    9c,12c,15c-18:3 .............. 7.13
    20:0......................... 0.91
    11c-20:1 ..................... 7.24
    22:0......................... 1.19
    13c-22:1 .................... 44.20

## Brassica Oleracea Seed Oil
*Brassica oleracea*

Specific Gravity (SG)
  15.5/15.5°C
  25/25°C
  Other SG
Refractive Index (RI)
  25°C
  40°C
  Other RI
Iodine Value
Saponification Value
Titer °C
% Unsaponifiable
Melting Point °C

**Fatty Acid Composition (%)**
    14:0......................... 0.04
    16:0......................... 3.64
    11c-16:1 ..................... 0.15
    18:0......................... 0.74
    9c-18:1 ..................... 16.55
    7c-18:1 ...................... 1.3
    9c,12c-18:2.................. 11.86
    9c,12c,15c-18:3 .............. 8.16
    20:0......................... 0.49
    11c-20:1 ..................... 9.08
    22:0......................... 0.41
    13c-22:1 .................... 42.05

22:2 .......................... 0.45
24:0 .......................... 0.29

Tocopherol Composition, mg/kg
   α-Tocopherol .................. 130
   β-Tocopherol..................... 1
   γ-Tocopherol................... 240
   δ-Tocopherol..................... 8
   Total, mg/kg

Tocotrienols Composition, mg/kg
   α-Tocotrienol .................... 2
   β-Tocotrienol
   γ-Tocotrienol
   δ-Tocotrienol
   Total Tocotrienols, mg/kg

**References** *JAOCS 80:* 1013–1020 (2003)

## Brazil Nut Oil
*Bertholletia excelsia, Nobilis, B. myrtaceae*

Specific Gravity (SG)
   15.5/15.5°C
   25/25°C................. 0.910–0.912
   Other SG
Refractive Index (RI)
   25°C .................. 1.464–1.468
   40°C .................. 1.458–1.462
   Other RI
Iodine Value .................. 97–106
Saponification Value ........... 192–202
Titer °C ....................... 29–32
% Unsaponifiable ................. 0–1
Melting Point °C

**Fatty Acid Composition (%)**
   14:0 .......................... 0.6
   16:0 ......................... 14–16
   9c-16:1 ....................... 0.3
   18:0 ......................... 6–10
   Total 18:1 .................... 29–48
   9c,12c-18:2................... 30–47
   20:0 .......................... 0.3

Sterol Composition, %
   Cholesterol ..................... 1
   Brassicasterol

Campesterol ...................... 2
Stigmasterol ..................... 9
Stigmasta-8,22-dien-3β-ol
5α-Stigmasta-7,22-dien-3β-ol
Δ7,25-Stigmastadienol
β-Sitosterol .................... 85
Δ5-Avenasterol
Δ7-Stigmasterol .................. 2
Δ7-Avenasterol
Δ7-Campesterol
Δ7-Ergosterol
Δ7,25-Stigmasterol
Sitostanol
Spinasterol
Squalene
24-Methylene Cholesterol
Other
% sterols in oil
Total Sterols, mg/kg

**References** *Riv. Ital. Sost. Grasse 52:* 79 (1975)
*J. Food Technol. 13:* 355 (1978)

## Brunfelsia Americana Seed Oil
*Brunfelsia americana*

Specific Gravity (SG)
   15.5/15.5°C
   25/25°C
   Other SG
Refractive Index (RI)
   25°C
   40°C
   Other RI
Iodine Value ..................... 124
Saponification Value ............. 198
Titer °C
% Unsaponifiable ................. 2.2
Melting Point °C

**Fatty Acid Composition (%)**
   14:0 .......................... 2.1
   16:0 .......................... 9.7
   18:0 ............................ 5
   9c-18:1 ...................... 16.9
   9c,12c-18:2................... 58.8
   Other: Ricinoleic, 5; Malvalic, 1.1;
      Sterculic, 1.4

**References** *JAOCS 68:* 608–609 (1991)

## Buchanania Lanzan Seed Oil
*Buchanania lanzan, B. latifolia*
Specific Gravity (SG)
  15.5/15.5°C
  25/25°C
  Other SG . . . . . . . . . . . . . (30/30) 0.9018
Refractive Index (RI)
  25°C
  40°C
  Other RI . . . . . . . . . . . . . . (30) 1.4620
Iodine Value . . . . . . . . . . . . . . . . . . . 57–63
Saponification Value . . . . . . . . . . . . . . . 193
Titer °C
% Unsaponifiable . . . . . . . . . . . . . . . . . . 0.7
Melting Point °C

**Fatty Acid Composition (%)**
  14:0 . . . . . . . . . . . . . . . . . . . . . . . . . . 0.6
  16:0 . . . . . . . . . . . . . . . . . . . . . . . . . . 33
  18:0 . . . . . . . . . . . . . . . . . . . . . . . . . . . 6
  Total 18:1 . . . . . . . . . . . . . . . . . . . . . 54
  9c,12c-18:2 . . . . . . . . . . . . . . . . . . . . . 6

**References** *J. Sci. Food Agric. 28:* 463 (1977)

## Buffalo Gourd Seed Oil
*Cucurbita foetidissima*
Specific Gravity (SG)
  15.5/15.5°C
  25/25°C . . . . . . . . . . . . . . . . . . . . . 0.9172
  Other SG
Refractive Index (RI)
  25°C . . . . . . . . . . . . . . . . 1.4692–1.4747
  40°C . . . . . . . . . . . . . . . . 1.4652–1.4686
  Other RI
Iodine Value . . . . . . . . . . . . . . . . 123–138
Saponification Value . . . . . . . . . . . 190–195
Titer °C
% Unsaponifiable
Melting Point °C

**Fatty Acid Composition (%)**
  16:0 . . . . . . . . . . . . . . . . . . . . . . . . . 6–24
  18:0 . . . . . . . . . . . . . . . . . . . . . . . . . 1–10
  Total 18:1 . . . . . . . . . . . . . . . . . . . 10–32
  Labellenic 5c,6c-18:2 (R)-form . . . . . 2–3
  9c,12c-18:2 . . . . . . . . . . . . . . . . . . 39–77

**References** *J. Am. Oil Chem. Soc. 57:* 310 (1980)
E.H. Pryde, *et al.*, eds., *New Sources Of Fats and Oils,* AOCS Press, Champaign, 1981, pp. 55

## Butternut Oil
*Juglans cinerea*
Specific Gravity (SG)
  15.5/15.5°C
  25/25°C
  Other SG
Refractive Index (RI)
  25°C
  40°C
  Other RI
Iodine Value
Saponification Value
Titer °C
% Unsaponifiable
Melting Point °C

**Fatty Acid Composition (%)**
  16:0 . . . . . . . . . . . . . . . . . . . . . . . . . . 1.6
  18:0 . . . . . . . . . . . . . . . . . . . . . . . . . . 0.8
  Total 18:1 . . . . . . . . . . . . . . . . . . . . . 19
  9c,12c-18:2 . . . . . . . . . . . . . . . . . . . . 62
  Undefined 18:3 . . . . . . . . . . . . . . . . . 16

**References** *J. Food Technol. 13:* 355 (1978)

## Calendula Seed Oil
*Calendula officinales*
Specific Gravity (SG)
  15.5/15.5°C
  25/25°C
  Other SG
Refractive Index (RI)
  25°C
  40°C
  Other RI

Iodine Value .................. 151–153
Saponification Value
Titer °C
% Unsaponifiable ................ 1–12
Melting Point °C

**Fatty Acid Composition (%)**
14:0 .......................... 0.5
16:0 .......................... 4.2
18:0 .......................... 2.0
Total 18:1 .................... 3.8
Undefined 18:2 ............... 28.5
Undefined 18:3 ................ 1.1
8t,10t, 12c-18:3 .............. 59.1
20:0 .......................... 0.4
Total 20:1 .................... 0.4

Tocopherol Composition, mg/kg
α-Tocopherol .................... 28
β-Tocopherol .................... 27
γ-Tocopherol .................. 1820
δ-Tocopherol .................... 36
Total, mg/kg .................. 1911

**References** *INFORM 12:* 468 (2001)

## California Laurel Seed Oil
*Umbellularia californica*
Specific Gravity (SG)
  15.5/15.5°C
  25/25°C
  Other SG
Refractive Index (RI)
  25°C ....................... 1.4533
  40°C
  Other RI
Iodine Value .................... 5–6
Saponification Value ............ 275
Titer °C
% Unsaponifiable ................. 2
Melting Point °C

**Fatty Acid Composition (%)**
10:0 ........................... 21
12:0 ........................... 70
14:0 ............................ 2
Total 18:1 ...................... 5

9c,12c-18:2 ..................... 2

**References** *Lipids 1:* 118 (1966)

## Cameline Oil (False Flax)
*Camelina sativa*
Specific Gravity (SG)
  15.5/15.5°C ............ 0.922–0.928
  25/25°C
  Other SG
Refractive Index (RI)
  25°C
  40°C
  Other RI ........... (20) 1.476–1.478
Iodine Value ................. 127–155
Saponification Value ......... 180–190
Titer °C
% Unsaponifiable
Melting Point °C

**Fatty Acid Composition (%)**
12:0 .......................... 0–0.1
14:0 .......................... 0–0.1
16:0 ............................ 5–6
9c-16:1 ......................... 0–2
18:0 ............................ 2–3
Total 18:1 .................... 12–24
9c,12c-18:2 ................... 15–16
Undefined 18:3 ................ 33–38
20:0 ............................. 1
Total 20:1 .................... 14–16
20:2 ............................. 2
Unidentified 20:3 ............... 1–2
22:0 .......................... 0.2–0.3
Unidentified 22:1 ................ 3
24:0 .......................... 0.2–0.3

**References** *INFORM 9:* 830 (1998)

## Camellia Oleifera Seed Oil
*Camellia oleifera*
Specific Gravity (SG)
  15.5/15.5°C
  25/25°C
  Other SG

Refractive Index (RI)
  25°C
  40°C
  Other RI
Iodine Value
Saponification Value
Titer °C
% Unsaponifiable
Melting Point °C

**Fatty Acid Composition (%)**
  14:0........................... 0.05
  16:0..................... 10–10.63
  9c-16:1 ....................... 0.11
  18:0...................... 2–3.48
  Total 18:1 ....................... 78
  9c-18:1 ....................... 77.89
  7c-18:1 ........................ 1.09
  9c,12c-18:2................ 5.03–9
  9c,12c,15c-18:3 ........... 0.17
  20:0...................... 0.07–9
  11c-20:1 ....................... 0.3
  22:0.......................... 0.43
  13c-22:1 ..................... 0.03
  22:2.......................... 0.32
  24:0.......................... 0.17

Tocopherol Composition, mg/kg
  α-Tocopherol ................... 107
  β-Tocopherol
  γ-Tocopherol
  δ-Tocopherol
  Total, mg/kg

**References**  *Acta Botanica Sinica 29:* 629 (1987)
  *JAOCS 80:* 1013–1020 (2003)

## Camellia Sinensis Seed Oil
*Camellia sinensis*
Specific Gravity (SG)
  15.5/15.5°C
  25/25°C
  Other SG
Refractive Index (RI)
  25°C
  40°C
  Other RI
Iodine Value
Saponification Value
Titer °C
% Unsaponifiable
Melting Point °C

**Fatty Acid Composition (%)**
  14:0............................ 14
  18:0............................. 2
  Total 18:1 ....................... 42
  9c,12c-18:2..................... 37
  20:0............................. 4

**References**  *Acta Botanica Sinica 29:* 629 (1987)

## Camphor Kernel Fat (Camphor Tree)
*Cinnamomum camphora*
Specific Gravity (SG)
  15.5/15.5°C
  25/25°C
  Other SG
Refractive Index (RI)
  25°C ....................... 1.4525
  40°C
  Other RI
Iodine Value ..................... 3–4
Saponification Value ............... 272
Titer °C
% Unsaponifiable
Melting Point °C

**Fatty Acid Composition (%)**
  10:0........................ 47–63
  12:0........................ 34–47
  14:0.......................... 0.7
  16:0.......................... 0.1
  Total 18:1 ..................... 1–3
  9c,12c-18:2................. 0.2–2

**References**  *Lipids 1:* 118 (1966)
  *Lipids 2:* 345 (1967)

## Canarium tramdenum
*Canarium tramdenum*
Specific Gravity (SG)
   15.5/15.5°C
   25/25°C
   Other SG
Refractive Index (RI)
   25°C
   40°C
   Other RI
Iodine Value
Saponification Value
Titer °C
% Unsaponifiable
Melting Point °C

**Fatty Acid Composition (%)**
   14:0 . . . . . . . . . . . . . . . . . . . . . . . 0.05
   16:0 . . . . . . . . . . . . . . . . . . . . . . . 25.19
   9c-16:1 . . . . . . . . . . . . . . . . . . . . 0.45
   18:0 . . . . . . . . . . . . . . . . . . . . . . . 5.69
   9c-18:1 . . . . . . . . . . . . . . . . . . . . 32.41
   7c-18:1 . . . . . . . . . . . . . . . . . . . . 0.64
   9c,12a-18:2 . . . . . . . . . . . . . . . . . 34
   9c,12c,15c-18:3 . . . . . . . . . . . . . . 0.43
   20:0 . . . . . . . . . . . . . . . . . . . . . . . 0.29
   11c-20:1 . . . . . . . . . . . . . . . . . . . 0.08
   22:0 . . . . . . . . . . . . . . . . . . . . . . . 0.13
   22:2 . . . . . . . . . . . . . . . . . . . . . . . 0.04
   24:0 . . . . . . . . . . . . . . . . . . . . . . . 0.09

Tocopherol Composition, mg/kg
   α-Tocopherol . . . . . . . . . . . . . . . . . . . . 51
   β-Tocopherol . . . . . . . . . . . . . . . . . . . . 45
   γ-Tocopherol . . . . . . . . . . . . . . . . . . . . 68
   δ-Tocopherol . . . . . . . . . . . . . . . . . . . . 939
   Total, mg/kg

**References** *JAOCS 80:* 1013–1020 (2003)

## Candlenut (Lumbang) Oil
*Aleurites moluccana*
Specific Gravity (SG)
   15.5/15.5°C . . . . . . . . . . . . . 0.924–0.929
   25/25°C
   Other SG
Refractive Index (RI)
   25°C . . . . . . . . . . . . . . . . . . . . 1.473–1.479
   40°C
   Other RI
Iodine Value . . . . . . . . . . . . . . . . . . 136–167
Saponification Value . . . . . . . . . . . 188–202
Titer °C
% Unsaponifiable . . . . . . . . . . . . . . . . . 0.3–1
Melting Point °C

**Fatty Acid Composition (%)**
   16:0 . . . . . . . . . . . . . . . . . . . . . . . . . . 5–9
   18:0 . . . . . . . . . . . . . . . . . . . . . . . . . . 3–7
   Total 18:1 . . . . . . . . . . . . . . . . . . . 11–35
   9c,12c-18:2 . . . . . . . . . . . . . . . . . . 37–49
   Undefined 18:3 . . . . . . . . . . . . . . . 24–35

**References**

## Cantaloupe Seed Oil
*Cucumis melo*
Specific Gravity (SG)
   15.5/15.5°C
   25/25°C
   Other SG
Refractive Index (RI)
   25°C . . . . . . . . . . . . . . . . . . . . . . . 1.4725
   40°C
   Other RI
Iodine Value . . . . . . . . . . . . . . . . . . 117–126
Saponification Value . . . . . . . . . . . 190–207
Titer °C
% Unsaponifiable . . . . . . . . . . . . . . . . . 0.5–1
Melting Point °C

**Fatty Acid Composition (%)**
   14:0 . . . . . . . . . . . . . . . . . . . . . . . . . . . . 2
   16:0 . . . . . . . . . . . . . . . . . . . . . . . . . . . . 2
   18:0 . . . . . . . . . . . . . . . . . . . . . . . . . . . . 5
   Total 18:1 . . . . . . . . . . . . . . . . . . . . . . 33
   9c,12c-18:2 . . . . . . . . . . . . . . . . . . . . . 56
   22:0 . . . . . . . . . . . . . . . . . . . . . . . . . . . . 1

**References**

## Cape Marigold Seed Oil
*Dimorphotheca pluvialis*

Specific Gravity (SG)
  15.5/15.5°C
  25/25°C...................... 0.905
  Other SG
Refractive Index (RI)
  25°C ....................... 1.4891
  40°C ....................... 1.4837
  Other RI
Iodine Value ..................... 167
Saponification Value
Titer °C
% Unsaponifiable
Melting Point °C

**Fatty Acid Composition (%)**
  16:0........................ 1.8–2
  18:0........................ 1.5–2
  Total 18:1 .................. 16–21
  9c,12c-18:2................. 11–12.4
  Undefined 18:3................ 0.6
  20:0......................... 0.9
  Total 20:1 .................. 0.4–1.1
  Other: D9-OH-10t,12t-octadecadienoic acid (dimorphecolic) 53–62

**References** *Indust. Crops Products 1:* 57 (1992)
*J. Am. Oil Chem. Soc. 74:* 277 (1997)

## Caraway Seed Oil
*Carum carvi*

Specific Gravity (SG)
  15.5/15.5°C
  25/25°C
  Other SG
Refractive Index (RI)
  25°C
  40°C
  Other RI ................. (35) 1.4710
Iodine Value ..................... 128
Saponification Value ................. 178
Titer °C
% Unsaponifiable .................. 2–3
Melting Point °C

**Fatty Acid Composition (%)**
  16:0............................ 3
  9c-18:1 ....................... 40
  6c-18:1 ....................... 26
  9c,12c-18:2.................... 30

**References**

## Carob Bean Oil
*Ceratonia siliqua*

Specific Gravity (SG)
  15.5/15.5°C................... 0.951
  25/25°C
  Other SG
Refractive Index (RI)
  25°C
  40°C ....................... 1.4691
  Other RI
Iodine Value .................. 98–99
Saponification Value ........... 198–205
Titer °C
% Unsaponifiable .................. 2.9
Melting Point °C

**Fatty Acid Composition (%)**
  16:0............................ 8
  18:0........................... 10
  Total 18:1 .................... 20
  9c,12c-18:2.................... 59
  Undefined 18:3................. 0.5
  24:0............................ 1

**References**

## Casca-de tatu Seed Oil
*Heisteria silvanii*

Specific Gravity (SG)
  15.5/15.5°C
  25/25°C
  Other SG
Refractive Index (RI)
  25°C
  40°C
  Other RI
Iodine Value

Saponification Value
Titer °C
% Unsaponifiable
Melting Point °C

**Fatty Acid Composition (%)**
16:0 . . . . . . . . . . . . . . . . . . . . . . . . . . . 3
9c-16:1 . . . . . . . . . . . . . . . . . . . . . . . 0.4
8a, 10t-17:2 . . . . . . . . . . . . . . . . . . . . 7
18:0 . . . . . . . . . . . . . . . . . . . . . . . . . . . 2
9,10 epoxy-18:0 . . . . . . . . . . . . . . . . 0.6
Total 18:1 . . . . . . . . . . . . . . . . . . . . . 47
Undefined 18:2 . . . . . . . . . . . . . . . . . . 1
9c,12c-18:2 . . . . . . . . . . . . . . . . . . . . 1
9a,11t-18:2 . . . . . . . . . . . . . . . . . . . . 3
Undefined 18:3 . . . . . . . . . . . . . . . . . . 2
7c, 9a,11t-18:3 . . . . . . . . . . . . . . . . . 23
20:0 . . . . . . . . . . . . . . . . . . . . . . . . . . . 1
Total 20:1 . . . . . . . . . . . . . . . . . . . . . . 1
24:0 . . . . . . . . . . . . . . . . . . . . . . . . . . . 1
26:0 . . . . . . . . . . . . . . . . . . . . . . . . . . . 4
28:0 . . . . . . . . . . . . . . . . . . . . . . . . . . . 1
30:0 . . . . . . . . . . . . . . . . . . . . . . . . . 0.3
Other: 8a,10t-17:2, 7: 9a;11t-18:2, 3;
  7c,9a,11t-18:3, 23; 9,10-epoxy-18:0,
  0.6; 9a,11a,13c-18:3, 0.4

**References** *Lipids 32:* 1189 (1997)

## Cashew Nut Oil
*Anarcadium occidentale*

Specific Gravity (SG)
  15.5/15.5°C . . . . . . . . . . . . . 0.911–0.918
  25/25°C
  Other SG
Refractive Index (RI)
  25°C
  40°C . . . . . . . . . . . . . . . . . . . 1.462–1.464
  Other RI
Iodine Value . . . . . . . . . . . . . . . . . . 79–89
Saponification Value . . . . . . . . . . . 180–196
Titer °C . . . . . . . . . . . . . . . . . . . . . 28–30
% Unsaponifiable . . . . . . . . . . . . . . 0.4–1.5
Melting Point °C

**Fatty Acid Composition (%)**
16:0 . . . . . . . . . . . . . . . . . . . . . . . 4–17
16:1 . . . . . . . . . . . . . . . . . . . . . . 0.3–0.4

9c-16:1 . . . . . . . . . . . . . . . . . . . . 0.3–0.5
17:0 . . . . . . . . . . . . . . . . . . . . . . . tr-0.2
18:0 . . . . . . . . . . . . . . . . . . . . . . . 2–11.6
Total 18:1 . . . . . . . . . . . . . . . . . . . 57–80
Undefined 18:2 . . . . . . . . . . . . 15.6–20.58
9c,12c-18:2 . . . . . . . . . . . . . . . . . . 16–22
Undefined 18:3 . . . . . . . . . . . . . . . . tr-0.3
20:0 . . . . . . . . . . . . . . . . . . . . . . . 0.3–0.8

Sterol Composition, %
  Cholesterol . . . . . . . . . . . . . . . . . 0.3–1.3
  Brassicasterol
  Campesterol . . . . . . . . . . . . . . . . . . 6–9
  Stigmasterol . . . . . . . . . . . . . . . . . . tr-2
  Stigmasta-8,22-dien-3β-ol
  5α-Stigmasta-7,22-dien-3β-ol
  Δ7,25-Stigmastadienol . . . . . . . . . . . 1.3
  β-Sitosterol . . . . . . . . . . . . . . . . . 75–83
  Δ5-Avenasterol . . . . . . . . . . . . . . 6–10.6
  Δ7-Stigmasterol . . . . . . . . . . . . . . . . 0.3
  Δ7-Avenasterol . . . . . . . . . . . . . . . . 0.4
  Δ7-Campesterol
  Δ7-Ergosterol
  Δ7,25-Stigmasterol
  Sitostanol
  Spinasterol
  Squalene
  24-Methylene Cholesterol
  Other: Fucosterol . . . . . . . . . . . . 0.6–0.8
  % sterols in oil
  Total Sterols, mg/kg . . . . . . . . . . . . 1840

Tocopherol Composition, mg/kg
  α-Tocopherol . . . . . . . . . . . . . . . . 28–75
  β-Tocopherol
  γ-Tocopherol . . . . . . . . . . . . . . . 450–835
  δ-Tocopherol . . . . . . . . . . . . . . . . 20–60
  Total, mg/kg . . . . . . . . . . . . . . . 600–950

**References** *Fat Sci. Technol. 91:* 23 (1989)
  *J. Am. Oil Chem. Soc. 70:* 1017 (1993)
  *JAOCS 74:* 375–380 (1997)
  *JAOCS 75:* 807–811 (1998)

## Cassia Alata (Ringworm Shrub)
*Cassia alata*

Specific Gravity (SG)
  15.5/15.5°C . . . . . . . . . . . . . . . . . 0.8898

25/25°C
Other SG
Refractive Index (RI)
25°C ........................ 1.4681
40°C
Other RI
Iodine Value ....................... 91
Saponification Value ............... 165
Titer °C
% Unsaponifiable ................... 4
Melting Point °C

**Fatty Acid Composition (%)**
12:0 ........................ 0.6–3
14:0 ........................... 2–4
16:0 ......................... 10–30
9c-16:1 ......................... 1
18:0 ............................ 5
Total 18:1 ................... 13–37
9c,12c-18:2 .................. 38–47
Undefined 18:3 .................. 1
20:0 ............................ 2
Total 20:1 .................... 0.2
22:0 ............................ 1
24:0 .......................... 0.6

Sterol Composition, %
  Cholesterol
  Brassicasterol
  Campesterol ...................... 6
  Stigmasterol .................... 21
  Stigmasta-8,22-dien-3β-ol
  5α-Stigmasta-7,22-dien-3β-ol
  Δ7,25-Stigmastadienol
  β-Sitosterol .................... 33
  Δ5-Avenasterol ................... 2
  Δ7-Stigmasterol
  Δ7-Avenasterol
  Δ7-Campesterol
  Δ7-Ergosterol
  Δ7,25-Stigmasterol
  Sitostanol
  Spinasterol
  Squalene
  24-Methylene Cholesterol
  Other Fucosterol, 3; 25(27)-dihydrochon-
    drillasterol, 3;22-dihydrospinasterol,
    20; 28-Isoavenasterol, 5
% sterols in oil
Total Sterols, mg/kg

**References** *Rev. Franc. Corps Gras 33:* 382 (1986)
*Food Chem. 30:* 205 (1988)

## Cassia Occidentalis (Wild Coffee)
*Cassia occidentalis*

Specific Gravity (SG)
  15.5/15.5°C
  25/25°C
  Other SG
Refractive Index (RI)
  25°C
  40°C
  Other RI
Iodine Value ..................... 114
Saponification Value ............. 179
Titer °C
% Unsaponifiable ................... 7
Melting Point °C

**Fatty Acid Composition (%)**
14:0 .......................... 0.3
16:0 ........................... 20
9c-16:1 ....................... 0.2
18:0 ........................... 1–2
Total 18:1 ...................... 16
9c,12c-18:2 ..................... 54
Undefined 18:3 ................... 5
20:0 .......................... 0.5
22:0 .......................... 0.7
24:0 .......................... 0.3

Sterol Composition, %
  Cholesterol
  Brassicasterol
  Campesterol .................... 11
  Stigmasterol ................... 32
  Stigmasta-8,22-dien-3β-ol
  5α-Stigmasta-7,22-dien-3β-ol
  Δ7,25-Stigmastadienol
  β-Sitosterol ................... 22
  Δ5-Avenasterol ................ 1–2
  Δ7-Stigmasterol
  Δ7-Avenasterol
  Δ7-Campesterol
  Δ7-Ergosterol
  Δ7,25-Stigmasterol

Sitostanol
Spinasterol
Squalene
24-Methylene Cholesterol
Other: Fucosterol, 1–2; 25(27)-dihydrochondrillasterol, 6;22-dihydrospinasterol, 16; 28-Isoavenasterol, 5
% sterols in oil
Total Sterols, mg/kg

**References** *Rev. Franc. Corps Gras 33:* 382 (1986)

## Cassia Siamea Seed Oil
*Cassia siamea*
Specific Gravity (SG)
  15.5/15.5°C
  25/25°C
  Other SG
Refractive Index (RI)
  25°C
  40°C
  Other RI
Iodine Value ...................... 101
Saponification Value ............... 197
Titer °C
% Unsaponifiable .................... 7
Melting Point °C

**Fatty Acid Composition (%)**
  16:0 ............................ 19
  18:0 ............................. 8
  Total 18:1 ...................... 12
  9c,12c-18:2 ..................... 43
  Other: Vernolic, 14; Malvalic, 2; Sterculic, 3

**References** *J. Am. Oil Chem. Soc. 65:* 952 (1993)

## Cassia Siberiana Seed Oil
*Cassia siberiana*
Specific Gravity (SG)
  15.5/15.5°C
  25/25°C
  Other SG
Refractive Index (RI)
  25°C
  40°C
  Other RI
Iodine Value
Saponification Value
Titer °C
% Unsaponifiable
Melting Point °C

**Fatty Acid Composition (%)**
  12:0 ........................... 0.7
  14:0 ............................. 1
  16:0 ............................ 16
  9c-16:1 ........................ 0.5
  18:0 ............................. 4
  Total 18:1 ...................... 32
  9c,12c-18:2 ..................... 43
  Undefined 18:3 ................... 1
  20:0 ........................... 1–2
  Total 20:1 ..................... 0.5
  22:0 ........................... 0.6
  24:0 ........................... 0.4

Sterol Composition, %
  Cholesterol
  Brassicasterol
  Campesterol .................... 11
  Stigmasterol ................... 22
  Stigmasta-8,22-dien-3β-ol
  5α-Stigmasta-7,22-dien-3β-ol
  Δ7,25-Stigmastadienol
  β-Sitosterol ................... 61
  Δ5-Avenasterol
  Δ7-Stigmasterol
  Δ7-Avenasterol
  Δ7-Campesterol
  Δ7-Ergosterol
  Δ7,25-Stigmasterol
  Sitostanol
  Spinasterol
  Squalene
  24-Methylene Cholesterol
  Other: Fucosterol, 4
% sterols in oil
Total Sterols, mg/kg

**References** *Rev. Franc. Corps Gras 33:* 382 (1986)

## Castor Oil
*Ricinus communis*
Specific Gravity (SG)
  15.5/15.5°C . . . . . . . . . . . . . 0.956–0.970
  25/25°C . . . . . . . . . . . . . . . . 0.945–0.965
  Other SG
Refractive Index (RI)
  25°C . . . . . . . . . . . . . . . . . . . 1.473–1.477
  40°C . . . . . . . . . . . . . . . . . . . 1.466–1.473
  Other RI
Iodine Value . . . . . . . . . . . . . . . . . . . . 81–91
Saponification Value . . . . . . . . . . . . 176–187
Titer °C
% Unsaponifiable
Melting Point °C

**Fatty Acid Composition (%)**
  16:0 . . . . . . . . . . . . . . . . . . . . . . . . . . . 1–2
  18:0 . . . . . . . . . . . . . . . . . . . . . . . . . 0.9–2
  Total 18:1 . . . . . . . . . . . . . . . . . . . . 2.9–6
  9c,12c-18:2 . . . . . . . . . . . . . . . . . . . . . 3–5
  Undefined 18:3 . . . . . . . . . . . . . . . . 0–0.5
  22:0 . . . . . . . . . . . . . . . . . . . . . . . . . . . 2.1
  Other: Ricinoleic, 88; dihydroxystearic, 1

Sterol Composition, %
  Cholesterol
  Brassicasterol
  Campesterol . . . . . . . . . . . . . . . . . . . . . 10
  Stigmasterol . . . . . . . . . . . . . . . . . . . . . 22
  Stigmasta-8,22-dien-3β-ol
  5α-Stigmasta-7,22-dien-3β-ol
  Δ7,25-Stigmastadienol
  β-Sitosterol . . . . . . . . . . . . . . . . . . . 44–56
  Δ5-Avenasterol . . . . . . . . . . . . . . . . 11–21
  Δ7-Stigmasterol . . . . . . . . . . . . . . . . . 0–2
  Δ7-Avenasterol . . . . . . . . . . . . . . . . . . . . 1
  Δ7-Campesterol
  Δ7-Ergosterol
  Δ7,25-Stigmasterol
  Sitostanol
  Spinasterol
  Squalene
  24-Methylene Cholesterol
  Other
  % sterols in oil
  Total Sterols, mg/kg

**References** *J. Am Oil Chem. Soc. 24:* 27 (1947)
*J. Am. Oil Chem. Soc. 34:* 513 (1962)
*Prog. Lipid Res. 22:* 161 (1983)
*Riv. Ital. Sost. Grasse 62:* 375 (1985)
*J. Am. Oil Chem. Soc. 74:* 277 (1997)

## Cay-Cay Fat
*Irvingia oliveri*
Specific Gravity (SG)
  15.5/15.5°C
  25/25°C
  Other SG . . . . . . . . . . . . . .(40/40) 0.9133
Refractive Index (RI)
  25°C
  40°C
  Other RI
Iodine Value . . . . . . . . . . . . . . . . . . . . . 6–7
Saponification Value . . . . . . . . . . . . . . . 235
Titer °C
% Unsaponifiable
Melting Point °C . . . . . . . . . . . . . . . . . . 40

**Fatty Acid Composition (%)**
  12:0 . . . . . . . . . . . . . . . . . . . . . . . . . . . . 39
  14:0 . . . . . . . . . . . . . . . . . . . . . . . . . . . . 56
  Total 18:1 . . . . . . . . . . . . . . . . . . . . . . . . 5

**References**

## Celastrus Orbiculatus Seed Oil
*Celastrus orbiculatus*
Specific Gravity (SG)
  15.5/15.5°C
  25/25°C
  Other SG
Refractive Index (RI)
  25°C
  40°C
  Other RI
Iodine Value
Saponification Value
Titer °C

% Unsaponifiable
Melting Point °C

**Fatty Acid Composition (%)**
14:0 . . . . . . . . . . . . . . . . . . . . . . . . . 0.2
16:0 . . . . . . . . . . . . . . . . . . . . . . . . . .21
9c-16:1 . . . . . . . . . . . . . . . . . . . . . . . 0.2
17:0 . . . . . . . . . . . . . . . . . . . . . . . . . 0.1
18:0 . . . . . . . . . . . . . . . . . . . . . . . . . . .4
Total 18:1 . . . . . . . . . . . . . . . . . . . . . .9
9c,12c-18:2 . . . . . . . . . . . . . . . . . . . .31
Undefined 18:3 . . . . . . . . . . . . . . . . .30
20:0 . . . . . . . . . . . . . . . . . . . . . . . . . 0.5
Total 20:1 . . . . . . . . . . . . . . . . . . . . 0.6
22:0 . . . . . . . . . . . . . . . . . . . . . . . . . .tr
Unidentified 22:1 . . . . . . . . . . . . . . . 1.6
24:0 . . . . . . . . . . . . . . . . . . . . . . . . . 0.2

**References**  *Lipids 9:* 928 (1974)

# Celery Seed Oil
*Apium graveolens*

Specific Gravity (SG)
 15.5/15.5°C
 25/25°C
 Other SG
Refractive Index (RI)
 25°C
 40°C
 Other RI . . . . . . . . . . . . . . . . . (35) 1.4783
Iodine Value . . . . . . . . . . . . . . . . . . . . 95
Saponification Value . . . . . . . . . . . . . . 178
Titer °C
% Unsaponifiable . . . . . . . . . . . . . . . . . . 0.8
Melting Point °C

**Fatty Acid Composition (%)**
16:0 . . . . . . . . . . . . . . . . . . . . . . . . . . . .3
9c-18:1 . . . . . . . . . . . . . . . . . . . . . . . .26
6c-18:1 . . . . . . . . . . . . . . . . . . . . . . . .50
9c,12c-18:2 . . . . . . . . . . . . . . . . . . . .20

**References**

# Cherry Kernel Oil
*Prunus avium*

Specific Gravity (SG)
 15.5/15.5°C
 25/25°C
 Other SG
Refractive Index (RI)
 25°C
 40°C
 Other RI
Iodine Value . . . . . . . . . . . . . . . . . . . . 113
Saponification Value . . . . . . . . . . . . . . 192
Titer °C
% Unsaponifiable . . . . . . . . . . . . . . . . . 0.66
Melting Point °C

**Fatty Acid Composition (%)**
12:0 . . . . . . . . . . . . . . . . . . . . . . . . . . . .tr
14:0 . . . . . . . . . . . . . . . . . . . . . . . . .tr-0.3
16:0 . . . . . . . . . . . . . . . . . . . . . . . 6.8–15
16:1 . . . . . . . . . . . . . . . . . . . . . . . 0.3–0.6
9c-16:1 . . . . . . . . . . . . . . . . . . . . .0.4–0.6
18:0 . . . . . . . . . . . . . . . . . . . . . . . . 1–2.6
Total 18:1 . . . . . . . . . . . . . . . . . . 37–52.9
9c-18:1 . . . . . . . . . . . . . . . . . . . 23.9–37.5
Undefined 18:2 . . . . . . . . . . . . . . . . 35–45
9c,12a-18:2 . . . . . . . . . . . . . . . . . 40–48.9
9c,12c-18:2 . . . . . . . . . . . . . . . . . . .40–49
Undefined 18:3 . . . . . . . . . . . . . . . . . . 0.5
9c,12c,15c-18:3 . . . . . . . . . . . . . . . . . tr-1
α-Eleostearic 9c,11t,13t-18:3 . . . 9.9–13.2
20:0 . . . . . . . . . . . . . . . . . . . . . . . . .tr-1.4
Total 20:1 . . . . . . . . . . . . . . . . . . . . tr-0.5

Sterol Composition, %
 Cholesterol . . . . . . . . . . . . . . . . . . . . 1.7
 Brassicasterol . . . . . . . . . . . . . . . . . . 0.6
 Campesterol . . . . . . . . . . . . . . . . . . . 2.8
 Stigmasterol . . . . . . . . . . . . . . . . . . . 6.1
 Stigmasta-8,22-dien-3β-ol
 5α-Stigmasta-7,22-dien-3β-ol
 Δ7,25-Stigmastadienol
 β-Sitosterol . . . . . . . . . . . . . . . . . . . 77.3
 Δ5-Avenasterol . . . . . . . . . . . . . . . . . . 7

Δ7-Stigmasterol ................. 2.5
Δ7-Avenasterol .................. 1.8
Δ7-Campesterol
Δ7-Ergosterol
Δ7,25-Stigmasterol .............. 0.3
Sitostanol
Spinasterol
Squalene
24-Methylene Cholesterol
Other
% sterols in oil ................. 0.2
Total Sterols, mg/kg

**References**  *J. Am Oil Chem. Soc. 69:* 1224 (1992)
*Riv. Ital. Sost. Grasse 75:* 405 (1998)
*JAOCS 69:* 492–494 (1992)

## Cherry Kernel Oil
*Prunus cerasus*
Specific Gravity (SG)
   15.5/15.5°C
   25/25°C ................ 0.916–0.925
   Other SG
Refractive Index (RI)
   25°C
   40°C ................... 1.466–1.471
   Other RI
Iodine Value .................. 110–118
Saponification Value .......... 190–198
Titer °C
% Unsaponifiable .............. 0.4–0.7
Melting Point °C

**Fatty Acid Composition (%)**
   14:0 ........................... 0.2
   16:0 ........................... 4–9
   18:0 ........................... 2–3
   Total 18:1 .................... 35–49
   Undefined 18:2 ................ 40–45
   9c,12c-18:2 ................... 42–45
   α-Eleostearic 9c,11t,13t-18:3 .... 3–10

Sterol Composition, %
   Cholesterol .................... 0.5
   Brassicasterol
   Campesterol ...................... 8
   Stigmasterol ..................... 7

Stigmasta-8,22-dien-3β-ol
5α-Stigmasta-7,22-dien-3β-ol
Δ7,25-Stigmastadienol
β-Sitosterol ..................... 69
Δ5-Avenasterol .................... 9
Δ7-Stigmasterol ................... 2
Δ7-Avenasterol .................... 1
Δ7-Campesterol .................... 3
Δ7-Ergosterol
Δ7,25-Stigmasterol ................ 1
Sitostanol
Spinasterol
Squalene
24-Methylene Cholesterol
Other
% sterols in oil ................. 0.8
Total Sterols, mg/kg

**References**  *Palm Oil Tech. Bull. 2:* 8 (1996)
*Riv. Ital. Sost. Grasse 75:* 405 (1998)

## Chestnut Oil
*Castanea mollisima*
Specific Gravity (SG)
   15.5/15.5°C
   25/25°C
   Other SG
Refractive Index (RI)
   25°C
   40°C
   Other RI
Iodine Value
Saponification Value
Titer °C
% Unsaponifiable
Melting Point °C

**Fatty Acid Composition (%)**
   16:0 ........................... 15
   9c-16:1 ........................ 0.7
   18:0 ............................ 1
   Total 18:1 ..................... 54
   9c,12c-18:2 .................... 25
   Undefined 18:3 ................ 2–3
   Total 20:1 ...................... 1
   22:0 ........................... 0.2

**References**  *J. Food Technol. 13:* 355 (1978)

## Chia Oil
*Salvia hispanica*

Specific Gravity (SG)
   15.5/15.5°C
   25/25°C ....................... 0.9330
   Other SG
Refractive Index (RI)
   25°C ......................... 1.4812
   40°C ......................... 1.4753
   Other RI
Iodine Value .................. 191–199
Saponification Value ................ 192
Titer °C ........................... -15
% Unsaponifiable ................... 1.2
Melting Point °C

**Fatty Acid Composition (%)**
   16:0 ........................... 7–8
   18:0 ............................. 3
   Total 18:1 ..................... 4–7
   9c,12c-18:2 ................... 24–26
   Undefined 18:3 ................ 54–59

**References** *Lipids 2:* 371 (1967)


## Chickling Vetch Seed Oil
*Lathyrus sativus*

Specific Gravity (SG)
   15.5/15.5°C
   25/25°C
   Other SG
Refractive Index (RI)
   25°C
   40°C
   Other RI
Iodine Value
Saponification Value
Titer °C
% Unsaponifiable
Melting Point °C

**Fatty Acid Composition (%)**
   14:0 ......................... 0.4–0.8
   16:0 ........................... 7–9
   9c-16:1 ...................... 0.3–0.5
   18:0 .......................... 13–15
   Total 18:1 .................... 57–60
   9c,12c-18:2 ................... 14–15
   Undefined 18:3 ..................... 1
   20:0 ......................... 0.3–0.8
   22:0 ......................... 0.3–0.5

Sterol Composition, %
   Cholesterol .................. 1.1–1.6
   Brassicasterol
   Campesterol ................... 11–13
   Stigmasterol .................. 20–26
   Stigmasta-8,22-dien-3β-ol
   5α-Stigmasta-7,22-dien-3β-ol
   Δ7,25-Stigmastadienol
   β-Sitosterol .................. 54–58
   Δ5-Avenasterol
   Δ7-Stigmasterol .................... 4
   Δ7-Avenasterol
   Δ7-Campesterol
   Δ7-Ergosterol
   Δ7,25-Stigmasterol
   Sitostanol
   Spinasterol
   Squalene
   24-Methylene Cholesterol
   Other
   % sterols in oil
   Total Sterols, mg/kg

**References** *Riv. Ital. Sost. Grasse 71:* 567 (1994)


## Chinese Melon Seed Oil (Bitter Gourd)
*Momordica charantia*

Specific Gravity (SG)
   15.5/15.5°C
   25/25°C
   Other SG
Refractive Index (RI)
   25°C
   40°C
   Other RI
Iodine Value
Saponification Value
Titer °C
% Unsaponifiable
Melting Point °C

**Fatty Acid Composition (%)**
 14:0 . . . . . . . . . . . . . . . . . . . . . . . . . 0.02
 16:0 . . . . . . . . . . . . . . . . . . . . . . . . 1.6–2.1
 9c-16:1 . . . . . . . . . . . . . . . . . . . . . 0.1–0.2
 18:0 . . . . . . . . . . . . . . . . . . . . . . . . . 22–27
 Total 18:1 . . . . . . . . . . . . . . . . . . . 2.6–4.0
 9c-18:1 . . . . . . . . . . . . . . . . . . . . . . . . 3.72
 7c-18:1 . . . . . . . . . . . . . . . . . . . . . . . . 0.11
 9c,12c-18:2 . . . . . . . . . . . . . . . . . . . . . 3–5
 Undefined 18:3 . . . . . . . . . . . . . . 0.5–60.6
 9c,12c,15c-18:3 . . . . . . . . . . . . . . . . 0.07
 α-Eleostearic 9c,11t,13t-18:3 . . . . . 63–68
 20:0 . . . . . . . . . . . . . . . . . . . . . . . . . . . 0.57
 11c-20:1 . . . . . . . . . . . . . . . . . . . . . . . 0.34
 22:0 . . . . . . . . . . . . . . . . . . . . . . . . . . . 1.12
 22:2 . . . . . . . . . . . . . . . . . . . . . . . . . . . 0.06
 24:0 . . . . . . . . . . . . . . . . . . . . . . . . . . . 0.03

Tocopherol Composition, mg/kg
 α-Tocopherol . . . . . . . . . . . . . . . . . . . 398
 β-Tocopherol . . . . . . . . . . . . . . . . . . . . . 1
 γ-Tocopherol . . . . . . . . . . . . . . . . . . . 492
 δ-Tocopherol
 Total, mg/kg

Tocotrienols Composition, mg/kg
 α-Tocotrienol
 β-Tocotrienol
 γ-Tocotrienol . . . . . . . . . . . . . . . . . . . . 30
 δ-Tocotrienol
 Total Tocotrienols, mg/kg

**References** *J. Am. Oil Chem. Soc. 73:* 263 (1996)
 *JAOCS 80:* 1013–1020 (2003)

# Chinese Soapberry Seed Oil
*Sapindus mukorossi*

Specific Gravity (SG)
 15.5/15.5°C
 25/25°C . . . . . . . . . . . . . . . . . . . . . 0.9040
 Other SG
Refractive Index (RI)
 25°C
 40°C . . . . . . . . . . . . . . . . . . . . . . . . 1.4632
 Other RI . . . . . . . . . . . . . . . . (28) 1.4680

Iodine Value . . . . . . . . . . . . . . . . . . . . 78–80
Saponification Value . . . . . . . . . . . . . . . 197
Titer °C
% Unsaponifiable
Melting Point °C

**Fatty Acid Composition (%)**
 16:0 . . . . . . . . . . . . . . . . . . . . . . . . . . . 4–6
 9c-16:1 . . . . . . . . . . . . . . . . . . . . . . . . 0.5
 18:0 . . . . . . . . . . . . . . . . . . . . . . . . . 0.2–1
 Total 18:1 . . . . . . . . . . . . . . . . . . . . 54–63
 9c,12c-18:2 . . . . . . . . . . . . . . . . . . . 5–14
 Undefined 18:3 . . . . . . . . . . . . . . . . . 1–6
 20:0 . . . . . . . . . . . . . . . . . . . . . . . . . . . 4–6
 Total 20:1 . . . . . . . . . . . . . . . . . . . . 15–22

**References** *Lipids 10:* 33 (1975)
 *Fette Seifen Anstrichm. 73:* 639 (1971)

# Chinese Vegetable Tallow (Mesocap fat; Chinese Tallow Tree)
*Sapium sebiferum*

Specific Gravity (SG)
 15.5/15.5°C
 25/25°C . . . . . . . . . . . . . . . . . . . . . 0.890
 Other SG
Refractive Index (RI)
 25°C
 40°C . . . . . . . . . . . . . . . . . . 1.455–1.457
 Other RI
Iodine Value . . . . . . . . . . . . . . . . . . . 16–29
Saponification Value . . . . . . . . . . . 200–218
Titer °C
% Unsaponifiable . . . . . . . . . . . . . . 0.5–1.3
Melting Point °C . . . . . . . . . . . . . . . . 42–45

**Fatty Acid Composition (%)**
 12:0 . . . . . . . . . . . . . . . . . . . . . . . . . 0–2.5
 14:0 . . . . . . . . . . . . . . . . . . . . . . . . 0.5–3.7
 16:0 . . . . . . . . . . . . . . . . . . . . . . . . . 58–72
 18:0 . . . . . . . . . . . . . . . . . . . . . . . . . . . 1–8
 Total 18:1 . . . . . . . . . . . . . . . . . . . . 20–35
 9c,12c-18:2 . . . . . . . . . . . . . . . . . . . . 0–2

**References**

## Chirongi Oil
*Buchanania latifolia*
Specific Gravity (SG)
   15.5/15.5°C
   25/25°C
   Other SG
Refractive Index (RI)
   25°C
   40°C
   Other RI
Iodine Value . . . . . . . . . . . . . . . . . . . . . . . . 63
Saponification Value . . . . . . . . . . . . . . . . 193
Titer °C
% Unsaponifiable . . . . . . . . . . . . . . . . . . . 0.6
Melting Point °C

**Fatty Acid Composition (%)**
   14:0 . . . . . . . . . . . . . . . . . . . . . . . . 0.1–0.2
   16:0 . . . . . . . . . . . . . . . . . . . . . . . . . 29–31
   18:0 . . . . . . . . . . . . . . . . . . . . . . . . . . . . . 8
   Total 18:1 . . . . . . . . . . . . . . . . . . . . . 55–58
   Undefined 18:2 . . . . . . . . . . . . . . . . . . . 5–6

**References** *INFORM 13:* 151 (2002)

## Chrysanthemum coronarium
*Chrysanthemum coronarium*
Specific Gravity (SG)
   15.5/15.5°C
   25/25°C
   Other SG
Refractive Index (RI)
   25°C
   40°C
   Other RI
Iodine Value
Saponification Value
Titer °C
% Unsaponifiable
Melting Point °C

**Fatty Acid Composition (%)**
   14:0 . . . . . . . . . . . . . . . . . . . . . . . . . . 0.07
   16:0 . . . . . . . . . . . . . . . . . . . . . . . . . . 9.40
   9c-16:1 . . . . . . . . . . . . . . . . . . . . . . . 0.11
   18:0 . . . . . . . . . . . . . . . . . . . . . . . . . . 2.25

   9c-18:1 . . . . . . . . . . . . . . . . . . . . . . . . 3.91
   7c-18:1 . . . . . . . . . . . . . . . . . . . . . . . . 0.51
   9c,12c-18:2 . . . . . . . . . . . . . . . . . . . 77.75
   9c,12c,15c-18:3 . . . . . . . . . . . . . . . . . 0.14
   20:0 . . . . . . . . . . . . . . . . . . . . . . . . . . . 0.49
   11c-20:1 . . . . . . . . . . . . . . . . . . . . . . . 0.11
   22:0 . . . . . . . . . . . . . . . . . . . . . . . . . . . 0.24
   13c-22:1 . . . . . . . . . . . . . . . . . . . . . . . 0.03
   22:2 . . . . . . . . . . . . . . . . . . . . . . . . . . . 0.09

Tocopherol Composition, mg/kg
   α-Tocopherol . . . . . . . . . . . . . . . . . . . 929
   β-Tocopherol . . . . . . . . . . . . . . . . . . . . 49
   γ-Tocopherol . . . . . . . . . . . . . . . . . . . . 31
   δ-Tocopherol . . . . . . . . . . . . . . . . . . . . 31
   Total, mg/kg

Tocotrienols Composition, mg/kg
   α-Tocotrienol
   β-Tocotrienol
   γ-Tocotrienol . . . . . . . . . . . . . . . . . . . . 35
   δ-Tocotrienol . . . . . . . . . . . . . . . . . . . . 28
   Total Tocotrienols, mg/kg

**References** *JAOCS 80:* 1013–1020 (2003)

## Cimicifuga racemosa
*Cimicifuga racemosa*
Specific Gravity (SG)
   15.5/15.5°C
   25/25°C
   Other SG
Refractive Index (RI)
   25°C
   40°C
   Other RI
Iodine Value
Saponification Value
Titer °C
% Unsaponifiable
Melting Point °C

**Fatty Acid Composition (%)**
   12:0 . . . . . . . . . . . . . . . . . . . . . . . . . . . . tr
   14:0 . . . . . . . . . . . . . . . . . . . . . . . . . . . 0.1
   15:0 . . . . . . . . . . . . . . . . . . . . . . . . . . . . tr
   16:0 . . . . . . . . . . . . . . . . . . . . . . . . . . . 5.4

9c-16:1 . . . . . . . . . . . . . . . . . . . . . . tr
11c-16:1 . . . . . . . . . . . . . . . . . . . . . 1.1
17:0 . . . . . . . . . . . . . . . . . . . . . . . . . . . tr
18:0 . . . . . . . . . . . . . . . . . . . . . . . . . . 2.6
Total 18:1 . . . . . . . . . . . . . . . . . . . . 1.9
9c-18:1 . . . . . . . . . . . . . . . . . . . . . . . .7
9c,12c-18:2 . . . . . . . . . . . . . . . . . . . . 29
9c,12c,15c-18:3 . . . . . . . . . . . . . . . 8.2
5,9c,12c,15c-18:4 . . . . . . . . . . . . . . 0.9
19:0 . . . . . . . . . . . . . . . . . . . . . . . . . . 0.1
20:0 . . . . . . . . . . . . . . . . . . . . . . . . . . 1.8
11c-20:1 . . . . . . . . . . . . . . . . . . . . . 18.6
15c-20:1 . . . . . . . . . . . . . . . . . . . . . 0.4
20:2 . . . . . . . . . . . . . . . . . . . . . . 0.8–4.8
Unidentified 20:3 . . . . . . . . . . . . . . 2.2
5,11c,14c-20:3 . . . . . . . . . . . . . . . . 5.8
5c,11c,14c, 17–20:4 . . . . . . . . . . . . .8
22:0 . . . . . . . . . . . . . . . . . . . . . . . . . . 0.3
13c-22:1 . . . . . . . . . . . . . . . . . . . . . . tr
24:0 . . . . . . . . . . . . . . . . . . . . . . . . . . 0.1

**References** *JAOCS 75:* 1761–1765 (1998)

## Citrullus Colocynthis
*Citrullus colocynthis*
Specific Gravity (SG)
  15.5/15.5°C
  25/25°C
  Other SG
Refractive Index (RI)
  25°C
  40°C
  Other RI
Iodine Value
Saponification Value
Titer °C
% Unsaponifiable
Melting Point °C

**Fatty Acid Composition (%)**
14:0 . . . . . . . . . . . . . . . . . . . . . . . . . . 0.1
15:0 . . . . . . . . . . . . . . . . . . . . . . . . . . . tr
16:0 . . . . . . . . . . . . . . . . . . . . . . . . . 11.9
9c-16:1 . . . . . . . . . . . . . . . . . . . . . . 0.1
18:0 . . . . . . . . . . . . . . . . . . . . . . . . . 10.6
9c-18:1 . . . . . . . . . . . . . . . . . . . . . 13.5
9c,12c-18:2 . . . . . . . . . . . . . . . . . . 63.4
9c,12c,15c-18:3 . . . . . . . . . . . . . . . 0.1

20:0 . . . . . . . . . . . . . . . . . . . . . . . . . . 0.3
11c-20:1 . . . . . . . . . . . . . . . . . . . . . . . tr

**References** *JAOCS 69:* 314–316 (1992)

## Cloudberry Seed Oil
*Rubus chamaemorus*
Specific Gravity (SG)
  15.5/15.5°C
  25/25°C
  Other SG
Refractive Index (RI)
  25°C
  40°C
  Other RI
Iodine Value
Saponification Value
Titer °C
% Unsaponifiable
Melting Point °C

**Fatty Acid Composition (%)**
14:0 . . . . . . . . . . . . . . . . . . . . . . . . . . 0.4
16:0 . . . . . . . . . . . . . . . . . . . . . . . . . . 3.0
9c-16:1 . . . . . . . . . . . . . . . . . . . . . . 0.4
7c-16:1 . . . . . . . . . . . . . . . . . . . . . . 0.5
18:0 . . . . . . . . . . . . . . . . . . . . . . . . . 1–2
18:1 . . . . . . . . . . . . . . . . . . . . . . . . . . 16
9c-18:1 . . . . . . . . . . . . . . . . . . . . . . 15
18:2 . . . . . . . . . . . . . . . . . . . . . . 40–46
9c,12c,15c-18:3 . . . . . . . . . . . . . . . 32
20:0 . . . . . . . . . . . . . . . . . . . . . . . . . . 0.1
Total 20:1 . . . . . . . . . . . . . . . . . . . . 0.7
20:2 . . . . . . . . . . . . . . . . . . . . . . . . 4–5
22:0 . . . . . . . . . . . . . . . . . . . . . . . . . . 0.5
22:2 . . . . . . . . . . . . . . . . . . . . . . . . . . . 2

**References**

## Cocoa Butter
*Theobroma cocoa*
Specific Gravity (SG)
  15.5/15.5°C
  25/25°C . . . . . . . . . . . . . . . 0.973–0.980
  Other SG
Refractive Index (RI)

25°C
40°C . . . . . . . . . . . . . . . . . . 1.456–1.458
Other RI
Iodine Value . . . . . . . . . . . . . . . . . . . . 32–40
Saponification Value . . . . . . . . . . . . 192–200
Titer °C . . . . . . . . . . . . . . . . . . . . . . . 45–50
% Unsaponifiable . . . . . . . . . . . . . . . 0.2–1.0
Melting Point °C . . . . . . . . . . . . . . . . . 31–35

**Fatty Acid Composition (%)**
   14:0 . . . . . . . . . . . . . . . . . . . . . . . . . . . 0.1
   16:0 . . . . . . . . . . . . . . . . . . . . . . . . . . 25–27
   9c-16:1 . . . . . . . . . . . . . . . . . . . . . . 0.1–0.3
   17:0 . . . . . . . . . . . . . . . . . . . . . . . . . . . 0.1
   18:0 . . . . . . . . . . . . . . . . . . . . . . . . . 31–37
   Total 18:1 . . . . . . . . . . . . . . . . . . . . . 31–35
   9c,12c-18:2 . . . . . . . . . . . . . . . . . . . . 2.8–4.0
   Undefined 18:3 . . . . . . . . . . . . . . . . . . . 0.1
   20:0 . . . . . . . . . . . . . . . . . . . . . . . . . 0.2–1.0

Sterol Composition, %
   Cholesterol . . . . . . . . . . . . . . . . . . . . . . . 1
   Brassicasterol
   Campesterol . . . . . . . . . . . . . . . . . . . . 8–11
   Stigmasterol . . . . . . . . . . . . . . . . . . . 24–31
   Stigmasta-8,22-dien-3β-ol
   5α-Stigmasta-7,22-dien-3β-ol
   Δ7,25-Stigmastadienol
   β-Sitosterol . . . . . . . . . . . . . . . . . . . . 58–63
   Δ5-Avenasterol . . . . . . . . . . . . . . . . . . . 3–5
   Δ7-Stigmasterol . . . . . . . . . . . . . . . . . . . . 1
   Δ7-Avenasterol
   Δ7-Campesterol
   Δ7-Ergosterol
   Δ7,25-Stigmasterol
   Sitostanol
   Spinasterol
   Squalene
   24-Methylene Cholesterol
   Other
   % sterols in oil
   Total Sterols, mg/kg

Tocopherol Composition, mg/kg
   α-Tocopherol . . . . . . . . . . . . . . . . . . . 1–19
   β-Tocopherol . . . . . . . . . . . . . . . . . . . . 0–10
   γ-Tocopherol . . . . . . . . . . . . . . . . . . 18–196
   δ-Tocopherol . . . . . . . . . . . . . . . . . . . . 0–17
   Total, mg/kg . . . . . . . . . . . . . . . . . . 25–220

**References** *J. Am. Oil Chem. Soc. 62:* 1047 (1985)
*Lebensmittelchem. Gerichtl. Chem. 36:* 53 (1982)
*J. Am. Oil Chem. Soc.53:* 732 (1976)
*J. Am. Oil Chem. Soc. 73:* 1217 (1996)
*Fette Seifen Anstrichm. 87:* 150 (1985)
*Deutsche Lebensm. Rundschau 72:* 6 (1976)
*J. Am. Oil Chem. Soc. 64:* 100 (1987)

## Coconut Oil
*Cocos nucifera*

Specific Gravity (SG)
   15.5/15.5°C
   25/25°C
   Other SG . . . . . . . . . . (40/20) 0.908–0.921
Refractive Index (RI)
   25°C
   40°C . . . . . . . . . . . . . . . . . . 1.448–1.450
   Other RI
Iodine Value . . . . . . . . . . . . . . . . . . . . . 5–13
Saponification Value . . . . . . . . . . . . 248–265
Titer °C
% Unsaponifiable . . . . . . . . . . . . . . . . . 0–1.5
Melting Point °C . . . . . . . . . . . . . . . . . 23–26

**Fatty Acid Composition (%)**
   6:0 . . . . . . . . . . . . . . . . . . . . . . . . . . 0–0.6
   8:0 . . . . . . . . . . . . . . . . . . . . . . . . 0.91–9.4
   10:0 . . . . . . . . . . . . . . . . . . . . . . . . 3.78–7.8
   12:0 . . . . . . . . . . . . . . . . . . . . . . 45.1–50.92
   14:0 . . . . . . . . . . . . . . . . . . . . . . 16.8–21.09
   16:0 . . . . . . . . . . . . . . . . . . . . . . . . 7.7–10.2
   18:0 . . . . . . . . . . . . . . . . . . . . . . . . 2.3–4.88
   Total 18:1 . . . . . . . . . . . . . . . . . . . . . 5.4–9.9
   Undefined 18:2 . . . . . . . . . . . . . . . . . . 0.56
   9c,12c-18:2 . . . . . . . . . . . . . . . . . . . 0.8–2.1
   Undefined 18:3 . . . . . . . . . . . . . . . . . . 0–0.2
   20:0 . . . . . . . . . . . . . . . . . . . . . . . . . 0–0.2
   Total 20:1 . . . . . . . . . . . . . . . . . . . . . . 0–0.2

Sterol Composition, %
   Cholesterol . . . . . . . . . . . . . . . . . . 0.6–3.0
   Brassicasterol . . . . . . . . . . . . . . . . . . 0–0.9
   Campesterol . . . . . . . . . . . . . . . . . 3.1–11.2
   Stigmasterol . . . . . . . . . . . . . . . . . 5.4–15.6

Stigmasta-8,22-dien-3β-ol
5α-Stigmasta-7,22-dien-3β-ol
Δ7,25-Stigmastadienol
β-Sitosterol . . . . . . . . . . . . . . . 19.7–50.7
Δ5-Avenasterol . . . . . . . . . . . . . 13–40.7
Δ7-Stigmasterol . . . . . . . . . . . . . . 0–3.0
Δ7-Avenasterol . . . . . . . . . . . . . . . 0–3.0
Δ7-Campesterol
Δ7-Ergosterol
Δ7,25-Stigmasterol
Sitostanol
Spinasterol
Squalene
24-Methylene Cholesterol
Other . . . . . . . . . . . . . . . . . . . . . . . 0–3.6
% sterols in oil
Total Sterols, mg/kg . . . . . . . . . 470–1140

Tocopherol Composition, mg/kg
  α-Tocopherol . . . . . . . . . . . . . . . . . 0–17
  β-Tocopherol. . . . . . . . . . . . . . . . . 0–11
  γ-Tocopherol . . . . . . . . . . . . . . . . . 0–14
  δ-Tocopherol
  Total, mg/kg

Tocotrienols Composition, mg/kg
  α-Tocotrienol . . . . . . . . . . . . . . . . . 0–44
  β-Tocotrienol
  γ-Tocotrienol
  δ-Tocotrienol
  Total Tocotrienols, mg/kg . . . . . . . . 0–44

References  *Codex* CX 1993/16
  *JAOCS* 75: 807–811 (1998)


## Coffee Bean Oil (Roasted)
*Coffea arabica*
Specific Gravity (SG)
  15.5/15.5°C . . . . . . . . . . . . . 0.928–0.952
  25/25°C
  Other SG
Refractive Index (RI)
  25°C . . . . . . . . . . . . . . . . . . 1.468–1.477
  40°C
  Other RI
Iodine Value . . . . . . . . . . . . . . . . . . 78–96
Saponification Value . . . . . . . . . . . 165–195
Titer °C

% Unsaponifiable . . . . . . . . . . . . . . . . 6–10
Melting Point °C

Fatty Acid Composition (%)
  16:0 . . . . . . . . . . . . . . . . . . . . . . . . 30–32
  18:0 . . . . . . . . . . . . . . . . . . . . . . . . . 7–8
  Total 18:1 . . . . . . . . . . . . . . . . . . . . . 23
  9c,12c-18:2 . . . . . . . . . . . . . . . . . . . . 32

References  *J. Am. Oil Chem. Soc.* 45: 577
  (1968)
  *J. Am. Oil Chem. Soc.* 50: 122 (1973)


## Coffee Bean Oil (Raw, Brazil)
*Coffea arabica*
Specific Gravity (SG)
  15.5/15.5°C
  25/25°C
  Other SG
Refractive Index (RI)
  25°C . . . . . . . . . . . . . . . . . . . . . . 1.4790
  40°C
  Other RI
Iodine Value . . . . . . . . . . . . . . . . . . . . 100
Saponification Value . . . . . . . . . . . 184–195
Titer °C
% Unsaponifiable . . . . . . . . . . . . . . . . 8–11
Melting Point °C

Fatty Acid Composition (%)
  16:0 . . . . . . . . . . . . . . . . . . . . . . . . 35–42
  18:0 . . . . . . . . . . . . . . . . . . . . . . . . 7–11
  Total 18:1 . . . . . . . . . . . . . . . . . . . 8–10
  9c,12c-18:2 . . . . . . . . . . . . . . . . . . 36–43
  20:0 . . . . . . . . . . . . . . . . . . . . . . . . . 4–7
  Total 20:1 . . . . . . . . . . . . . . . . . . . . 4–7
  22:0 . . . . . . . . . . . . . . . . . . . . . . . . . 4–7
  Unidentified 22:1 . . . . . . . . . . . . . . . 4–7
  24:0 . . . . . . . . . . . . . . . . . . . . . . . . . 4–7
  15c-24:1 . . . . . . . . . . . . . . . . . . . . . 4–7

Sterol Composition, %
  Cholesterol
  Brassicasterol
  Campesterol . . . . . . . . . . . . . . . . . . . . 19
  Stigmasterol . . . . . . . . . . . . . . . . . . . . 20
  Stigmasta-8,22-dien-3β-ol
  5α-Stigmasta-7,22-dien-3β-ol

Δ7,25-Stigmastadienol
β-Sitosterol . . . . . . . . . . . . . . . . . . . . . . 54
Δ5-Avenasterol . . . . . . . . . . . . . . . . . . . . 6
Δ7-Stigmasterol . . . . . . . . . . . . . . . . . . . 1
Δ7-Avenasterol
Δ7-Campesterol
Δ7-Ergosterol
Δ7,25-Stigmasterol
Sitostanol
Spinasterol
Squalene
24-Methylene Cholesterol
Other
% sterols in oil
Total Sterols, mg/kg

**References**

## Cohune Nut Oil (Palm Oil)
*Specific Gravity (SG)*
15.5/15.5°C
25/25°C . . . . . . . . . . . . . . . . . 0.916–0.918
Other SG . . . . . . . . . (199/15) 0.868–0.871
Refractive Index (RI)
25°C
40°C . . . . . . . . . . . . . . . . . . . 1.449–1.450
Other RI
Iodine Value . . . . . . . . . . . . . . . . . . . 9–14
Saponification Value . . . . . . . . . . . 251–260
Titer °C
% Unsaponifiable . . . . . . . . . . . . . . . 0.2–0.5
Melting Point °C

**Fatty Acid Composition (%)**
8:0 . . . . . . . . . . . . . . . . . . . . . . . . . . . 7–9
10:0 . . . . . . . . . . . . . . . . . . . . . . . . . . 6–8
12:0 . . . . . . . . . . . . . . . . . . . . . . . . 44–48
14:0 . . . . . . . . . . . . . . . . . . . . . . . . 16–17
16:0 . . . . . . . . . . . . . . . . . . . . . . . . . 7–10
18:0 . . . . . . . . . . . . . . . . . . . . . . . . . . 3–4
Total 18:1 . . . . . . . . . . . . . . . . . . . . . 8–10
9c,12c-18:2 . . . . . . . . . . . . . . . . . . . . . . 1

**References** *J. Am. Dietetic Assn. 68:* 224 (1976)

## Coincya Transtagana
*Coincya transtagana*
Specific Gravity (SG)
15.5/15.5°C
25/25°C
Other SG
Refractive Index (RI)
25°C
40°C
Other RI
Iodine Value
Saponification Value
Titer °C
% Unsaponifiable
Melting Point °C

**Fatty Acid Composition (%)**
16:0 . . . . . . . . . . . . . . . . . . . . . . . . . . . . 5
16:1 . . . . . . . . . . . . . . . . . . . . . . . . . . . 0.4
18:0 . . . . . . . . . . . . . . . . . . . . . . . . . . . 1.4
Total 18:1 . . . . . . . . . . . . . . . . . . . . . 12.3
Undefined 18:2 . . . . . . . . . . . . . . . . . 17.5
Undefined 18:3 . . . . . . . . . . . . . . . . . 25.4
20:0 . . . . . . . . . . . . . . . . . . . . . . . . . . . 0.9
Total 20:1 . . . . . . . . . . . . . . . . . . . . . . 5.2
20:2 . . . . . . . . . . . . . . . . . . . . . . . . . . . . 1
21:1 . . . . . . . . . . . . . . . . . . . . . . . . . . . 0.6
22:0 . . . . . . . . . . . . . . . . . . . . . . . . . . . 0.8
Unidentified 22:1 . . . . . . . . . . . . . . . 28.6
22:2 . . . . . . . . . . . . . . . . . . . . . . . . . . . 0.8

**References** *JAOCS 70:* 1157–1158 (1993)

## Coincya Longirostra
*Coincya longirostra*
Specific Gravity (SG)
15.5/15.5°C
25/25°C
Other SG
Refractive Index (RI)
25°C
40°C
Other RI
Iodine Value

Saponification Value
Titer °C
% Unsaponifiable
Melting Point °C

**Fatty Acid Composition (%)**
   16:0 . . . . . . . . . . . . . . . . . . . . . . . . 3.8
   16:1 . . . . . . . . . . . . . . . . . . . . . . . . 0.3
   18:0 . . . . . . . . . . . . . . . . . . . . . . . . 1.4
   Total 18:1 . . . . . . . . . . . . . . . . . . 14.2
   Undefined 18:2 . . . . . . . . . . . . . . . 14.8
   Undefined 18:3 . . . . . . . . . . . . . . . 27.7
   20:0 . . . . . . . . . . . . . . . . . . . . . . . . 1.1
   Total 20:1 . . . . . . . . . . . . . . . . . . . 6.0
   20:2 . . . . . . . . . . . . . . . . . . . . . . . . 1.2
   21:1 . . . . . . . . . . . . . . . . . . . . . . . . 0.6
   22:0 . . . . . . . . . . . . . . . . . . . . . . . . 0.9
   Unidentified 22:1 . . . . . . . . . . . . . . 27.6
   22:2 . . . . . . . . . . . . . . . . . . . . . . . . 0.9

**References** *JAOCS 70:* 1157–1158 (1993)

## Coincya Rupestris
*Coincya rupestris*

Specific Gravity (SG)
   15.5/15.5°C
   25/25°C
   Other SG
Refractive Index (RI)
   25°C
   40°C
   Other RI
Iodine Value
Saponification Value
Titer °C
% Unsaponifiable
Melting Point °C

**Fatty Acid Composition (%)**
   16:0 . . . . . . . . . . . . . . . . . . . . . . . . 3–4.2
   16:1 . . . . . . . . . . . . . . . . . . . . . . . . 0.4–0.7
   18:0 . . . . . . . . . . . . . . . . . . . . . . . . 1.5–1.6
   Total 18:1 . . . . . . . . . . . . . . . . . . 13.5–14.4
   Undefined 18:2 . . . . . . . . . . . . . . . 14.3–16.3
   Undefined 18:3 . . . . . . . . . . . . . . . 24.6–25.7
   20:0 . . . . . . . . . . . . . . . . . . . . . . . . 0.9–1.1
   Total 20:1 . . . . . . . . . . . . . . . . . . . 5.9–6.1
   20:2 . . . . . . . . . . . . . . . . . . . . . . . . 0.5–1.2
   21:1 . . . . . . . . . . . . . . . . . . . . . . . . 0.5–0.8
   22:0 . . . . . . . . . . . . . . . . . . . . . . . . 0.7–0.9
   Unidentified 22:1 . . . . . . . . . . . . . . 29.8–30.5
   22:2 . . . . . . . . . . . . . . . . . . . . . . . . 0.4–0.9

**References** *JAOCS 70:* 1157–1158 (1993)

## Coincya Monensis
*Coincya monensis*

Specific Gravity (SG)
   15.5/15.5°C
   25/25°C
   Other SG
Refractive Index (RI)
   25°C
   40°C
   Other RI
Iodine Value
Saponification Value
Titer °C
% Unsaponifiable
Melting Point °C

**Fatty Acid Composition (%)**
   16:0 . . . . . . . . . . . . . . . . . . . . . . . . 3.5–4.3
   16:1 . . . . . . . . . . . . . . . . . . . . . . . . 0.4–0.8
   18:0 . . . . . . . . . . . . . . . . . . . . . . . . 1.7–1.9
   Total 18:1 . . . . . . . . . . . . . . . . . . 14.5–21.8
   Undefined 18:2 . . . . . . . . . . . . . . . 13.9–24.6
   Undefined 18:3 . . . . . . . . . . . . . . . 17.7–24.4
   20:0 . . . . . . . . . . . . . . . . . . . . . . . . 0.9–1.2
   Total 20:1 . . . . . . . . . . . . . . . . . . . 6.2–8.7
   20:2 . . . . . . . . . . . . . . . . . . . . . . . . 0.6–1.2
   21:1 . . . . . . . . . . . . . . . . . . . . . . . . 0.3–0.6
   22:0 . . . . . . . . . . . . . . . . . . . . . . . . 0.5–1.7
   Unidentified 22:1 . . . . . . . . . . . . . . 24.6–28.6
   22:2 . . . . . . . . . . . . . . . . . . . . . . . . 0.3–0.7

**References** *JAOCS 70:* 1157–1158 (1993)

## Comphrey Seed Oil
*Symphytum officinale*

Specific Gravity (SG)
   15.5/15.5°C
   25/25°C
   Other SG
Refractive Index (RI)

25°C
40°C
Other RI
Iodine Value
Saponification Value
Titer °C
% Unsaponifiable
Melting Point °C

**Fatty Acid Composition (%)**
14:0 .......................... 0.4
16:0 .......................... 6.7
18:0 .......................... 1.5
Total 18:1 .................... 17.1
9c,12c-18:2 ................... 44.1
6c,9c,12c-18:3 ................ 25.8
9c,12c,15c-18:3 ............... 2.4
6c,9c,12c,15c-18:4 ............ 1.2

**References** *J. Sci. Food Agric. 54:* 309 (1991)

## Connarus paniculatus
*Connarus paniculatus*
Specific Gravity (SG)
15.5/15.5°C
25/25°C
Other SG
Refractive Index (RI)
25°C
40°C
Other RI
Iodine Value
Saponification Value
Titer °C
% Unsaponifiable
Melting Point °C

**Fatty Acid Composition (%)**
14:0 .......................... 0.2
16:0 .......................... 25.21
9c-16:1 ....................... 0.10
18:0 .......................... 4.01
9c-18:1 ....................... 30.05
7c-18:1 ....................... 0.62
9c,12c-18:2 ................... 37.87
9c,12c,15c-18:3 ............... 0.5
20:0 .......................... 0.23

11c-20:1 ...................... 0.29
22:0 .......................... 0.25

Tocopherol Composition, mg/kg
α-Tocopherol .................. 355
β-Tocopherol .................. 65
γ-Tocopherol .................. 61
δ-Tocopherol
Total, mg/kg

**References** *JAOCS 80:* 1013–1020 (2003)

## Corchorus olitorius
*Corchorus olitorius*
Specific Gravity (SG)
15.5/15.5°C
25/25°C
Other SG
Refractive Index (RI)
25°C
40°C
Other RI
Iodine Value
Saponification Value
Titer °C
% Unsaponifiable
Melting Point °C

**Fatty Acid Composition (%)**
14:0 .......................... 0.08
16:0 .......................... 14.08
9c-16:1 ....................... 0.18
18:0 .......................... 2.82
9c-18:1 ....................... 9.58
7c-18:1 ....................... 1.18
9c,12c-18:2 ................... 66.39
9c,12c,15c-18:3 ............... 1.96
20:0 .......................... 0.88
11c-20:1 ...................... 0.26
22:0 .......................... 1.25

Tocopherol Composition, mg/kg
α-Tocopherol .................. 397
β-Tocopherol .................. 38
γ-Tocopherol .................. 1237
δ-Tocopherol .................. 32
Total, mg/kg

**References** *JAOCS 80:* 1013–1020 (2003)

## Cordia Rothii Seed Oil

*Specific Gravity (SG)*
  15.5/15.5°C
  25/25°C
  Other SG
Refractive Index (RI)
  25°C
  40°C
  Other RI
Iodine Value . . . . . . . . . . . . . . . . . . . . . . 90
Saponification Value
Titer °C
% Unsaponifiable . . . . . . . . . . . . . . . . . . . 2
Melting Point °C

**Fatty Acid Composition (%)**
  14:0. . . . . . . . . . . . . . . . . . . . . . . . . . . . 2
  16:0. . . . . . . . . . . . . . . . . . . . . . . . . . . 33
  18:0. . . . . . . . . . . . . . . . . . . . . . . . . . . . 2
  Total 18:1 . . . . . . . . . . . . . . . . . . . . . . 8
  Undefined 18:2 . . . . . . . . . . . . . . . . . 40
  Other: Ricinoleic, 11; Malvalic 2–3;
    Sterculic 1–2

**References** *J. Sci. Food Agric. 58:* 285 (1992)


## Coriander Seed Oil
*Coriandrum sativum*

Specific Gravity (SG)
  15.5/15.5°C
  25/25°C. . . . . . . . . . . . . . . . . . . . . 0.9110
  Other SG
Refractive Index (RI)
  25°C . . . . . . . . . . . . . . . . . . . . . . . 1.4635
  40°C
  Other RI . . . . . . . . . . . . . . . . (30) 1.4704
Iodine Value . . . . . . . . . . . . . . . . . . 86–100
Saponification Value . . . . . . . . . . . 182–191
Titer °C
% Unsaponifiable . . . . . . . . . . . . . . . . 1–2
Melting Point °C

**Fatty Acid Composition (%)**
  12:0. . . . . . . . . . . . . . . . . . . . . . . . . . 0.1
  14:0. . . . . . . . . . . . . . . . . . . . . 0.05–0.8
  16:0. . . . . . . . . . . . . . . . . . . . . . . 2.91–8

  9c-16:1 . . . . . . . . . . . . . . . . . . . . 0.19–0.4
  18:0. . . . . . . . . . . . . . . . . . . . . . . . 0.51–2
  Total 18:1 . . . . . . . . . . . . . . . . . . . . . 32
  6c-18:1 . . . . . . . . . . . . . . . . . . . . . . . 53
  7c-18:1 . . . . . . . . . . . . . . . . . . . . . . 0.82
  9c,12c-18:2 . . . . . . . . . . . . . . . 7–14.23
  9c,12c,15c-18:3 . . . . . . . . . . . . . . . 0.2
  20:0. . . . . . . . . . . . . . . . . . . . . . . . 0.07
  11c-20:1 . . . . . . . . . . . . . . . . . . . . 0.31
  22:0. . . . . . . . . . . . . . . . . . . . . . . . 0.06
  22:2. . . . . . . . . . . . . . . . . . . . . . . . 0.06
  24:0. . . . . . . . . . . . . . . . . . . . . . . . 0.07

Tocopherol Composition, mg/kg
  α-Tocopherol . . . . . . . . . . . . . . . . . . 46
  β-Tocopherol
  γ-Tocopherol . . . . . . . . . . . . . . . . . . 31
  δ-Tocopherol
  Total, mg/kg

Tocotrienols Composition, mg/kg
  α-Tocotrienol . . . . . . . . . . . . . . . . . . 96
  β-Tocotrienol
  γ-Tocotrienol. . . . . . . . . . . . . . . . . 231
  δ-Tocotrienol. . . . . . . . . . . . . . . . . . 41
  Total Tocotrienols, mg/kg

**References** *Bangladesh J. Sci. Ind. Res. 17:* 172 (1982)
*Bangladesh J. Sci. Ind. Res 26:* 33 (1993)
*JAOCS 80:* 1013–1020 (2003)


## Corn Oil (High Oleic)

Specific Gravity (SG)
  15.5/15.5°C
  25/25°C
  Other SG
Refractive Index (RI)
  25°C
  40°C
  Other RI
Iodine Value
Saponification Value
Titer °C
% Unsaponifiable
Melting Point °C

**Fatty Acid Composition (%)**
  16:0. . . . . . . . . . . . . . . . . . . . . . . . 10–16

18:0 . . . . . . . . . . . . . . . . . . . . . . . . . . . 2
Total 18:1 . . . . . . . . . . . . . . . . . . . . 44–64
9c,12c-18:2 . . . . . . . . . . . . . . . . . . . 20–38
Undefined 18:3 . . . . . . . . . . . . . . . 0.8–1.0
20:0 . . . . . . . . . . . . . . . . . . . . . . . . . . . 1

**References** *J. Am. Oil Chem. Soc. 72:* 989 (1995)

## Corn Oil (Maize)
*Zea mays*

Specific Gravity (SG)
   15.5/15.5°C
   25/25°C
   Other SG . . . . . . . . . . (20/20) 0.917–0.925
Refractive Index (RI)
   25°C . . . . . . . . . . . . . . . . . . . 1.470–1.473
   40°C . . . . . . . . . . . . . . . . . . . 1.465–1.468
   Other RI
Iodine Value . . . . . . . . . . . . . . . . . . 107–135
Saponification Value . . . . . . . . . . . 187–195
Titer °C
% Unsaponifiable . . . . . . . . . . . . . . . . . . 1–3
Melting Point °C

**Fatty Acid Composition (%)**
   12:0 . . . . . . . . . . . . . . . . . . . . . . . . . 0–0.3
   14:0 . . . . . . . . . . . . . . . . . . . . . . . . . 0–0.3
   16:0 . . . . . . . . . . . . . . . . . . . . . . . 9.2–16.5
   16:1 . . . . . . . . . . . . . . . . . . . . . . . . . . . . 0
   9c-16:1 . . . . . . . . . . . . . . . . . . . . . . . 0–0.4
   18:0 . . . . . . . . . . . . . . . . . . . . . . . . . 0–3.3
   Total 18:1 . . . . . . . . . . . . . . . . . . . 20–42.2
   Undefined 18:2 . . . . . . . . . . . . . . . . . . 44.7
   9c,12c-18:2 . . . . . . . . . . . . . . . . . . 39.4–65.6
   Undefined 18:3 . . . . . . . . . . . . . . . . 0.5–1.5
   20:0 . . . . . . . . . . . . . . . . . . . . . . . . 0.3–0.7
   Total 20:1 . . . . . . . . . . . . . . . . . . . . . 0–0.4
   20:2 . . . . . . . . . . . . . . . . . . . . . . . . . 0–0.1
   22:0 . . . . . . . . . . . . . . . . . . . . . . . . . 0–0.5
   Unidentified 22:1 . . . . . . . . . . . . . . . 0–0.1
   24:0 . . . . . . . . . . . . . . . . . . . . . . . . . 0–0.4

Sterol Composition, %
   Cholesterol . . . . . . . . . . . . . . . . . . 0.2–0.6
   Brassicasterol . . . . . . . . . . . . . . . . . 0–0.2
   Campesterol . . . . . . . . . . . . . . 18.6–24.1
   Stigmasterol . . . . . . . . . . . . . . . . 4.3–7.7
   Stigmasta-8,22-dien-3β-ol
   5α-Stigmasta-7,22-dien-3β-ol
   Δ7,25-Stigmastadienol
   β-Sitosterol . . . . . . . . . . . . . . . . 54.8–66.6
   Δ5-Avenasterol . . . . . . . . . . . . . . . 4.2–8.2
   Δ7-Stigmasterol . . . . . . . . . . . . . . 1.0–4.2
   Δ7-Avenasterol . . . . . . . . . . . . . . . 0.7–2.7
   Δ7-Campesterol
   Δ7-Ergosterol
   Δ7,25-Stigmasterol
   Sitostanol
   Spinasterol
   Squalene
   24-Methylene Cholesterol
   Other . . . . . . . . . . . . . . . . . . . . . . . . 0–2.4
% sterols in oil
Total Sterols, mg/kg . . . . . . . 7950–22150

Tocopherol Composition, mg/kg
   α-Tocopherol . . . . . . . . . . . . . . . . 23–573
   β-Tocopherol . . . . . . . . . . . . . . . . . . 0–356
   γ-Tocopherol . . . . . . . . . . . . . . . . 268–2468
   δ-Tocopherol . . . . . . . . . . . . . . . . . . 23–75
   Total, mg/kg . . . . . . . . . . . . . . . 331–3716

Tocotrienols Composition, mg/kg
   α-Tocotrienol . . . . . . . . . . . . . . . . . 0–239
   β-Tocotrienol
   γ-Tocotrienol . . . . . . . . . . . . . . . . . 0–450
   δ-Tocotrienol . . . . . . . . . . . . . . . . . . 0–20
   Total Tocotrienols, mg/kg . . . . . . . . 0–709

**References** *Codex* 1997/17
*JAOCS 74:* 375–380 (1997)

## Corn Oil (Low Saturate)

Specific Gravity (SG)
   15.5/15.5°C
   25/25°C
   Other SG
Refractive Index (RI)
   25°C
   40°C
   Other RI
Iodine Value
Saponification Value
Titer °C
% Unsaponifiable
Melting Point °C

**Fatty Acid Composition (%)**
  16:0 . . . . . . . . . . . . . . . . . . . . . . . . . 6–8
  18:0 . . . . . . . . . . . . . . . . . . . . . . . . . . 1
  Total 18:1 . . . . . . . . . . . . . . . . . . . . 25–31
  9c,12c-18:2 . . . . . . . . . . . . . . . . . . . 58–64
  Undefined 18:3 . . . . . . . . . . . . . . . 0.8–0.9
  20:0 . . . . . . . . . . . . . . . . . . . . . . . . . 0.5

**References** *J. Am. Oil Chem. Soc.* 72: 989 (1995)

# Cottonseed Oil
*Gossypium spp.*

Specific Gravity (SG)
  15.5/15.5°C
  25/25°C
  Other SG . . . . . . . . . . (20/20) 0.918–0.926
Refractive Index (RI)
  25°C
  40°C . . . . . . . . . . . . . . . . . . . 1.458–1.466
  Other RI
Iodine Value . . . . . . . . . . . . . . . . . . . 96–115
Saponification Value . . . . . . . . . . . . 189–198
Titer °C
% Unsaponifiable . . . . . . . . . . . . . . . . . . 0–2
Melting Point -1°C

**Fatty Acid Composition (%)**
  12:0 . . . . . . . . . . . . . . . . . . . . . . . . 0–0.2
  14:0 . . . . . . . . . . . . . . . . . . . . . . . . 0.6–1.0
  16:0 . . . . . . . . . . . . . . . . . . . . . . 18.1–26.4
  16:1 . . . . . . . . . . . . . . . . . . . . . . . . 0–0.7
  9c-16:1 . . . . . . . . . . . . . . . . . . . . . . 0–1.2
  18:0 . . . . . . . . . . . . . . . . . . . . . . . . 2.1–3.3
  Total 18:1 . . . . . . . . . . . . . . . . . . 14.2–21.7
  Undefined 18:2 . . . . . . . . . . . . . . 55.6–61.6
  9c,12c-18:2 . . . . . . . . . . . . . . . . . 46.7–58.3
  Undefined 18:3 . . . . . . . . . . . . . . . . 0–0.4
  20:0 . . . . . . . . . . . . . . . . . . . . . . . . 0.2–0.5
  Total 20:1 . . . . . . . . . . . . . . . . . . . . 0–0.1
  20:2 . . . . . . . . . . . . . . . . . . . . . . . . 0–0.1
  22:0 . . . . . . . . . . . . . . . . . . . . . . . . 0–0.6
  Unidentified 22:1 . . . . . . . . . . . . . . 0–0.3
  22:2 . . . . . . . . . . . . . . . . . . . . . . . . 0–0.1
  24:0 . . . . . . . . . . . . . . . . . . . . . . . . 0–0.1

Sterol Composition, %
  Cholesterol . . . . . . . . . . . . . . . . . . 0.7–2.3
  Brassicasterol . . . . . . . . . . . . . . . . 0.1–0.9
  Campesterol . . . . . . . . . . . . . . . . 6.4–14.5
  Stigmasterol . . . . . . . . . . . . . . . . . 2.1–6.8
  Stigmasta-8,22-dien-3β-ol
  5α-Stigmasta-7,22-dien-3β-ol
  Δ7,25-Stigmastadienol
  β-Sitosterol . . . . . . . . . . . . . . . . 76.0–87.1
  Δ5-Avenasterol . . . . . . . . . . . . . . . 1.8–7.3
  Δ7-Stigmasterol . . . . . . . . . . . . . . . 0–1.4
  Δ7-Avenasterol . . . . . . . . . . . . . . . 0.8–3.3
  Δ7-Campesterol
  Δ7-Ergosterol
  Δ7,25-Stigmasterol
  Sitostanol
  Spinasterol
  Squalene
  24-Methylene Cholesterol
  Other . . . . . . . . . . . . . . . . . . . . . . . . 0–1.5
% sterols in oil
Total Sterols, mg/kg . . . . . . . . 2690–6430

Tocopherol Composition, mg/kg
  α-Tocopherol . . . . . . . . . . . . . . . 136–674
  β-Tocopherol. . . . . . . . . . . . . . . . . . 0–30
  γ-Tocopherol. . . . . . . . . . . . . . . . 138–750
  δ-Tocopherol. . . . . . . . . . . . . . . . . . 0–20
  Total, mg/kg . . . . . . . . . . . . . . 390–1430

Tocotrienols Composition, mg/kg
  α-Tocotrienol . . . . . . . . . . . . . . . . . 0–30
  β-Tocotrienol
  γ-Tocotrienol. . . . . . . . . . . . . . . . . . 0–30
  δ-Tocotrienol
  Total Tocotrienols, mg/kg

**References** *Codex* 1997/17
  *JAOCS* 74: 375–380 (1997)
  *JAOCS* 68: 518–519 (1991)

# Couepia longipendula
*Couepia longipendula*

Specific Gravity (SG)
  15.5/15.5°C
  25/25°C
  Other SG

Refractive Index (RI)
 25°C
 40°C
 Other RI
Iodine Value
Saponification Value
Titer °C
% Unsaponifiable
Melting Point °C

**Fatty Acid Composition (%)**
 16:0 . . . . . . . . . . . . . . . . . . . . . . . . . 25.2
 9c-16:1 . . . . . . . . . . . . . . . . . . . . . . . 0.9
 18:0 . . . . . . . . . . . . . . . . . . . . . . . . . . 6.2
 9c-18:1 . . . . . . . . . . . . . . . . . . . . . . . 26.5
 11c-18:1 . . . . . . . . . . . . . . . . . . . . . . 0.4
 9c,12c-18:2 . . . . . . . . . . . . . . . . . . . . 7.4
 α-Eleostearic 9c,11t,13t-18:3 . . 11.3–21.8
 20:0 . . . . . . . . . . . . . . . . . . . . . . . . . . 0.3

**References** *JAOCS 68:* 440 (1991)

# Cowpea Oil
*Vigna unguiculata*
Specific Gravity (SG)
 15.5/15.5°C
 25/25°C
 Other SG
Refractive Index (RI)
 25°C
 40°C
 Other RI
Iodine Value
Saponification Value
Titer °C
% Unsaponifiable
Melting Point °C

**Fatty Acid Composition (%)**
 16:0 . . . . . . . . . . . . . . . . . . . . . . . 25–37
 18:0 . . . . . . . . . . . . . . . . . . . . . . . 5.9–8.0
 9c-18:1 . . . . . . . . . . . . . . . . . . . . 7.3–16.4
 9c,12c-18:2 . . . . . . . . . . . . . . . . 18.5–25.5
 Undefined 18:3 . . . . . . . . . . . . . 20–29.7

**References** *J. Sci. Food Agric. 78:* 1 (1998)

# Crambe Oil
*Crambe abyssinica*
Specific Gravity (SG)
 15.5/15.5°C
 25/25°C . . . . . . . . . . . . . . . . . 0.908–0.910
 Other SG
Refractive Index (RI)
 25°C . . . . . . . . . . . . . . . . . . . . . . . 1.4700
 40°C . . . . . . . . . . . . . . . . . . . 1.4648–1.466
 Other RI
Iodine Value . . . . . . . . . . . . . . . . . . . 87–113
Saponification Value
Titer °C
% Unsaponifiable
Melting Point °C

**Fatty Acid Composition (%)**
 16:0 . . . . . . . . . . . . . . . . . . . . . . . . . . . . 2
 9c-16:1 . . . . . . . . . . . . . . . . . . . . . . . 0.3
 18:0 . . . . . . . . . . . . . . . . . . . . . . . . . . . 1
 Total 18:1 . . . . . . . . . . . . . . . . . . . 12–15
 9c,12c-18:2 . . . . . . . . . . . . . . . . . . . 8–10
 Undefined 18:3 . . . . . . . . . . . . . . . . . 6–7
 20:0 . . . . . . . . . . . . . . . . . . . . . . . . . 1–2
 Total 20:1 . . . . . . . . . . . . . . . . . . . . 3–4
 22:0 . . . . . . . . . . . . . . . . . . . . . . . . . 0.2
 13c-22:1 . . . . . . . . . . . . . . . . . . . . 55–60
 22:2 . . . . . . . . . . . . . . . . . . . . . . . . . . . 1
 15c-24:1 . . . . . . . . . . . . . . . . . . . . . . . 1

**References** *J. Am. Oil Chem. Soc. 43:* 330 (1966)
*Indust. Crops Prod. 1:* 57 (1992)

# Crepis Alpina Seed Oil
*Crepis alpina*
Specific Gravity (SG)
 15.5/15.5°C
 25/25°C
 Other SG
Refractive Index (RI)
 25°C
 40°C
 Other RI
Iodine Value

Saponification Value
Titer °C
% Unsaponifiable
Melting Point °C

**Fatty Acid Composition (%)**
14:0 . . . . . . . . . . . . . . . . . . . . . . . . . . . 1
16:0 . . . . . . . . . . . . . . . . . . . . . . . . . . . 4
18:0 . . . . . . . . . . . . . . . . . . . . . . . . . . . 1
Total 18:1 . . . . . . . . . . . . . . . . . . . . . 2
Undefined 18:2 . . . . . . . . . . . . . . . . . 14
9c,12t-18:2 . . . . . . . . . . . . . . . . . . . . . 2
Other . . . . . . . . . . . . . . . . crepenynic, 74

**References** *JAOCS 70:* 817 (1993)
*J. Liq. Chromatogr. 18:* 4165 (1995)

## Crotalaria Juncea Seed Oil
*Crotalaria juncea*
Specific Gravity (SG)
15.5/15.5°C
25/25°C
Other SG
Refractive Index (RI)
25°C
40°C
Other RI
Iodine Value
Saponification Value
Titer °C
% Unsaponifiable
Melting Point °C

**Fatty Acid Composition (%)**
8:0 . . . . . . . . . . . . . . . . . . . . . . . . . . . . 6
10:0 . . . . . . . . . . . . . . . . . . . . . . . . . . . 6
12:0 . . . . . . . . . . . . . . . . . . . . . . . . . . . 6
13:0 . . . . . . . . . . . . . . . . . . . . . . . . . . . 6
14:0 . . . . . . . . . . . . . . . . . . . . . . . . . . . 6
15:0 . . . . . . . . . . . . . . . . . . . . . . . . . . . 4
16:0 . . . . . . . . . . . . . . . . . . . . . . . . . . 19
9c-16:1 . . . . . . . . . . . . . . . . . . . . . . . . . 4
17:0 . . . . . . . . . . . . . . . . . . . . . . . . . . . 1
18:0 . . . . . . . . . . . . . . . . . . . . . . . . . . . 5
Total 18:1 . . . . . . . . . . . . . . . . . . . . . 12
9c,12c-18:2 . . . . . . . . . . . . . . . . . . . . . 6
Undefined 18:3 . . . . . . . . . . . . . . . . . 3

20:0 . . . . . . . . . . . . . . . . . . . . . . . . . . 0.7
22:0 . . . . . . . . . . . . . . . . . . . . . . . . . . . 7
24:0 . . . . . . . . . . . . . . . . . . . . . . . . . . . 5
Other . . . . . . . . . . . . . . . . . . . . . . . . . 2.5

**References** *Fat Sci. Technol. 97:* 457 (1995)

## Croton Seed Oil
*Croton tiglium*
Specific Gravity (SG)
15.5/15.5°C
25/25°C
Other SG . . . . . . . . . . (15/15) 0.935–0.960
Refractive Index (RI)
25°C
40°C . . . . . . . . . . . . . . . . 1.4760–1.4730
Other RI
Iodine Value . . . . . . . . . . . . . . . . . 102–115
Saponification Value . . . . . . . . . . 200–215
Titer °C
% Unsaponifiable . . . . . . . . . . . . . . 0.3–0.9
Melting Point °C

**Fatty Acid Composition (%)**
16:0 . . . . . . . . . . . . . . . . . . . . . . . . . . . 1
18:0 . . . . . . . . . . . . . . . . . . . . . . . . . . 0.5
Total 18:1 . . . . . . . . . . . . . . . . . . . . . 56
9c,12c-18:2 . . . . . . . . . . . . . . . . . . . . 29

**References**

## Cryptolepis Buchnani
*Cryptolepis buchnani*
Specific Gravity (SG)
15.5/15.5°C
25/25°C
Other SG
Refractive Index (RI)
25°C
40°C
Other RI
Iodine Value . . . . . . . . . . . . . . . . . . 53.28
Saponification Value . . . . . . . . . . . . 190.44
Titer °C
% Unsaponifiable . . . . . . . . . . . . . . . . 1.7
Melting Point °C

**Fatty Acid Composition (%)**
    16:0 .......................... 30.9
    18:0 ........................... 6.5
    9c-18:1 ........................ 5.5
    Undefined 18:3 ................. 7.4
    22:0 ........................... 0.8
    24:0 ............................. 3
    Other ............... keto acid, 45.9

**References** *JAOCS 69:* 188 (1992)

## Cryptostegia Grandiflora Seed Oil
*Cryptostegia grandiflora*
Specific Gravity (SG)
    15.5/15.5°C
    25/25°C
    Other SG
Refractive Index (RI)
    25°C
    40°C
    Other RI
Iodine Value ..................... 114
Saponification Value ............. 280
Titer °C
% Unsaponifiable ................... 1
Melting Point °C

**Fatty Acid Composition (%)**
    16:0 ............................. 6
    18:0 ............................. 4
    Total 18:1 ...................... 52
    9c,12c-18:2 ..................... 34
    Undefined 18:3 ................... 2
    20:0 ........................... 0.6
    22:0 ........................... 0.9

**References** *Fette Seifen Anstrichm. 86:* 167 (1984)

## Cucumeropsis Edulis Seed Oil
*Cucumeropsis edulis*
Specific Gravity (SG)
    15.5/15.5°C
    25/25°C ...................... 0.9080
    Other SG
Refractive Index (RI)
    25°C ......................... 1.4622
    40°C
    Other RI
Iodine Value ...................... 95
Saponification Value ............. 186
Titer °C
% Unsaponifiable ................... 1
Melting Point °C

**Fatty Acid Composition (%)**
    16:0 ............................ 15
    18:0 ............................. 9
    Total 18:1 ....................... 1
    9c,12c-18:2 ..................... 12
    Undefined 18:3 .................. 62
    20:0 ........................... 0.1

**References** *Riv. Ital. Sost. Grasse 75:* 191 (1998)

## Cucumeropsis Manni Seed Oil
*Cucumeropsis manni*
Specific Gravity (SG)
    15.5/15.5°C
    25/25°C
    Other SG
Refractive Index (RI)
    25°C
    40°C
    Other RI
Iodine Value ...................... 92
Saponification Value ............. 198
Titer °C
% Unsaponifiable
Melting Point °C

**Fatty Acid Composition (%)**
    16:0 ............................ 19
    18:0 ............................ 11
    9c,12c-18:2 ..................... 50
    Undefined 18:3 .................. 20

**References** *Riv. Ital Sost. Grasse 67:* 259 (1990)

## Cucumis Sativus Seed Oil
*Cucumis sativus*

Specific Gravity (SG)
　15.5/15.5°C
　25/25°C
　Other SG
Refractive Index (RI)
　25°C
　40°C
　Other RI
Iodine Value
Saponification Value
Titer °C
% Unsaponifiable
Melting Point °C

**Fatty Acid Composition (%)**
　14:0........................... 0.07
　16:0........................... 13.65
　9c-16:1 ....................... 0.11
　18:0........................... 10.41
　9c-18:1 ....................... 18.01
　7c-18:1 ....................... 0.61
　9c,12c-18:2................... 54.32
　9c,12c,15c-18:3 .............. 0.37
　20:0........................... 0.37
　11c-20:1 ..................... 0.07
　22:0........................... 0.06
　22:2........................... 0.07
　24:0........................... 0.16

Tocopherol Composition, mg/kg
　α-Tocopherol ....................... 4
　β-Tocopherol........................ 4
　γ-Tocopherol....................... 75
　δ-Tocopherol..................... 913
　Total, mg/kg

Tocotrienols Composition, mg/kg
　α-Tocotrienol ..................... 73
　β-Tocotrienol
　γ-Tocotrienol
　δ-Tocotrienol
　Total Tocotrienols, mg/kg

**References**  *JAOCS 80:* 1013–1020 (2003)

## Cucurbita Pepo Seed Oil
*Cucurbita pepo*

Specific Gravity (SG)
　15.5/15.5°C
　25/25°C
　Other SG
Refractive Index (RI)
　25°C
　40°C
　Other RI
Iodine Value
Saponification Value
Titer °C
% Unsaponifiable
Melting Point °C

**Fatty Acid Composition (%)**
　14:0........................... 0.08
　16:0........................... 17.79
　9c-16:1 ....................... 0.06
　18:0........................... 7.98
　9c-18:1 ....................... 15.46
　7c-18:1 ....................... 0.50
　9c,12c-18:2................... 56.19
　9c,12c,15c-18:3 .............. 0.23
　20:0........................... 0.36
　11c-20:1 ..................... 0.09
　22:0........................... 0.10

**References**

## Cupania Anacardioides Seed Oil
*Sapindaceae (Soapberry) Family*

Specific Gravity (SG)
　15.5/15.5°C
　25/25°C
　Other SG
Refractive Index (RI)
　25°C
　40°C
　Other RI

Iodine Value
Saponification Value
Titer °C
% Unsaponifiable
Melting Point °C

**Fatty Acid Composition (%)**
 16:0 .......................... 12
 16:1 ........................... 8
 18:0 ........................... 6
 Total 18:1 ..................... 10
 Undefined 18:2 ................. 16
 20:0 ........................... 2
 Total 20:1 ..................... 46

**References**

## Cuphea Seed Oil (Capric acid rich)
*Heterodon koehneana*

Specific Gravity (SG)
 15.5/15.5°C
 25/25°C
 Other SG
Refractive Index (RI)
 25°C
 40°C
 Other RI
Iodine Value
Saponification Value
Titer °C
% Unsaponifiable
Melting Point °C

**Fatty Acid Composition (%)**
 8:0 ........................... 0.1
 10:0 .......................... 84–92
 12:0 .......................... 1.5–3
 14:0 .......................... 0.6–2
 16:0 .......................... 1.3–3
 18:0 .......................... 0.3
 Total 18:1 .................... 1–4
 9c,12c-18:2 ................... 2–4
 Undefined 18:3 ................ 0.1–0.2
 20:0 .......................... 0–0.1

**References** *Crit. Rev. Food Sci. Nutr. 28:* 139 (1989)

*J. Am. Oil Chem. Soc. 65:* 139 (1988)
*Lipids 2:* 345 (1967)

## Cuphea Seed Oil (Caprylic acid rich)
*Diploptychia painteri*

Specific Gravity (SG)
 15.5/15.5°C
 25/25°C
 Other SG
Refractive Index (RI)
 25°C
 40°C
 Other RI
Iodine Value
Saponification Value
Titer °C
% Unsaponifiable
Melting Point °C

**Fatty Acid Composition (%)**
 8:0 .......................... 65–78
 10:0 ......................... 19–24
 12:0 ......................... 0.1–0.2
 14:0 ......................... 0.4
 16:0 ......................... 0.6–3.0
 18:0 ......................... 0.1–0.4
 Total 18:1 ................... 0.5–3
 9c,12c-18:2 .................. 1–4
 Undefined 18:3 ............... 0.1–0.2

**References** *Lipids 2:* 345 (1967)
*Crit. Rev. Food Sci. Nutr. 28:* 139 (1989)

## Cuphea Seed Oil (Lauric acid rich)
*Cuphea wrightii*

Specific Gravity (SG)
 15.5/15.5°C
 25/25°C
 Other SG
Refractive Index (RI)
 25°C
 40°C

Other RI
Iodine Value
Saponification Value
Titer °C
% Unsaponifiable
Melting Point °C

**Fatty Acid Composition (%)**
- 10:0 . . . . . . . . . . . . . . . . . . . . . . . . 31–39
- 12:0 . . . . . . . . . . . . . . . . . . . . . . . . 49–57
- 14:0 . . . . . . . . . . . . . . . . . . . . . . . . . 3–4
- 16:0 . . . . . . . . . . . . . . . . . . . . . . . . 1–1.6
- Total 18:1 . . . . . . . . . . . . . . . . . . . 1–2.7
- 9c,12c-18:2 . . . . . . . . . . . . . . . . . . 3.6–4.9
- Undefined 18:3 . . . . . . . . . . . . . . . . 0.1
- 20:0 . . . . . . . . . . . . . . . . . . . . . . . . . 0.1

**References** *J. Am. Oil Chem. Soc. 65:* 139 (1988)
*Crit. Rev. Food Sci. Nutr. 28:* 139 (1989)

## Cuphea Seed Oil (Lauric acid rich)
*Heteranthus epilobiifolia*

Specific Gravity (SG)
  15.5/15.5°C
  25/25°C
  Other SG
Refractive Index (RI)
  25°C
  40°C
  Other RI
Iodine Value
Saponification Value
Titer °C
% Unsaponifiable
Melting Point °C

**Fatty Acid Composition (%)**
- 10:0 . . . . . . . . . . . . . . . . . . . . . . . . . 0.3
- 12:0 . . . . . . . . . . . . . . . . . . . . . . . . 32–39
- 14:0 . . . . . . . . . . . . . . . . . . . . . . . . 49–55
- 16:0 . . . . . . . . . . . . . . . . . . . . . . . . . 3–5
- 18:0 . . . . . . . . . . . . . . . . . . . . . . . . . . 1
- Total 18:1 . . . . . . . . . . . . . . . . . . . . . 3
- 9c,12c-18:2 . . . . . . . . . . . . . . . . . . . . . 5

Undefined 18:3 . . . . . . . . . . . . . . . . 0.1
20:0 . . . . . . . . . . . . . . . . . . . . . . . . . 0.1

**References** *Crit. Rev. Food Sci. Nutr. 28:* 139 (1989)

## Cuphea Seed Oil (Linoleic acid rich)
*Cuphea fruiticosa*

Specific Gravity (SG)
  15.5/15.5°C
  25/25°C
  Other SG
Refractive Index (RI)
  25°C
  40°C
  Other RI
Iodine Value
Saponification Value
Titer °C
% Unsaponifiable
Melting Point °C

**Fatty Acid Composition (%)**
- 14:0 . . . . . . . . . . . . . . . . . . . . . . . . . 0–3
- 16:0 . . . . . . . . . . . . . . . . . . . . . . . . 17–18
- 18:0 . . . . . . . . . . . . . . . . . . . . . . . . . . 2
- Total 18:1 . . . . . . . . . . . . . . . . . . . 12–14
- 9c,12c-18:2 . . . . . . . . . . . . . . . . . . 62–67
- Undefined 18:3 . . . . . . . . . . . . . . . 0–0.5
- 20:0 . . . . . . . . . . . . . . . . . . . . . . . . . 0–2
- Total 20:1 . . . . . . . . . . . . . . . . . . . . 0–1
- 22:0 . . . . . . . . . . . . . . . . . . . . . . . . . 0–1

**References** *J. Am. Oil Chem. Soc. 62:* 81 (1985)
*Crit. Rev. Food Sci. Nutr. 28:* 139 (1989)

## Cuphea Viscosissima Seed Oil
*Cuphea viscosissima*

Specific Gravity (SG)
  15.5/15.5°C
  25/25°C
  Other SG

Refractive Index (RI)
   25°C
   40°C
   Other RI
Iodine Value
Saponification Value
Titer °C
% Unsaponifiable
Melting Point °C

**Fatty Acid Composition (%)**
   6:0 . . . . . . . . . . . . . . . . . . . . . . . . 0.7–1.1
   8:0 . . . . . . . . . . . . . . . . . . . . . . . 15.9–21.2
   10:0 . . . . . . . . . . . . . . . . . . . . . . 65.9–71.5
   12:0 . . . . . . . . . . . . . . . . . . . . . . . . 2.5–3.6
   14:0 . . . . . . . . . . . . . . . . . . . . . . . . 0.8–1.3
   16:0 . . . . . . . . . . . . . . . . . . . . . . . . 1.4–1.9
   18:0 . . . . . . . . . . . . . . . . . . . . . . . . 0.1–0.2
   Total 18:1 . . . . . . . . . . . . . . . . . . . 1.4–2.6
   Undefined 18:2 . . . . . . . . . . . . . . . . . 3–4
   Undefined 18:3 . . . . . . . . . . . . . . 0.2–0.3

**References** *JAOCS 68:* 515–517 (1991)

## Cupu Assu Kernel Oil
*Theobroma grandiflora*

Specific Gravity (SG)
   15.5/15.5°C
   25/25°C
   Other SG
Refractive Index (RI)
   25°C
   40°C . . . . . . . . . . . . . . . . . . . . . . . 1.4563
   Other RI
Iodine Value . . . . . . . . . . . . . . . . . . . . 44–45
Saponification Value . . . . . . . . . . . . . . . . 189
Titer °C
% Unsaponifiable
Melting Point °C

**Fatty Acid Composition (%)**
   14:0 . . . . . . . . . . . . . . . . . . . . . . . . . . . 0.8
   16:0 . . . . . . . . . . . . . . . . . . . . . . . . . . 6–12
   18:0 . . . . . . . . . . . . . . . . . . . . . . . . . 22–35
   Total 18:1 . . . . . . . . . . . . . . . . . . . . 39–47
   9c,12c-18:2 . . . . . . . . . . . . . . . . . . . . . 3–9
   Undefined 18:3 . . . . . . . . . . . . . . . . . . 0–1
   20:0 . . . . . . . . . . . . . . . . . . . . . . . . . 10–12

Sterol Composition, %
   Cholesterol
   Brassicasterol
   Campesterol . . . . . . . . . . . . . . . . . . . . . . 4
   Stigmasterol . . . . . . . . . . . . . . . . . . . . . . 9
   Stigmasta-8,22-dien-3β-ol
   5α-Stigmasta-7,22-dien-3β-ol
   Δ7,25-Stigmastadienol
   β-Sitosterol . . . . . . . . . . . . . . . . . . . . . . 80
   Δ5-Avenasterol
   Δ7-Stigmasterol
   Δ7-Avenasterol
   Δ7-Campesterol
   Δ7-Ergosterol
   Δ7,25-Stigmasterol
   Sitostanol
   Spinasterol
   Squalene
   24-Methylene Cholesterol
   Other . . . . . . . . . . . . . . . . . . . . . . . . . . . 7
% sterols in oil
Total Sterols, mg/kg . . . . . . . . . . . . . . 245

Tocopherol Composition, mg/kg
   α-Tocopherol
   β-Tocopherol
   γ-Tocopherol . . . . . . . . . . . . . . . . . . . 122
   δ-Tocopherol . . . . . . . . . . . . . . . . . . . . . 6
   Total, mg/kg . . . . . . . . . . . . . . . . . . . 128

**References** *J. Am. Dietetic Assn. 68:* 224 (1976)
*J. Am. Oil Chem. Soc. 71:* 845 (1994)

## Daniellia Ogea Seed Oil
*Daniella ogea*

Specific Gravity (SG)
   15.5/15.5°C
   25/25°C
   Other SG
Refractive Index (RI)
   25°C
   40°C
   Other RI
Iodine Value
Saponification Value
Titer °C

% Unsaponifiable
Melting Point °C

**Fatty Acid Composition (%)**
16:0 .......................... 7.5
9c-16:1 ........................ 5
16:2 .......................... 1.2
18:0 .......................... 1.4
9c-18:1 ....................... 6.6
11c-18:1 ...................... 0.2
9c,12c-18:2 .................. 27.7
6c,9c,12c-18:3 ................ 0.3
9c,12c,15c-18:3 ............... 0.7
20:0 .......................... 0.6
Total 20:1 .................... 0.8
22:0 .......................... 3.3
24:0 .......................... 9.2

**References** *J. Am. Oil Chem. Soc. 75:* 1031 (1998)

## Delavaya Toxocarpa Seed Oil
*Delavaya toxocarpa*
Specific Gravity (SG)
  15.5/15.5°C
  25/25°C
  Other SG
Refractive Index (RI)
  25°C
  40°C
  Other RI
Iodine Value
Saponification Value
Titer °C
% Unsaponifiable
Melting Point °C

**Fatty Acid Composition (%)**
14:0 ......................... 0.01
16:0 ......................... 4.20
9c-16:1 ...................... 0.05
18:0 ......................... 2.12
9c-18:1 ..................... 39.10
7c-18:1 ...................... 0.54
9c,12c-18:2 .................. 2.72
9c,12c,15c-18:3 .............. 0.62

20:0 ......................... 9.65
11c-20:1 .................... 37.49
22:0 ......................... 0.78
13c-22:1 ..................... 0.91
24:0 ......................... 0.16

Tocopherol Composition, mg/kg
  α-Tocopherol ..................... 2
  β-Tocopherol
  γ-Tocopherol ..................... 1
  δ-Tocopherol
  Total, mg/kg

Tocotrienols Composition, mg/kg
  α-Tocotrienol .................... 29
  β-Tocotrienol .................... 2
  γ-Tocotrienol
  δ-Tocotrienol
  Total Tocotrienols, mg/kg

**References** *JAOCS 80:* 1013–1020 (2003)

## Delphinium ajacis
*Delphinium ajacis*
Specific Gravity (SG)
  15.5/15.5°C
  25/25°C
  Other SG
Refractive Index (RI)
  25°C
  40°C
  Other RI
Iodine Value
Saponification Value
Titer °C
% Unsaponifiable
Melting Point °C

**Fatty Acid Composition (%)**
14:0 ......................... 0.04
16:0 ......................... 4.44
9c-16:1 ...................... 0.08
18:0 ......................... 2.17
9c-18:1 ..................... 46.46
7c-18:1 ...................... 0.71
9c,12c-18:2 ................. 15.39

9c,12c,15c-18:3 ............... 1.68
20:0 ......................... 0.22
11c-20:1 .................... 26.92
22:0 ......................... 0.22
13c-22:1 .................... 0.05
22:2 ......................... 0.12
24:0 ......................... 0.24

Tocopherol Composition, mg/kg
α-Tocopherol .................. 120
β-Tocopherol ................... 78
γ-Tocopherol ................... 83
δ-Tocopherol
Total, mg/kg

Tocotrienols Composition, mg/kg
α-Tocotrienol .................. 566
β-Tocotrienol .................. 153
γ-Tocotrienol
δ-Tocotrienol
Total Tocotrienols, mg/kg

**References**  *JAOCS 80:* 1013–1020 (2003)

## Dhupa Fat (Malabar Tallow)
*Valeria indica*
Specific Gravity (SG)
  15.5/15.5°C
  25/25°C
  Other SG
Refractive Index (RI)
  25°C
  40°C ................... 1.456–1.459
  Other RI
Iodine Value ..................... 36–43
Saponification Value ............ 187–192
Titer °C
% Unsaponifiable ................ 0.5–2.5
Melting Point °C

**Fatty Acid Composition (%)**
  16:0 ........................ 9–15
  18:0 ........................ 38–45
  Total 18:1 .................. 38–50
  9c,12c-18:2 ................. 0–2
  20:0 ........................ 0.5–5.0

**References**  *Indian Standard IS:* 8879–8978 (1979)

## Dimocarpus Longan Seed Oil
*Dimocarpus longan*
Specific Gravity (SG)
  15.5/15.5°C
  25/25°C
  Other SG
Refractive Index (RI)
  25°C
  40°C
  Other RI
Iodine Value
Saponification Value
Titer °C
% Unsaponifiable
Melting Point °C

**Fatty Acid Composition (%)**
  14:0 ........................ 0.26
  16:0 ....................... 12.15
  9c-16:1 ..................... 0.18
  18:0 ........................ 8.04
  7c-18:1 ..................... 0.66
  11c-18:1 ................... 36.87
  9c,12c-18:2 ................. 8.4
  9c,12c,15c-18:3 ............. 2.65
  20:0 ........................ 4.27
  11c-20:1 .................... 1.90
  22:0 ........................ 2.74
  24:0 ........................ 2.41

Tocopherol Composition, mg/kg
  α-Tocopherol ................ 139
  β-Tocopherol .................. 2
  γ-Tocopherol ................. 92
  δ-Tocopherol .................. 3
  Total, mg/kg

Tocotrienols Composition, mg/kg
  α-Tocotrienol ................. 2
  β-Tocotrienol
  γ-Tocotrienol
  δ-Tocotrienol
  Total Tocotrienols, mg/kg

**References** *JAOCS 80:* 1013–1020 (2003)

## Diospyros Mespiliformis Seed Oil
*Diospyros mespiliformis*

Specific Gravity (SG)
   15.5/15.5°C
   25/25°C
   Other SG
Refractive Index (RI)
   25°C
   40°C
   Other RI
Iodine Value
Saponification Value
Titer °C
% Unsaponifiable
Melting Point °C

**Fatty Acid Composition (%)**
   16:0 . . . . . . . . . . . . . . . . . . . . . . . . . 11.8
   9c-16:1 . . . . . . . . . . . . . . . . . . . . . . . 0.4
   16:2 . . . . . . . . . . . . . . . . . . . . . . . . . . 0.5
   18:0 . . . . . . . . . . . . . . . . . . . . . . . . . . 3.2
   9c-18:1 . . . . . . . . . . . . . . . . . . . . . . . 7.7
   11c-18:1 . . . . . . . . . . . . . . . . . . . . . . 0.5
   9c,12c-18:2 . . . . . . . . . . . . . . . . . . . . 8.0
   9c,12c,15c-18:3 . . . . . . . . . . . . . . . . 0.6
   20:0 . . . . . . . . . . . . . . . . . . . . . . . . . . 0.6
   22:0 . . . . . . . . . . . . . . . . . . . . . . . . . . 0.5
   4c,7c,10c,13c,16c,19c-22:6 . . . . . . . . 4.7
   24:0 . . . . . . . . . . . . . . . . . . . . . . . . . . 0.6

**References** *J. Am. Oil Chem. Soc. 75:* 1031 (1998)

## Diploptychea Painteri Seed Oil
*Cuphea lythracea*

Specific Gravity (SG)
   15.5/15.5°C
   25/25°C
   Other SG
Refractive Index (RI)
   25°C
   40°C
   Other RI
Iodine Value
Saponification Value
Titer °C
% Unsaponifiable
Melting Point °C

**Fatty Acid Composition (%)**
   8:0 . . . . . . . . . . . . . . . . . . . . . . . . . . . . 65
   10:0 . . . . . . . . . . . . . . . . . . . . . . . . . . . 24
   12:0 . . . . . . . . . . . . . . . . . . . . . . . . . . 0.2
   14:0 . . . . . . . . . . . . . . . . . . . . . . . . . . 0.4
   16:0 . . . . . . . . . . . . . . . . . . . . . . . . . . . . 3
   18:0 . . . . . . . . . . . . . . . . . . . . . . . . . . 0.4
   Total 18:1 . . . . . . . . . . . . . . . . . . . . . . . 3
   Undefined 18:2 . . . . . . . . . . . . . . . . . . . 4

**References** *Crit. Rev. Food Sci. Nutr. 28:* 139 (1989)

## Domba Fat
*Calophyllum inophyllum*

Specific Gravity (SG)
   15.5/15.5°C
   25/25°C
   Other SG
Refractive Index (RI)
   25°C
   40°C
   Other RI
Iodine Value . . . . . . . . . . . . . . . . . . . 82–98
Saponification Value . . . . . . . . . . . 192–201
Titer °C
% Unsaponifiable
Melting Point °C . . . . . . . . . . . . . . . . . . . 8

**Fatty Acid Composition (%)**
   14:0 . . . . . . . . . . . . . . . . . . . . . . . . . 0.02
   16:0 . . . . . . . . . . . . . . . . . . . . . . . . . . . 14
   9c-16:1 . . . . . . . . . . . . . . . . . . . . . . . 0.24
   18:0 . . . . . . . . . . . . . . . . . . . . . . . . 14.78
   Total 18:1 . . . . . . . . . . . . . . . . . . . 36–53
   9c-18:1 . . . . . . . . . . . . . . . . . . . . . . 44.99
   7c-18:1 . . . . . . . . . . . . . . . . . . . . . . . 0.86
   Undefined 18:2 . . . . . . . . . . . . . . . 16–29
   9c,12c-18:2 . . . . . . . . . . . . . . . . . . 20.96

9c,12c,15c-18:3 .............. 0.17
20:0 ....................... 0.89
9c-20:1 .................... 0.25
22:0 ....................... 0.26
13c-22:1 ................... 0.06
24:0 ....................... 0.85

Tocopherol Composition, mg/kg
α-Tocopherol .................. 58
β-Tocopherol .................. 36
γ-Tocopherol .................. 42
δ-Tocopherol .................. 42
Total, mg/kg

Tocotrienols Composition, mg/kg
α-Tocotrienol ................. 49
β-Tocotrienol
γ-Tocotrienol ................. 57
δ-Tocotrienol ................. 94
Total Tocotrienols, mg/kg

**References** *INFORM 13:* 151 (2002)
*JAOCS 80:* 1013–1020 (2003)

## Dukudu Seed Oil
*Celastrus parniculatus*
Specific Gravity (SG)
  15.5/15.5°C
  25/25°C ..................... 0.9586
  Other SG
Refractive Index (RI)
  25°C
  40°C
  Other RI
Iodine Value .................... 104
Saponification Value ........... 239–258
Titer °C
% Unsaponifiable ................... 3
Melting Point °C

**Fatty Acid Composition (%)**
16:0 ........................... 20
18:0 ............................ 4
Total 18:1 ..................... 15
9c,12c-18:2 .................... 39
Undefined 18:3 ................. 12
Other ................... Benzoic, 2

**References** *Lipids 9:* 928 (1974)

## Egusi Seed Oil
*Colocynthis citrullus*
Specific Gravity (SG)
  15.5/15.5°C
  25/25°C
  Other SG
Refractive Index (RI)
  25°C
  40°C
  Other RI
Iodine Value
Saponification Value
Titer °C
% Unsaponifiable
Melting Point °C

**Fatty Acid Composition (%)**
14:0 .......................... 0.02
16:0 ............................ 10
9c-16:1 ........................ 0.1
18:0 ............................ 10
Total 18:1 ...................... 16
9c,12c-18:2 ..................... 62
Undefined 18:3 ................. 0.4

**References** *J. Food Sci. 47:* 829 (1982)

## Elm Seed Oil
*Ulmus americana*
Specific Gravity (SG)
  15.5/15.5°C
  25/25°C ..................... 0.9305
  Other SG
Refractive Index (RI)
  25°C .................. 1.4535–1.4574
  40°C
  Other RI
Iodine Value .................. 24–25
Saponification Value ......... 273–275
Titer °C
% Unsaponifiable ............... 1–1.5
Melting Point °C

**Fatty Acid Composition (%)**
- 8:0 .......................... 10
- 10:0 .......................... 70
- 12:0 .......................... 4
- 14:0 .......................... 2
- 16:0 .......................... 4
- 18:0 .......................... 1
- Total 18:1 ...................... 5
- 9c,12c-18:2 ..................... 4
- Undefined 18:3 .................. 1
- Total 20:1 ..................... 0.1

**References** *Lipids 2:* 345 (1967)

## Entandrophragma angolense Seed Oil
*Entandrophragma angolense*

Specific Gravity (SG)
  15.5/15.5°C
  25/25°C
  Other SG
Refractive Index (RI)
  25°C
  40°C
  Other RI
Iodine Value
Saponification Value
Titer °C
% Unsaponifiable
Melting Point °C

**Fatty Acid Composition (%)**
- 16:0 .......................... 3.8
- 9c-16:1 ....................... 10.8
- 16:2 .......................... 2.8
- 18:0 .......................... 10.4
- 9c-18:1 ........................ 1.6
- 11c-18:1 ...................... 31.7
- 9c,12c-18:2 .................... 11.1
- 9c,12c,15c-18:3 ................ 0.2
- 20:0 .......................... 1.18
- 22:0 .......................... 0.3

**References** *J. Am. Oil Chem. Soc. 75:* 1031 (1998)

## Enterolobium Cyclocarpium Seed Oil
*Enterolobium cyclocarpium*

Specific Gravity (SG)
  15.5/15.5°C
  25/25°C
  Other SG
Refractive Index (RI)
  25°C
  40°C
  Other RI
Iodine Value
Saponification Value
Titer °C
% Unsaponifiable
Melting Point °C

**Fatty Acid Composition (%)**
- 16:0 .......................... 5.5
- 9c-16:1 ....................... 0.5
- 16:2 .......................... 0.8
- 18:0 .......................... 4.4
- 9c-18:1 ....................... 10.5
- 11c-18:1 ...................... 1.6
- 9c,12c-18:2 .................... 18.1
- 6c,9c,12c-18:3 ................. 0.2
- 9c,12c,15c-18:3 ................ 5.5
- 20:0 .......................... 1.4
- Total 20:1 .................... 0.3
- 22:0 .......................... 2.4
- 24:0 .......................... 1.1

**References** *J. Am. Oil Chem. Soc. 75:* 1031 (1998)

## Ephedra Gerardiana
*Ephedra gerardiana*

Specific Gravity (SG)
  15.5/15.5°C
  25/25°C
  Other SG
Refractive Index (RI)
  25°C
  40°C

Other RI
Iodine Value
Saponification Value
Titer °C
% Unsaponifiable
Melting Point °C

**Fatty Acid Composition (%)**
10:0 .......................... 0.1
12:0 .......................... tr
14:0 .......................... 0.1
9c-14:1 ....................... 0.1
15:0 .......................... 0.1
16:0 .......................... 6.6
9c-16:1 ....................... 0.6
7c-16:1 ....................... 0.1
17:0 .......................... 0.1
18:0 .......................... 2.5
9c-18:1 ....................... 17.1
11c-18:1 ...................... 11.2
5,11–18:2 ..................... 1.7
5,9–18:2 ...................... 0.4
9c,12c-18:2 ................... 8.7
9c,12c,15c-18:3 ............... 10.5
5,9c,12c,15c-18:4 ............. 0.5
20:0 .......................... 0.4
11c-20:1 ...................... 0.5
13c-20:1 ...................... 0.5
20:2 .......................... 1.5
Unidentified 20:3 ............. 3.3
5,11c,14c-20:3 ................ 7.5
5c,11c,14c, 17–20:4 ........... 19.2
22:0 .......................... 0.2
Unidentified 22:1 ............. tr
24:0 .......................... 0.1
26:0 .......................... 0.1

**References** *JAOCS 75:* 1761–1765 (1998)

# Erythrophleum fordii
*Erythrophleum fordii*

Specific Gravity (SG)
  15.5/15.5°C
  25/25°C
  Other SG
Refractive Index (RI)
  25°C

  40°C
Other RI
Iodine Value
Saponification Value
Titer °C
% Unsaponifiable
Melting Point °C

**Fatty Acid Composition (%)**
14:0 .......................... 0.03
16:0 .......................... 11.18
9c-16:1 ....................... 5.07
18:0 .......................... 6.44
9c-18:1 ....................... 20.60
7c-18:1 ....................... 14.30
9c,12c-18:2 ................... 37
9c,12c,15c-18:3 ............... 0.23
20:0 .......................... 1.39
11c-20:1 ...................... 0.13
22:0 .......................... 0.34
24:0 .......................... 0.33

Tocopherol Composition, mg/kg
  α-Tocopherol .................. 599
  β-Tocopherol .................. 45
  γ-Tocopherol .................. 159
  δ-Tocopherol .................. 32
  Total, mg/kg

Tocotrienols Composition, mg/kg
  α-Tocotrienol ................. 45
  β-Tocotrienol
  γ-Tocotrienol
  δ-Tocotrienol
  Total Tocotrienols, mg/kg

**References** *JAOCS 80:* 1013–1020 (2003)

# Euphorbia Lagascae Seed Oil
*Euphorbia lagascae*

Specific Gravity (SG)
  15.5/15.5°C
  25/25°C ....................... 0.955
  Other SG
Refractive Index (RI)
  25°C .......................... 1.4731
  40°C .......................... 1.4680
  Other RI

Iodine Value ..................... 102
Saponification Value
Titer °C
% Unsaponifiable
Melting Point °C

**Fatty Acid Composition (%)**
  14:0............................. 1
  16:0............................. 4
  18:0............................ 1.5
  Total 18:1 ...................... 19
  9c,12c-18:2....................... 9
  Undefined 18:3 ................. 0.3
  Total 20:1 ....................... 1
  Other................... Vernolic, 64

**References** *Indust. Crops Prod. 1:* 135 (1992)

# Evening Primrose Oil
*Oenothera biennis*

Specific Gravity (SG)
  15.5/15.5°C
  25/25°C
  Other SG
Refractive Index (RI)
  25°C
  40°C
  Other RI .................. (20) 1.4791
Iodine Value .................. 147–155
Saponification Value ............ 193–198
Titer °C
% Unsaponifiable ................ 1.5–2.0
Melting Point °C

**Fatty Acid Composition (%)**
  12:0........................... 0.03
  14:0........................... 0.07
  16:0............................6–10
  9c-16:1 ....................... 0.04
  18:0........................... 1.5–3.5
  Total 18:1 .....................5–12
  9c,12c-18:2.................... 65–80
  6c,9c,12c-18:3 .................8–14
  9c,12c,15c-18:3 ................ 0.2
  20:0........................... 0.3

Total 20:1 ....................... 0.2
22:0............................ 0.1
24:0............................ 0.1

Sterol Composition, %
  Cholesterol
  Brassicasterol
  Campesterol .................... 8–9
  Stigmasterol
  Stigmasta-8,22-dien-3β-ol
  5α-Stigmasta-7,22-dien-3β-ol
  Δ7,25-Stigmastadienol
  β-Sitosterol.................. 87–90
  Δ5-Avenasterol .................. 4
  Δ7-Stigmasterol ................. 2
  Δ7-Avenasterol
  Δ7-Campesterol
  Δ7-Ergosterol
  Δ7,25-Stigmasterol
  Sitostanol
  Spinasterol
  Squalene
  24-Methylene Cholesterol
  Other
  % sterols in oil
  Total Sterols, mg/kg

Tocopherol Composition, mg/kg
  α-Tocopherol ............... 76–356
  β-Tocopherol
  γ-Tocopherol................ 187–358
  δ-Tocopherol.................. 0–19
  Total, mg/kg ............... 263–661

**References** *J. Am. Oil Chem. Soc. 61:* 540 (1984)
*Food Res. Intl. 26:* 181 (1993)
*Rev. Franc. Corps Gras 39:* 339 (1992)
*J. Am. Oil Chem. Soc. 60:* 1858 (1993)
*Riv. Ital. Sost. Grasse. 53:* 25 (1976)
*Lipids 31:* 1311 (1996)

# Fennel Seed Oil
*Foeniculum officinale*

Specific Gravity (SG)
  15.5/15.5°C

25/25°C
Other SG
Refractive Index (RI)
  25°C
  40°C
  Other RI ................ (35) 1.4795
Iodine Value ...................... 99
Saponification Value ............... 181
Titer °C
% Unsaponifiable ................. 3–4
Melting Point °C

**Fatty Acid Composition (%)**
  16:0 ............................ 4
  9c-18:1 ........................ 22
  6c-18:1 ........................ 60
  9c,12c-18:2 .................... 14

**References**

## Fenugreek Seed Oil
*Trigonella foenum-graecum*
Specific Gravity (SG)
  15.5/15.5°C
  25/25°C ................. 0.910–0.922
  Other SG
Refractive Index (RI)
  25°C .................. 1.4741–1.4789
  40°C
  Other RI
Iodine Value ................ 115–158
Saponification Value ........... 178–185
Titer °C
% Unsaponifiable ................ 3.5–4
Melting Point °C

**Fatty Acid Composition (%)**
  16:0 ......................... 8–10
  18:0 ......................... 2.5–5
  Total 18:1 .................... 24–35
  9c,12c-18:2 .................. 33–42
  Undefined 18:3 ............... 14–22
  20:0 .......................... 1–2
  22:0 ......................... 0.5–1

**References**

## Fig Seed Oil
*Ficus carica*
Specific Gravity (SG)
  15.5/15.5°C
  25/25°C
  Other SG
Refractive Index (RI)
  25°C
  40°C
  Other RI
Iodine Value
Saponification Value ............... 167
Titer °C
% Unsaponifiable
Melting Point °C

**Fatty Acid Composition (%)**
  16:0 .......................... 6–7
  18:0 .......................... 2–3
  Total 18:1 .................... 15–20
  9c,12c-18:2 .................. 30–35
  Undefined 18:3 ............... 34–45

**References** *Fette Seifen Anstrichm. 85:* 23 (1983)

## Finger Millet
*Eleusine coracana*
Specific Gravity (SG)
  15.5/15.5°C
  25/25°C
  Other SG
Refractive Index (RI)
  25°C
  40°C
  Other RI
Iodine Value
Saponification Value
Titer °C
% Unsaponifiable
Melting Point °C

**Fatty Acid Composition (%)**
  16:0 .......................... 23
  18:0 ........................... 2

Total 18:1 . . . . . . . . . . . . . . . . . . . . . . 48
9c,12c-18:2 . . . . . . . . . . . . . . . . . . . . . 22
Undefined 18:3 . . . . . . . . . . . . . . . . . . . 4
20:0 . . . . . . . . . . . . . . . . . . . . . . . . . . 0.5

**References** *Cereal Chem. 71:* 355 (1994)

## Fokienia hodginsii
*Fokienia hodginsii*
Specific Gravity (SG)
  15.5/15.5°C
  25/25°C
  Other SG
Refractive Index (RI)
  25°C
  40°C
  Other RI
Iodine Value
Saponification Value
Titer °C
% Unsaponifiable
Melting Point °C

**Fatty Acid Composition (%)**
16:0 . . . . . . . . . . . . . . . . . . . . . . . . . . 6.07
16:1 . . . . . . . . . . . . . . . . . . . . . . . . . . 0.05
17:0 . . . . . . . . . . . . . . . . . . . . . . . . . . 0.08
18:0 . . . . . . . . . . . . . . . . . . . . . . . . . . 3.09
9c-18:1 . . . . . . . . . . . . . . . . . . . . . . . 10.55
11c-18:1 . . . . . . . . . . . . . . . . . . . . . . . 0.23
5,9–18:2 . . . . . . . . . . . . . . . . . . . . . . . 0.19
9c,12a-18:2 . . . . . . . . . . . . . . . . . . . . 33.18
5c,9c,12c-18:3 . . . . . . . . . . . . . . . . . . . 0.85
5,9c,12c,15c-18:4 . . . . . . . . . . . . . . . . 2.82
20:2 . . . . . . . . . . . . . . . . . . . . . . . . . . 0.27
5,11c,14c-20:3 . . . . . . . . . . . . . . . . . . 0.28
5c,11c,14c, 17–20:4 . . . . . . . . . . . . . . . 0.8

**References** *JAOCS 76:* 535–536 (1999)

## Foxtail Millet
*Setaria italica*
Specific Gravity (SG)
  15.5/15.5°C
  25/25°C . . . . . . . . . . . . . . . . . . . . . 0.9156

Other SG
Refractive Index (RI)
  25°C
  40°C . . . . . . . . . . . . . . . . . . . . . . . 1.4710
  Other RI
Iodine Value . . . . . . . . . . . . . . . . . 105–132
Saponification Value . . . . . . . . . . . . 160–193
Titer °C
% Unsaponifiable . . . . . . . . . . . . . . . . . 2–3
Melting Point °C

**Fatty Acid Composition (%)**
16:0 . . . . . . . . . . . . . . . . . . . . . . . . . . . . 9
18:0 . . . . . . . . . . . . . . . . . . . . . . . . . . . . 1
Total 18:1 . . . . . . . . . . . . . . . . . . . . . . . 20
9c,12c-18:2 . . . . . . . . . . . . . . . . . . . . . 66
Undefined 18:3 . . . . . . . . . . . . . . . . . . 2.5
9c,12c,15c-18:3 . . . . . . . . . . . . . . . . 40.07
20:0 . . . . . . . . . . . . . . . . . . . . . . . . . . 0.15
11c-20:1 . . . . . . . . . . . . . . . . . . . . . . 0.18
20:2 . . . . . . . . . . . . . . . . . . . . . . . . . . 0.06
5c,8c,11c,14c-20:4 . . . . . . . . . . . . . . . 0.2
22:0 . . . . . . . . . . . . . . . . . . . . . . . . . . 0.5
Unidentified 22:1 . . . . . . . . . . . . . . . . 0.1

**References** *Cereal Chem. 71:* 355 (1994)

## Fucus Serratus
Specific Gravity (SG)
  15.5/15.5°C
  25/25°C
  Other SG
Refractive Index (RI)
  25°C
  40°C
  Other RI
Iodine Value
Saponification Value
Titer °C
% Unsaponifiable
Melting Point °C

**Fatty Acid Composition (%)**
16:0 . . . . . . . . . . . . . . . . . . . . . . . . 19–30
16:1 . . . . . . . . . . . . . . . . . . . . . . . . . 0–10
Total 18:1 . . . . . . . . . . . . . . . . . . . . 11–34
Undefined 18:2 . . . . . . . . . . . . . . . . . 8–14
Undefined 18:3 . . . . . . . . . . . . . . . . . 5–10

18:4 . . . . . . . . . . . . . . . . . . . . . . . . . 4–11
20:4 . . . . . . . . . . . . . . . . . . . . . . . . 13–16
20:5 . . . . . . . . . . . . . . . . . . . . . . . . . 6–16
Total lipids (dry wt basis) . . . . . . . . 0.4–2

**References** *Phytochemistry 43:* 49 (1996)

## Fungal Oil
*Mortierella alpina*
Specific Gravity (SG)
 15.5/15.5°C
 25/25°C
 Other SG
Refractive Index (RI)
 25°C
 40°C
 Other RI
Iodine Value
Saponification Value
Titer °C
% Unsaponifiable
Melting Point °C

**Fatty Acid Composition (%)**
 16:0 . . . . . . . . . . . . . . . . . . . . . . . . . 8–14
 18:0 . . . . . . . . . . . . . . . . . . . . . . . . . 6–13
 Total 18:1 . . . . . . . . . . . . . . . . . . . 13–15
 Undefined 18:2 . . . . . . . . . . . . . . . . 7–20
 Unidentified 20:3 . . . . . . . . . . . . . . . 2–3
 20:4 . . . . . . . . . . . . . . . . . . . . . . . 29–46
 22:0 . . . . . . . . . . . . . . . . . . . . . . . . . 1–2
 24:0 . . . . . . . . . . . . . . . . . . . . . . . . . 1–5

**References** *JAOCS 75:* 507 (1998)

## Gamboge Butter (Kernel Fat)
*Garcinia morella*
Specific Gravity (SG)
 15.5/15.5°C
 25/25°C
 Other SG
Refractive Index (RI)
 25°C
 40°C
 Other RI
Iodine Value . . . . . . . . . . . . . . . . . . . 48–55
Saponification Value . . . . . . . . . . . . . . . 191
Titer °C
% Unsaponifiable
Melting Point °C . . . . . . . . . . . . . . . . 29–37

**Fatty Acid Composition (%)**
 14:0 . . . . . . . . . . . . . . . . . . . . . . . . 0–0.3
 16:0 . . . . . . . . . . . . . . . . . . . . . . . . 0.7–7.2
 18:0 . . . . . . . . . . . . . . . . . . . . . . . . 42–46
 Total 18:1 . . . . . . . . . . . . . . . . . . . . 43–50
 9c,12c-18:2 . . . . . . . . . . . . . . . . . . . . . 1–6
 20:0 . . . . . . . . . . . . . . . . . . . . . . . . 0.3–2.5

**References**

## Garlic Oil
*Allium sativum*
Specific Gravity (SG)
 15.5/15.5°C
 25/25°C
 Other SG
Refractive Index (RI)
 25°C
 40°C
 Other RI . . . . . . . . . . . . . . . (30) 1.4525
Iodine Value . . . . . . . . . . . . . . . . . . . . 96.5
Saponification Value . . . . . . . . . . . . . . . 198
Titer °C . . . . . . . . . . . . . . . . . . . . . . . 43–44
% Unsaponifiable
Melting Point °C

**Fatty Acid Composition (%)**
 12:0 . . . . . . . . . . . . . . . . . . . . . . . . . . 0.6
 14:0 . . . . . . . . . . . . . . . . . . . . . . . . . . 0.5
 16:0 . . . . . . . . . . . . . . . . . . . . . . . . . . . 26
 18:0 . . . . . . . . . . . . . . . . . . . . . . . . . . . . 3
 Total 18:1 . . . . . . . . . . . . . . . . . . . . . . 13
 9c,12c-18:2 . . . . . . . . . . . . . . . . . . . . . 46
 Undefined 18:3 . . . . . . . . . . . . . . . . . . . 1
 20:0 . . . . . . . . . . . . . . . . . . . . . . . . . . 4.5

**References** *Bangladesh J. Sci. Ind. Res. 26:* 41 (1993)

## Glyricidia Sepium Seed Oil
*Glyricidia sepium*

Specific Gravity (SG)
 15.5/15.5°C
 25/25°C
 Other SG
Refractive Index (RI)
 25°C
 40°C
 Other RI
Iodine Value
Saponification Value
Titer °C
% Unsaponifiable
Melting Point °C

**Fatty Acid Composition (%)**
 16:0 . . . . . . . . . . . . . . . . . . . . . . . . 15.1
 9c-16:1 . . . . . . . . . . . . . . . . . . . . . 0.4
 16:2 . . . . . . . . . . . . . . . . . . . . . . . . . 2
 18:0 . . . . . . . . . . . . . . . . . . . . . . . . 16.2
 9c-18:1 . . . . . . . . . . . . . . . . . . . . . 24
 11c-18:1 . . . . . . . . . . . . . . . . . . . . 0.7
 9c,12c-18:2 . . . . . . . . . . . . . . . . . 28.5
 6c,9c,12c-18:3 . . . . . . . . . . . . . . . 0.3
 9c,12c,15c-18:3 . . . . . . . . . . . . . . 1.4
 20:0 . . . . . . . . . . . . . . . . . . . . . . . . 3.2
 Total 20:1 . . . . . . . . . . . . . . . . . . . 0.2
 22:0 . . . . . . . . . . . . . . . . . . . . . . . . 1.7
 24:0 . . . . . . . . . . . . . . . . . . . . . . . . 1.3

**References** *J. Am. Oil Chem. Soc. 75:* 1031 (1998)


## Gnetum sp.
*Gnetum sp.*

Specific Gravity (SG)
 15.5/15.5°C
 25/25°C
 Other SG
Refractive Index (RI)
 25°C
 40°C
 Other RI
Iodine Value
Saponification Value
Titer °C
% Unsaponifiable
Melting Point °C

**Fatty Acid Composition (%)**
 14:0 . . . . . . . . . . . . . . . . . . . . . . . . 0.13
 16:0 . . . . . . . . . . . . . . . . . . . . . . . . 8.11
 9c-16:1 . . . . . . . . . . . . . . . . . . . . . 0.14
 18:0 . . . . . . . . . . . . . . . . . . . . . . . . 9.18
 11c-18:1 . . . . . . . . . . . . . . . . . . . . 16.36
 9c,12c-18:2 . . . . . . . . . . . . . . . . . 3.39
 9c,12c,15c-18:3 . . . . . . . . . . . . . . 3.68
 20:0 . . . . . . . . . . . . . . . . . . . . . . . . 1.85
 11c-20:1 . . . . . . . . . . . . . . . . . . . . 0.57
 22:0 . . . . . . . . . . . . . . . . . . . . . . . . 1.18
 13c-22:1 . . . . . . . . . . . . . . . . . . . . 0.39
 24:0 . . . . . . . . . . . . . . . . . . . . . . . . 0.44

Tocopherol Composition, mg/kg
 α-Tocopherol . . . . . . . . . . . . . . . . . . . . 23
 β-Tocopherol
 γ-Tocopherol . . . . . . . . . . . . . . . . . . . . 11
 δ-Tocopherol
 Total, mg/kg

Tocotrienols Composition, mg/kg
 α-Tocotrienol . . . . . . . . . . . . . . . . . . . . 29
 β-Tocotrienol
 γ-Tocotrienol . . . . . . . . . . . . . . . . . . . . 17
 δ-Tocotrienol
 Total Tocotrienols, mg/kg

**References** *JAOCS 80:* 1013–1020 (2003)


## Gomphera Globosa Seed Oil
*Gomphera globosa*

Specific Gravity (SG)
 15.5/15.5°C
 25/25°C
 Other SG
Refractive Index (RI)
 25°C
 40°C
 Other RI
Iodine Value . . . . . . . . . . . . . . . . . . . . . . 83
Saponification Value . . . . . . . . . . . . . . . 276
Titer °C

% Unsaponifiable . . . . . . . . . . . . . . . . . . . 0.6
Melting Point °C

**Fatty Acid Composition (%)**
14:0. . . . . . . . . . . . . . . . . . . . . . . . . . . 1
16:0. . . . . . . . . . . . . . . . . . . . . . . . . . 19
18:0. . . . . . . . . . . . . . . . . . . . . . . . . . 10
Total 18:1 . . . . . . . . . . . . . . . . . . . . . 46
9c,12c-18:2. . . . . . . . . . . . . . . . . . . . 24
20:0. . . . . . . . . . . . . . . . . . . . . . . . . . . 1
22:0. . . . . . . . . . . . . . . . . . . . . . . . . 0.4

**References** *Fette Seifen Anstrichm. 86:* 165 (1984)

## Gooseberry Seed Oil
*Ribes grossularia*
Specific Gravity (SG)
  15.5/15.5°C
  25/25°C
  Other SG
Refractive Index (RI)
  25°C
  40°C
  Other RI
Iodine Value . . . . . . . . . . . . . . . . . . . . . 171
Saponification Value . . . . . . . . . . . . . . . 188
Titer °C
% Unsaponifiable
Melting Point °C

**Fatty Acid Composition (%)**
16:0. . . . . . . . . . . . . . . . . . . . . . . . . 7–8
18:0. . . . . . . . . . . . . . . . . . . . . . . . . . . 1
Total 18:1 . . . . . . . . . . . . . . . . . . . 15–18
9c,12c-18:2. . . . . . . . . . . . . . . . . . 39–41
6c,9c,12c-18:3 . . . . . . . . . . . . . . . 10–12
9c,12c,15c-18:3 . . . . . . . . . . . . . . 19–20
6c,9c,12c,15c-18:4 . . . . . . . . . . . . . . 4–5
20:0. . . . . . . . . . . . . . . . . . . . . . . . . 0–2
Total 20:1 . . . . . . . . . . . . . . . . . . . . . 0–2
22:0. . . . . . . . . . . . . . . . . . . . . . . . . 0–2

**References** *J. Am. Oil Chem. Soc. 65:* 755 (1988)

## Grape Seed Oil
*Vitis vinifera*
Specific Gravity (SG)
  15.5/15.5°C
  25/25°C
  Other SG . . . . . . . . . . (20/20) 0.923–0.926
Refractive Index (RI)
  25°C
  40°C . . . . . . . . . . . . . . . . . . 1.473–1.477
  Other RI
Iodine Value . . . . . . . . . . . . . . . . . . 130–138
Saponification Value . . . . . . . . . . . . 188–194
Titer °C
% Unsaponifiable . . . . . . . . . . . . . . . . . 0–2
Melting Point °C

**Fatty Acid Composition (%)**
12:0. . . . . . . . . . . . . . . . . . . . . . . . 0–0.5
14:0. . . . . . . . . . . . . . . . . . . . . . . . 0–0.3
16:0. . . . . . . . . . . . . . . . . . . . . . . 5.5–11
9c-16:1 . . . . . . . . . . . . . . . . . . . . . 0–1.2
18:0. . . . . . . . . . . . . . . . . . . . . . . . . 3–6
Total 18:1 . . . . . . . . . . . . . . . . . . . 12–28
9c,12c-18:2. . . . . . . . . . . . . . . . . . 58–78
6c,9c,12c-18:3 . . . . . . . . . . . . . . . . . 0–1
20:0. . . . . . . . . . . . . . . . . . . . . . . . . 0–1
22:0. . . . . . . . . . . . . . . . . . . . . . . . 0–0.3
24:0. . . . . . . . . . . . . . . . . . . . . . . . 0–0.4

Sterol Composition, %
  Cholesterol . . . . . . . . . . . . . . . . . . 0–0.5
  Brassicasterol . . . . . . . . . . . . . . . . 0–0.2
  Campesterol . . . . . . . . . . . . . . . . . 9–14
  Stigmasterol . . . . . . . . . . . . . . . . . 9–17
  Stigmasta-8,22-dien-3β-ol
  5α-Stigmasta-7,22-dien-3β-ol
  Δ7,25-Stigmastadienol
  β-Sitosterol
  Δ5-Avenasterol . . . . . . . . . . . . . . . . . 1–3
  Δ7-Stigmasterol . . . . . . . . . . . . . . . . 1–3
  Δ7-Avenasterol . . . . . . . . . . . . . . . . . 0–1
  Δ7-Campesterol
  Δ7-Ergosterol
  Δ7,25-Stigmasterol
  Sitostanol
  Spinasterol

Squalene
24-Methylene Cholesterol
Other: Sitostanol, 64–70; Δ5,24-
   Stigmastadienol, 1
% sterols in oil
Total Sterols, mg/kg . . . . . . . . . . . . . 5800

Tocopherol Composition, mg/kg
   α-Tocopherol . . . . . . . . . . . . . . . . . 16–38
   β-Tocopherol. . . . . . . . . . . . . . . . . . 0–89
   γ-Tocopherol . . . . . . . . . . . . . . . . . . 0–73
   δ-Tocopherol. . . . . . . . . . . . . . . . . . . 0–4
   Total, mg/kg . . . . . . . . . . . . . . . . 16–204

Tocotrienols Composition, mg/kg
   α-Tocotrienol . . . . . . . . . . . . . . . . 18–107
   β-Tocotrienol
   γ-Tocotrienol. . . . . . . . . . . . . . . . 115–205
   δ-Tocotrienol. . . . . . . . . . . . . . . . . . . 0–3
   Total Tocotrienols, mg/kg. . . . . . 133–313

**References** *Codex* CX *1993/16*
*Riv. Ital. Sost. Grasse 65:* 227 (1988)
*Riv. Ital. Sost. Grasse 73:* 287 (1996)
*Riv. Ital. Sost. Grasse 70:* 601 (1993)
*J. Am. Dietetic Assn. 73:* 41 (1988)

## Grapefruit Seed Oil
*Citrus grandis (paradisi)*
Specific Gravity (SG)
   15.5/15.5°C
   25/25°C. . . . . . . . . . . . . . . . . 0.917–0.920
   Other SG
Refractive Index (RI)
   25°C . . . . . . . . . . . . . . . . . . . 1.469–1.470
   40°C
   Other RI
Iodine Value . . . . . . . . . . . . . . . . . . 92–106
Saponification Value . . . . . . . . . . . 178–197
Titer °C
% Unsaponifiable . . . . . . . . . . . . . . . 0.3–0.7
Melting Point °C

**Fatty Acid Composition (%)**
   12:0. . . . . . . . . . . . . . . . . . . . . . . . . . . 0.5
   14:0. . . . . . . . . . . . . . . . . . . . . . . . . . . . 1
   16:0. . . . . . . . . . . . . . . . . . . . . . . . . 18–29

9c-16:1 . . . . . . . . . . . . . . . . . . . . . . . 0–1
18:0. . . . . . . . . . . . . . . . . . . . . . . . . . 2–8
Total 18:1 . . . . . . . . . . . . . . . . . . . 20–28
9c,12c-18:2 . . . . . . . . . . . . . . . . . . 36–51
Undefined 18:3 . . . . . . . . . . . . . . . . 5–6
20:0. . . . . . . . . . . . . . . . . . . . . . . . 0.5–2

Sterol Composition, %
   Cholesterol
   Brassicasterol
   Campesterol
   Stigmasterol
   Stigmasta-8,22-dien-3β-ol
   5α-Stigmasta-7,22-dien-3β-ol
   Δ7,25-Stigmastadienol
   β-Sitosterol . . . . . . . . . . . . . . . . . . . . . 7
   Δ5-Avenasterol . . . . . . . . . . . . . . . . . . 3
   Δ7-Stigmasterol . . . . . . . . . . . . . . . . 90
   Δ7-Avenasterol
   Δ7-Campesterol
   Δ7-Ergosterol
   Δ7,25-Stigmasterol
   Sitostanol
   Spinasterol
   Squalene
   24-Methylene Cholesterol
   Other
% sterols in oil
Total Sterols, mg/kg

**References** *J. Am. Oil Chem. Soc. 49:* 85 (1972)
*Pakistan J. Sci Ind. Res. 34:* 238 (1991)

## Green Gram
*Vigna radiata*
Specific Gravity (SG)
   15.5/15.5°C
   25/25°C
   Other SG
Refractive Index (RI)
   25°C
   40°C
   Other RI
Iodine Value
Saponification Value

Titer °C
% Unsaponifiable
Melting Point °C

**Fatty Acid Composition (%)**
   16:0 .......................... 24.8
   18:0 ............................. 6
   9c-18:1 ........................ 5.4
   9c,12c-18:2 ................... 37.1
   9c,12c,15c-18:3 ............. 21.8
   20:0 ........................... 1.2
   22:0 ........................... 2.2
   24:0 ........................... 1.4

Tocopherol Composition, mg/kg
   α-Tocopherol ................... 0.9
   β-Tocopherol ................... 0.1
   γ-Tocopherol ................. 116.6
   δ-Tocopherol ................... 7.8
   Total, mg/kg ................. 125.4

Tocotrienols Composition, mg/kg
   α-Tocotrienol
   β-Tocotrienol
   γ-Tocotrienol
   δ-Tocotrienol
   Total Tocotrienols, mg/kg .......... 0.6

**References** *J. Am. Oil Chem. Soc. 74:* 1603 (1997)

## Guava Seed Oil
*Psidium guajava*

Specific Gravity (SG)
   15.5/15.5°C
   25/25°C
   Other SG .............(30/30) 0.9207
Refractive Index (RI)
   25°C
   40°C ........................ 1.4772
   Other RI
Iodine Value ...................... 134
Saponification Value ................ 196
Titer °C
% Unsaponifiable ................... 0.5
Melting Point °C

**Fatty Acid Composition (%)**
   14:0 ........................... 0.1
   16:0 ........................... 6.6
   18:0 ........................... 4.6
   Total 18:1 .................... 10.8
   9c,12c-18:2 ................... 76.4
   Undefined 18:3 ................. 0.1
   20:0 ........................... 0.3
   22:0 ........................... 0.1
   24:0 ........................... 0.1
   Other .......................... 0.9

**References** *J. Am. Oil Chem. Soc. 71:* 457 (1994)

## Hannoa Undulata Seed Oil
*Hannoa undulata (simarubacea)*

Specific Gravity (SG)
   15.5/15.5°C
   25/25°C
   Other SG
Refractive Index (RI)
   25°C
   40°C
   Other RI
Iodine Value ....................... 66
Saponification Value ................ 191
Titer °C
% Unsaponifiable ..................... 1
Melting Point °C

**Fatty Acid Composition (%)**
   9c-16:1 ........................ 8
   Total 18:1 ..................... 20
   9c,12c-18:2 .................... 61
   Undefined 18:3 ................. 7.6
   20:0 ........................... 0.4
   22:0 ............................ 3

Sterol Composition, %
   Cholesterol
   Brassicasterol
   Campesterol
   Stigmasterol ..................... 7
   Stigmasta-8,22-dien-3β-ol
   5α-Stigmasta-7,22-dien-3β-ol
   Δ7,25-Stigmastadienol
   β-Sitosterol .................... 70
   Δ5-Avenasterol .................. 12

Δ7-Stigmasterol
Δ7-Avenasterol
Δ7-Campesterol
Δ7-Ergosterol
Δ7,25-Stigmasterol
Sitostanol
Spinasterol
Squalene
24-Methylene Cholesterol
Other: 24-methylene-cholesterol, 10 % sterols in oil
Total Sterols, mg/kg . . . . . . . . . . . . . . 618

Tocopherol Composition, mg/kg
α-Tocopherol . . . . . . . . . . . . . . . . . . . 71
β-Tocopherol. . . . . . . . . . . . . . . . . . . . . 6
γ-Tocopherol . . . . . . . . . . . . . . . . . . . . 20
δ-Tocopherol
Total, mg/kg . . . . . . . . . . . . . . . . . . . 97

**References** *Rev. Franc. Corp Gras 39:* 195 (1992)

## Hazelnut Oil (Chilean)
*Gevuina avellana*
Specific Gravity (SG)
 15.5/15.5°C
 25/25°C
 Other SG
Refractive Index (RI)
 25°C
 40°C
 Other RI
Iodine Value
Saponification Value
Titer °C
% Unsaponifiable
Melting Point °C

**Fatty Acid Composition (%)**
 16:0. . . . . . . . . . . . . . . . . . . . . . . . . . 1.9
 9c-16:1 . . . . . . . . . . . . . . . . . . . . . . 22.7
 18:0. . . . . . . . . . . . . . . . . . . . . . . . . . 0.5
 9c-18:1 . . . . . . . . . . . . . . . . . . . . . . 39.4
 11c-18:1 . . . . . . . . . . . . . . . . . . . . . . 6.2
 9c,12c-18:2 . . . . . . . . . . . . . . . . . . . . 5.6
 Undefined 18:3 . . . . . . . . . . . . . . . . . 0.1

20:0. . . . . . . . . . . . . . . . . . . . . . . . . . . 1.4
Total 20:1 . . . . . . . . . . . . . . . . . . . . . 9.7
22:0. . . . . . . . . . . . . . . . . . . . . . . . . . . 2.2
Unidentified 22:1 . . . . . . . . . . . . . . . . 9.5
24:0. . . . . . . . . . . . . . . . . . . . . . . . . . . 0.5

Tocopherol Composition, mg/kg
α-Tocopherol . . . . . . . . . . . . . . . . . . 0.4
β-Tocopherol
γ-Tocopherol . . . . . . . . . . . . . . . . . . . 0.6
δ-Tocopherol
Total, mg/kg

Tocotrienols Composition, mg/kg
α-Tocotrienol . . . . . . . . . . . . . . . . . . 130
β-Tocotrienol . . . . . . . . . . . . . . . . . . . 1.3
γ-Tocotrienol. . . . . . . . . . . . . . . . . . . . 0.9
δ-Tocotrienol. . . . . . . . . . . . . . . . . . . . 0.1
Total Tocotrienols, mg/kg . . . . . . . . . 132

**References** *J. Am. Oil Chem. Soc. 75:* 1037 (1998)

## Hazelnut Oil (Filbert)
*Corylus avellana*
Specific Gravity (SG)
 15.5/15.5°C . . . . . . . . . . . . . . 0.914–0.920
 25/25°C . . . . . . . . . . . . . . . . . 0.908–0.915
 Other SG
Refractive Index (RI)
 25°C . . . . . . . . . . . . . . . . . . . 1.469–1.476
 40°C . . . . . . . . . . . . . . . . . . . 1.456–1.463
 Other RI
Iodine Value . . . . . . . . . . . . . . . . . . 83–90
Saponification Value . . . . . . . . . . . 188–197
Titer °C
% Unsaponifiable . . . . . . . . . . . . . . 0.2–0.3
Melting Point °C

**Fatty Acid Composition (%)**
 16:0. . . . . . . . . . . . . . . . . . . . . . . 4.1–7.2
 9c-16:1 . . . . . . . . . . . . . . . . . . . . 0.1–0.3
 17:0. . . . . . . . . . . . . . . . . . . . . . . . 0–0.2
 18:0. . . . . . . . . . . . . . . . . . . . . . . 1.5–2.4
 Total 18:1 . . . . . . . . . . . . . . . . . 71.9–84.0
 11c-18:1 . . . . . . . . . . . . . . . . . . . 0.9–1.2
 9c,12c-18:2 . . . . . . . . . . . . . . . . . 5.7–22.2

Undefined 18:3 . . . . . . . . . . . . . . . . 0–0.2
20:0. . . . . . . . . . . . . . . . . . . . . . . . . . 0.1
Total 20:1 . . . . . . . . . . . . . . . . . . 0.1–0.3
22:0. . . . . . . . . . . . . . . . . . . . . . . . . . 0.1
Unidentified 22:1 . . . . . . . . . . . . 0.1–0.2
Other 17:1, 0.1

Sterol Composition, %
    Cholesterol . . . . . . . . . . . . . . . . . . . 0–0.7
    Brassicasterol
    Campesterol . . . . . . . . . . . . . . . . . . . . 5–6
    Stigmasterol . . . . . . . . . . . . . . . . . . . . . . 1
    Stigmasta-8,22-dien-3β-ol
    5α-Stigmasta-7,22-dien-3β-ol
    Δ7,25-Stigmastadienol
    β-Sitosterol . . . . . . . . . . . . . . . . . . . . 82–93
    Δ5-Avenasterol . . . . . . . . . . . . . . . . . . 2–8
    Δ7-Stigmasterol . . . . . . . . . . . . . . . . . 1–3
    Δ7-Avenasterol . . . . . . . . . . . . . . . . . . 2–3
    Δ7-Campesterol
    Δ7-Ergosterol
    Δ7,25-Stigmasterol
    Sitostanol
    Spinasterol
    Squalene
    24-Methylene Cholesterol
    Other
    % sterols in oil
    Total Sterols, mg/kg . . . . . . . . 1200–2000

Tocopherol Composition, mg/kg
    α-Tocopherol . . . . . . . . . . . . . . . 200–409
    β-Tocopherol. . . . . . . . . . . . . . . . . . . 6–17
    γ-Tocopherol. . . . . . . . . . . . . . . . . 18–150
    δ-Tocopherol. . . . . . . . . . . . . . . . . . . 1–7
    Total, mg/kg . . . . . . . . . . . . . . . . 225–583

**References** *Riv. Ital Sost. Grasse 68:* 411 (1993)
*J. Am. Dietetic Assn. 73:* 39 (1978)
*Food Chem. 50:* 245 (1994)
*Food Chem. 48:* 411 (1993)
*J. Food Technol. 13:* 355 (1978)
*J. Am. Oil Chem. Soc. 74:* 755 (1997)

# Hempseed Oil
*Cannabis sativa*

Specific Gravity (SG)
    15.5/15.5°C
    25/25°C. . . . . . . . . . . . . . . . . 0.923–0.925
    Other SG
Refractive Index (RI)
    25°C
    40°C . . . . . . . . . . . . . . . . . . . 1.470–1.473
    Other RI
Iodine Value . . . . . . . . . . . . . . . . . . 145–166
Saponification Value . . . . . . . . . . . 190–195
Titer °C
% Unsaponifiable . . . . . . . . . . . . . . . 1.0–1.5
Melting Point °C

**Fatty Acid Composition (%)**
    16:0. . . . . . . . . . . . . . . . . . . . . . . . . 6–12
    18:0. . . . . . . . . . . . . . . . . . . . . . . . . . 1–2
    Total 18:1 . . . . . . . . . . . . . . . . . . . 11–16
    9c,12c-18:2. . . . . . . . . . . . . . . . . . 45–65
    Undefined 18:3 . . . . . . . . . . . . . . . 15–30
    20:0. . . . . . . . . . . . . . . . . . . . . . . . . . . . 2

Sterol Composition, %
    Cholesterol
    Brassicasterol
    Campesterol . . . . . . . . . . . . . . . . . . . . . 17
    Stigmasterol . . . . . . . . . . . . . . . . . . . . . 15
    Stigmasta-8,22-dien-3β-ol
    5α-Stigmasta-7,22-dien-3β-ol
    Δ7,25-Stigmastadienol
    β-Sitosterol . . . . . . . . . . . . . . . . . . . . . . 44
    Δ5-Avenasterol . . . . . . . . . . . . . . . . . . . . 2
    Δ7-Stigmasterol . . . . . . . . . . . . . . . . . . . 2
    Δ7-Avenasterol . . . . . . . . . . . . . . . . . . . . 1
    Δ7-Campesterol
    Δ7-Ergosterol
    Δ7,25-Stigmasterol
    Sitostanol
    Spinasterol
    Squalene
    24-Methylene Cholesterol

Other
% sterols in oil
Total Sterols, mg/kg . . . . . . . . . . . . . 3720

**References** *J. Am. Dietetic Assn. 73:* 41 (1978)
K.A. Williams, *Oils, Fats and Fatty Foods*, 4th edn., Elsevier, NY, 1966, pp. 288

## Heteranthus Epilobiifolia Seed Oil
*Heteranthus epilobiifolia*
Specific Gravity (SG)
  15.5/15.5°C
  25/25°C
  Other SG
Refractive Index (RI)
  25°C
  40°C
  Other RI
Iodine Value
Saponification Value
Titer °C
% Unsaponifiable
Melting Point °C

**Fatty Acid Composition (%)**
  10:0. . . . . . . . . . . . . . . . . . . . . . . . . . 0.3
  12:0. . . . . . . . . . . . . . . . . . . . . . . . . . 32
  14:0. . . . . . . . . . . . . . . . . . . . . . . . . . 55
  16:0. . . . . . . . . . . . . . . . . . . . . . . . . . 5
  18:0. . . . . . . . . . . . . . . . . . . . . . . . . . 1
  Total 18:1 . . . . . . . . . . . . . . . . . . . . 1
  Undefined 18:2 . . . . . . . . . . . . . . . . 5
  Undefined 18:3 . . . . . . . . . . . . . . . . 0.1
  20:0. . . . . . . . . . . . . . . . . . . . . . . . . . 0.1

**References** *Crit. Rev. Food Sci. Nutr. 28:* 139 (1989)

## Hibiscus Cannabinus
*Hibiscus cannabinus*
Specific Gravity (SG)
  15.5/15.5°C
  25/25°C
  Other SG

Refractive Index (RI)
  25°C
  40°C
  Other RI
Iodine Value
Saponification Value
Titer °C
% Unsaponifiable
Melting Point °C

**Fatty Acid Composition (%)**

**References**

## Hibiscus Coatesii
*Hibiscus coatesii*
Specific Gravity (SG)
  15.5/15.5°C
  25/25°C
  Other SG
Refractive Index (RI)
  25°C
  40°C
  Other RI
Iodine Value
Saponification Value
Titer °C
% Unsaponifiable
Melting Point °C

**Fatty Acid Composition (%)**
  14:0. . . . . . . . . . . . . . . . . . . . . . . . . . 0.2
  16:0. . . . . . . . . . . . . . . . . . . . . . . . . . 17.1
  16:1. . . . . . . . . . . . . . . . . . . . . . . . . . 0.7
  18:0. . . . . . . . . . . . . . . . . . . . . . . . . . 3.5
  Total 18:1 . . . . . . . . . . . . . . . . . . . . 9.9
  Undefined 18:2 . . . . . . . . . . . . . . . . 62.1
  Undefined 18:3 . . . . . . . . . . . . . . . . 0.5
  20:0. . . . . . . . . . . . . . . . . . . . . . . . . . 0.4

**References** *JAOCS 68:* 518–519 (1991)

## Hibiscus sabdariffa
*Hibiscus sabdariffa*
Specific Gravity (SG)
  15.5/15.5°C

25/25°C
Other SG
Refractive Index (RI)
    25°C
    40°C
    Other RI
Iodine Value
Saponification Value
Titer °C
% Unsaponifiable
Melting Point °C

**Fatty Acid Composition (%)**
    14:0 . . . . . . . . . . . . . . . . . . . . . . . . . 0.15
    16:0 . . . . . . . . . . . . . . . . . . . . . . . . 17.17
    9c-16:1 . . . . . . . . . . . . . . . . . . . . . . 0.47
    18:0 . . . . . . . . . . . . . . . . . . . . . . . . . 2.77
    9c-18:1 . . . . . . . . . . . . . . . . . . . . . 27.38
    7c-18:1 . . . . . . . . . . . . . . . . . . . . . . 1.01
    9c,12c-18:2 . . . . . . . . . . . . . . . . . . 42.49
    9c,12c,15c-18:3 . . . . . . . . . . . . . . . 0.28
    20:0 . . . . . . . . . . . . . . . . . . . . . . . . . 0.39
    11c-20:1 . . . . . . . . . . . . . . . . . . . . . 0.03
    22:0 . . . . . . . . . . . . . . . . . . . . . . . . . 0.33
    24:0 . . . . . . . . . . . . . . . . . . . . . . . . . 0.19

Tocopherol Composition, mg/kg
    α-Tocopherol . . . . . . . . . . . . . . . . . . 135
    β-Tocopherol . . . . . . . . . . . . . . . . . . . 38
    γ-Tocopherol . . . . . . . . . . . . . . . . . . 159
    δ-Tocopherol . . . . . . . . . . . . . . . . . . . 40
    Total, mg/kg

**References** *JAOCS 80:* 1013–1020 (2003)

## Hickory Nut Oil
*Caryaovata ovata*
Specific Gravity (SG)
    15.5/15.5°C
    25/25°C
    Other SG
Refractive Index (RI)
    25°C
    40°C
    Other RI
Iodine Value
Saponification Value
Titer °C

% Unsaponifiable
Melting Point °C

**Fatty Acid Composition (%)**
    16:0 . . . . . . . . . . . . . . . . . . . . . . . . . . . . 9
    9c-16:1 . . . . . . . . . . . . . . . . . . . . . . . 0.5
    18:0 . . . . . . . . . . . . . . . . . . . . . . . . . . . . 2
    Total 18:1 . . . . . . . . . . . . . . . . . . . . . 52
    9c,12c-18:2 . . . . . . . . . . . . . . . . . . . 34
    Undefined 18:3 . . . . . . . . . . . . . . . 1–2
    20:0 . . . . . . . . . . . . . . . . . . . . . . . . . . 0.2

**References** *J. Food Technol. 13:* 355 (1978)

## Hollyhock
*Althea rosea*
Specific Gravity (SG)
    15.5/15.5°C
    25/25°C
    Other SG
Refractive Index (RI)
    25°C
    40°C
    Other RI
Iodine Value
Saponification Value
Titer °C
% Unsaponifiable
Melting Point °C

**Fatty Acid Composition (%)**

**References**

## Horse Chestnut Oil
*Aesculus hippocastanum*
Specific Gravity (SG)
    15.5/15.5°C
    25/25°C
    Other SG
Refractive Index (RI)
    25°C . . . . . . . . . . . . . . . . . . . . 1.467–1.473
    40°C
    Other RI
Iodine Value . . . . . . . . . . . . . . . . . . . 90–109
Saponification Value . . . . . . . . . . . . 180–194

Titer °C
% Unsaponifiable .................. 1–4
Melting Point °C

**Fatty Acid Composition (%)**
    16:0 .......................... 4–6
    18:0 .......................... 1–4
    Total 18:1 .................... 67–72
    9c,12c-18:2 ................... 21–23
    Undefined 18:3 ................ 0–2

**References**

## Horsegram
*Dolichos biflorus*

Specific Gravity (SG)
    15.5/15.5°C
    25/25°C
    Other SG
Refractive Index (RI)
    25°C
    40°C
    Other RI
Iodine Value
Saponification Value
Titer °C
% Unsaponifiable
Melting Point °C

**Fatty Acid Composition (%)**
    16:0 ......................... 19.6
    18:0 .......................... 2.4
    9c-18:1 ....................... 14.9
    9c,12c-18:2 ................... 37.8
    9c,12c,15c-18:3 ............... 13.0
    20:0 .......................... 1.0
    Total 20:1 .................... 0.4
    22:0 .......................... 4.7
    22:2 .......................... 0.8
    24:0 .......................... 2.9

Tocopherol Composition, mg/kg
    α-Tocopherol .................. 0.3
    β-Tocopherol
    γ-Tocopherol ................. 66.3
    δ-Tocopherol .................. 6.9
    Total, mg/kg .... 73.5 (original material)

**References** *J. Am. Oil Chem. Soc.* 74: 1603 (1997)

## Illipe (mowrah) Butter
*Madhuca latiflora/longiflora (Bassia latiflora)*

Specific Gravity (SG)
    15.5/15.5°C
    25/25°C
    Other SG ......... (100/15) 0.856–0.870
Refractive Index (RI)
    25°C
    40°C .................. 1.458–1.462
    Other RI
Iodine Value .................... 53–70
Saponification Value ........... 188–207
Titer °C
% Unsaponifiable ................. 1–3
Melting Point °C ................. 25–29

**Fatty Acid Composition (%)**
    8:0 ........................... 0.2
    10:0 .......................... 0.1
    12:0 .......................... 0.2
    14:0 .......................... 0.3
    16:0 ........................... 23
    9c-16:1 ....................... 0.2
    18:0 ........................... 23
    Total 18:1 ..................... 34
    9c,12c-18:2 .................... 14
    Undefined 18:3 ................ 0.2
    20:0 .......................... 0.2

Sterol Composition, %
    Cholesterol
    Brassicasterol
    Campesterol ..................... 16
    Stigmasterol ..................... 7
    Stigmasta-8,22-dien-3β-ol
    5α-Stigmasta-7,22-dien-3β-ol
    Δ7,25-Stigmastadienol
    β-Sitosterol .................... 70
    Δ5-Avenasterol ................... 6
    Δ7-Stigmasterol .................. 1
    Δ7-Avenasterol
    Δ7-Campesterol

Δ7-Ergosterol
Δ7,25-Stigmasterol
Sitostanol
Spinasterol
Squalene
24-Methylene Cholesterol
Other
% sterols in oil
Total Sterols, mg/kg . . . . . . . . . . . . . . 550

**References** *J. Am. Dietetic Assn. 68:* 224 (1976)
*J. Am. Dietetic Assn. 73:* 39 (1978)
*JAOCS 76:* 1431–1436 (1999)

## Ipomoea aquatica
*Ipomoea aquatica*
Specific Gravity (SG)
  15.5/15.5°C
  25/25°C
  Other SG
Refractive Index (RI)
  25°C
  40°C
  Other RI
Iodine Value
Saponification Value
Titer °C
% Unsaponifiable
Melting Point °C

**Fatty Acid Composition (%)**
  14:0 . . . . . . . . . . . . . . . . . . . . . . . . . 0.23
  16:0 . . . . . . . . . . . . . . . . . . . . . . . . 20.92
  9c-16:1 . . . . . . . . . . . . . . . . . . . . . . 0.19
  18:0 . . . . . . . . . . . . . . . . . . . . . . . . . 9.87
  9c-18:1 . . . . . . . . . . . . . . . . . . . . . 31.82
  9c,12c-18:2 . . . . . . . . . . . . . . . . . . 27.66
  9c,12c,15c-18:3 . . . . . . . . . . . . . . . 1.05
  20:0 . . . . . . . . . . . . . . . . . . . . . . . . . 2.22
  11c-20:1 . . . . . . . . . . . . . . . . . . . . . 0.14
  22:0 . . . . . . . . . . . . . . . . . . . . . . . . . 0.87
  22:2 . . . . . . . . . . . . . . . . . . . . . . . . . 0.65
  24:0 . . . . . . . . . . . . . . . . . . . . . . . . . 1.54

Tocopherol Composition, mg/kg
  α-Tocopherol . . . . . . . . . . . . . . . . . . . . 63
  β-Tocopherol
  γ-Tocopherol . . . . . . . . . . . . . . . . . . . 680
  δ-Tocopherol . . . . . . . . . . . . . . . . . . . . 36
  Total, mg/kg

Tocotrienols Composition, mg/kg
  α-Tocotrienol
  β-Tocotrienol
  γ-Tocotrienol . . . . . . . . . . . . . . . . . . . . 35
  δ-Tocotrienol
  Total Tocotrienols, mg/kg

**References** *JAOCS 80:* 1013–1020 (2003)

## Ironwood/Nahar Fat (Indian Rose Chestnut)
*Mesua ferrea*
Specific Gravity (SG)
  15.5/15.5°C
  25/25°C
  Other SG
Refractive Index (RI)
  25°C
  40°C
  Other RI
Iodine Value . . . . . . . . . . . . . . . . . . . 73–93
Saponification Value . . . . . . . . . . . . 193–205
Titer °C
% Unsaponifiable
Melting Point °C

**Fatty Acid Composition (%)**
  14:0 . . . . . . . . . . . . . . . . . . . . . . . . . . 0–3
  16:0 . . . . . . . . . . . . . . . . . . . . . . . . . 8–14
  18:0 . . . . . . . . . . . . . . . . . . . . . . . . 10–16
  Total 18:1 . . . . . . . . . . . . . . . . . . . . 55–66
  9c,12c-18:2 . . . . . . . . . . . . . . . . . . 10–20
  20:0 . . . . . . . . . . . . . . . . . . . . . . . . . . 0–2

**References** *INFORM 13:* 151 (2002)

## Irvingia Gabonensis Kernel Fat (Dika Fat)
*Irvingia gabonensis*

Specific Gravity (SG)
  15.5/15.5°C
  25/25°C
  Other SG
Refractive Index (RI)
  25°C
  40°C
  Other RI
Iodine Value . . . . . . . . . . . . . . . . . . . . . . . 2
Saponification Value . . . . . . . . . . . . . . . . 252
Titer °C
% Unsaponifiable . . . . . . . . . . . . . . . . . . . 0.4
Melting Point °C

**Fatty Acid Composition (%)**
  8:0 . . . . . . . . . . . . . . . . . . . . . . . . . . . . . . 3
  12:0 . . . . . . . . . . . . . . . . . . . . . . . . . 35–59
  14:0 . . . . . . . . . . . . . . . . . . . . . . . . . 33–59
  16:0 . . . . . . . . . . . . . . . . . . . . . . . . . . . 2–5
  18:0 . . . . . . . . . . . . . . . . . . . . . . . . . 0.4–1
  Total 18:1 . . . . . . . . . . . . . . . . . . . . 0.6–2

**References** *Rev. Franc. Corps Gras 39:* 147 (1992)

## Isano (boleko) Seed Oil
*Oneguekoa gore E.*

Specific Gravity (SG)
  15.5/15.5°C
  25/25°C
  Other SG . . . . . . . . . (20/4) 0.973–0.9838
Refractive Index (RI)
  25°C . . . . . . . . . . . . . . . . 1.5060–1.5079
  40°C
  Other RI
Iodine Value
Saponification Value . . . . . . . . . . . . 187–194
Titer °C
% Unsaponifiable . . . . . . . . . . . . . . . . . . 1–3
Melting Point °C

**Fatty Acid Composition (%)**
  14:0 . . . . . . . . . . . . . . . . . . . . . . . . . . . . . 1
  16:0 . . . . . . . . . . . . . . . . . . . . . . . . . . . . . 4
  18:0 . . . . . . . . . . . . . . . . . . . . . . . . . . . . . 1
  Total 18:1 . . . . . . . . . . . . . . . . . . . . . . . 14
  9c,12c–18:2 . . . . . . . . . . . . . . . . . . . . . . . 5
  Other: 9a,11c–18:2, 10; 9a,11a–18:2, 10; 9a,11a,17c–18:3, 32; 9a,11a,13c–18:3, 2; 9a,11a,13c,17c–18:4, 6; 8–OH, 9a,11a–18:2, 4; 8–OH,9a,11a,17c–18:3, 15; 8–OH, 9a,11a, 13c, 17c–18:4, 2; 8–OH, 9a, 11a, 13c–18:3, 1; threo–9,10–dihydroxy–18:0, 2.

**References**

## Isotoma Longiflora Seed Oil
*Isotoma longiflora*

Specific Gravity (SG)
  15.5/15.5°C
  25/25°C
  Other SG
Refractive Index (RI)
  25°C
  40°C
  Other RI
Iodine Value . . . . . . . . . . . . . . . . . . . . . . 84
Saponification Value . . . . . . . . . . . . . . . 278
Titer °C
% Unsaponifiable . . . . . . . . . . . . . . . . . . 0.5
Melting Point °C

**Fatty Acid Composition (%)**
  14:0 . . . . . . . . . . . . . . . . . . . . . . . . . . . 0.2
  16:0 . . . . . . . . . . . . . . . . . . . . . . . . . . . . 20
  18:0 . . . . . . . . . . . . . . . . . . . . . . . . . . . . 11
  Total 18:1 . . . . . . . . . . . . . . . . . . . . . . . 35
  9c,12c-18:2 . . . . . . . . . . . . . . . . . . . . . . 25
  Undefined 18:3 . . . . . . . . . . . . . . . . . . . . 3
  20:0 . . . . . . . . . . . . . . . . . . . . . . . . . . . 3.6
  22:0 . . . . . . . . . . . . . . . . . . . . . . . . . . . 2.2

**References** *Fette Seifen Anstrichm. 86:* 165 (1984)

## Jaboty Tallow (Fat, Butter)
*Erisma calcaratum*

Specific Gravity (SG)
  15.5/15.5°C
  25/25°C
  Other SG . . . . . . . . . . . . . . . .(78/4) 0.8764
Refractive Index (RI)
  25°C
  40°C . . . . . . . . . . . . . . . . . . . 1.449–1.452
  Other RI . . . . . . . . . . . . . . . . (77) 1.4366
Iodine Value . . . . . . . . . . . . . . . . . . . . 4–23
Saponification Value . . . . . . . . . . . . 228–236
Titer °C
% Unsaponifiable . . . . . . . . . . . . . . . 0.3–1.6
Melting Point °C

**Fatty Acid Composition (%)**
  12:0. . . . . . . . . . . . . . . . . . . . . . . . . . . . 24
  14:0. . . . . . . . . . . . . . . . . . . . . . . . . . . . 53
  16:0. . . . . . . . . . . . . . . . . . . . . . . . . . . . 19
  Total 18:1 . . . . . . . . . . . . . . . . . . . . . . . 3

**References** K.A.Williams, *Oils, Fats and Fatty Foods*, 4th edn., Elsevier, NY, 1966, pp. 288


## Jack Bean Oil
*Canavalia ensiformis*

Specific Gravity (SG)
  15.5/15.5°C
  25/25°C
  Other SG
Refractive Index (RI)
  25°C
  40°C
  Other RI
Iodine Value . . . . . . . . . . . . . . . . . . . . . . 36
Saponification Value . . . . . . . . . . . . 381–385
Titer °C
% Unsaponifiable
Melting Point °C

**Fatty Acid Composition (%)**
  12:0. . . . . . . . . . . . . . . . . . . . . . . . . . . . 48
  14:0. . . . . . . . . . . . . . . . . . . . . 0.44–19
  16:0. . . . . . . . . . . . . . . . . . . . . . 6–14.99

9c-16:1 . . . . . . . . . . . . . . . . . . . . . . 0.13–8
18:0. . . . . . . . . . . . . . . . . . . . . . . . . . 2.17–4
Total 18:1 . . . . . . . . . . . . . . . . . . . . . . . . 18
9c-18:1 . . . . . . . . . . . . . . . . . . . . . . . . 37.21
7c-18:1 . . . . . . . . . . . . . . . . . . . . . . . . . 4.05
9c,12c-18:2 . . . . . . . . . . . . . . . . . . . . . 21.17
9c,12c,15c-18:3 . . . . . . . . . . . . . . . . . . 8.25
20:0. . . . . . . . . . . . . . . . . . . . . . . . . . . . 0.74
11c-20:1 . . . . . . . . . . . . . . . . . . . . . . . 0.64
22:0. . . . . . . . . . . . . . . . . . . . . . . . . . . . 0.39
13c-22:1 . . . . . . . . . . . . . . . . . . . . . . . 0.28
22:2 . . . . . . . . . . . . . . . . . . . . . . . . . . . 0.26
24:0. . . . . . . . . . . . . . . . . . . . . . . . . . . . 1.27

Tocopherol Composition, mg/kg
  α-Tocopherol . . . . . . . . . . . . . . . . . . . 58
  β-Tocopherol. . . . . . . . . . . . . . . . . . . . 34
  γ-Tocopherol . . . . . . . . . . . . . . . . . . . 186
  δ-Tocopherol. . . . . . . . . . . . . . . . . . . 608
  Total, mg/kg

Tocotrienols Composition, mg/kg
  α-Tocotrienol
  β-Tocotrienol
  γ-Tocotrienol. . . . . . . . . . . . . . . . . . . . 29
  δ-Tocotrienol. . . . . . . . . . . . . . . . . . . . 33
  Total Tocotrienols, mg/kg

**References** *Riv. Ital. Sost. Grasse 71:* 421 (1994)
*JAOCS 80:* 1013–1020 (2003)


## Japan Tallow (Wax)
*Rhus succedanea*

Specific Gravity (SG)
  15.5/15.5°C . . . . . . . . . . . . . . 0.975–1.00
  25/25°C . . . . . . . . . . . . . . . . . 0.965–0.990
  Other SG
Refractive Index (RI)
  25°C
  40°C . . . . . . . . . . . . . . . . . . . 1.450–1.458
  Other RI
Iodine Value . . . . . . . . . . . . . . . . . . . . 5–17
Saponification Value . . . . . . . . . . . . 209–227
Titer °C
% Unsaponifiable
Melting Point °C

**Fatty Acid Composition (%)**
Other Dibasic acids, 5–7

**References**

## Jatropha Oil (See also Physic Nut Oil)
*Jatropha curcas*
Specific Gravity (SG)
  15.5/15.5°C
  25/25°C
  Other SG ............... (20/20) 0.916
Refractive Index (RI)
  25°C
  40°C
  Other RI .................. (20) 1.471
Iodine Value ................... 95–110
Saponification Value ........... 185–210
Titer °C
% Unsaponifiable .................. 0.9
Melting Point °C

**Fatty Acid Composition (%)**
  14:0 ........................ 0–1.4
  16:0 ........................ 3–17
  9c-16:1 ..................... 0.7–0.9
  17:0 ........................ 0.1
  18:0 ........................ 5–10
  Total 18:1 .................. 34–64
  9c,12c-18:2 ................. 18–45
  Undefined 18:3 .............. 0.2
  20:0 ........................ 0.2
  9c-20:1 ..................... 0.1
  22:0 ........................ 0.4–0.7
  24:0 ........................ 0.1
  15c-24:1 .................... 0.1

**References** www.jatropha.de/oil.htm

## Java Almond Fat
*Dacryodes rostrata*
Specific Gravity (SG)
  15.5/15.5°C
  25/25°C
  Other SG
Refractive Index (RI)
  25°C
  40°C
  Other RI
Iodine Value ....................... 53
Saponification Value ............... 185
Titer °C
% Unsaponifiable
Melting Point °C

**Fatty Acid Composition (%)**
  14:0 ........................ 0–1
  16:0 ........................ 12–13
  18:0 ........................ 31–46
  Total 18:1 .................. 38–50
  9c,12c-18:2 ................. 2–3
  Undefined 18:3 .............. 0.3
  20:0 ........................ 1–3
  22:0 ........................ 0.1

**References** *Fat Sci. Technol. 95:* 367 (1993)

## Java Olive Oil
*Sterculia foetida*
Specific Gravity (SG)
  15.5/15.5°C
  25/25°C
  Other SG
Refractive Index (RI)
  25°C
  40°C ........................ 1.4615
  Other RI
Iodine Value ................... 76–85
Saponification Value ........... 191–201
Titer °C
% Unsaponifiable ............... 0.5–1
Melting Point °C

**Fatty Acid Composition (%)**
  16:0 ........................ 21–27
  18:0 ........................ 1–3.7
  Total 18:1 .................. 9
  9c-18:1 ..................... 5.6
  Undefined 18:2 .............. 6.3
  9c,12c-18:2 ................. 9
  19:1 ........................ 45
  Other: Malvalic, 6; Sterculic, 49

**References** *J. Am. Oil Chem. Soc. 45:* 585 (1968)
*JAOCS 75:* 1757–1760 (1998)

## Jojoba Oil
*Simmondsia chinensis*
Specific Gravity (SG)
  15.5/15.5°C
  25/25°C
  Other SG
Refractive Index (RI)
  25°C
  40°C
  Other RI
Iodine Value
Saponification Value
Titer °C
% Unsaponifiable
Melting Point °C

**Fatty Acid Composition (%)**
  16:0 . . . . . . . . . . . . . . . . . . . . . . . . 0.5–3
  9c-16:1 . . . . . . . . . . . . . . . . . . . . . 0.3–0.5
  18:0 . . . . . . . . . . . . . . . . . . . . . . . . 0.1–0.2
  Total 18:1 . . . . . . . . . . . . . . . . . . . 5–12
  20:0 . . . . . . . . . . . . . . . . . . . . . . . . 0.1
  Total 20:1 . . . . . . . . . . . . . . . . . . . 66–74
  22:0 . . . . . . . . . . . . . . . . . . . . . . . . 0.2–1
  Unidentified 22:1 . . . . . . . . . . . . . . 9–19
  24:0 . . . . . . . . . . . . . . . . . . . . . . . . 0–0.5
  15c-24:1 . . . . . . . . . . . . . . . . . . . . 1–5
  26:0 . . . . . . . . . . . . . . . . . . . . . . . . 0–0.1
  Other: 26:1, 0–0.4

**References** *J. Am. Oil Chem. Soc. 54:* 187 (1977)
*J. Am. Oil Chem. Soc. 61:* 1061 (1984)

## Kaiphal Oil
*Myristica malabarica*
Specific Gravity (SG)
  15.5/15.5°C
  25/25°C
  Other SG
Refractive Index (RI)
  25°C
  40°C . . . . . . . . . . . . . . . . . 1.4580–1.4593
  Other RI
Iodine Value . . . . . . . . . . . . . . . . . . . 50–54
Saponification Value . . . . . . . . . . . . 189–191
Titer °C
% Unsaponifiable
Melting Point °C . . . . . . . . . . . . . . . . 31–32

**Fatty Acid Composition (%)**
  14:0 . . . . . . . . . . . . . . . . . . . . . . . . . 39
  16:0 . . . . . . . . . . . . . . . . . . . . . . . . . 13
  Total 18:1 . . . . . . . . . . . . . . . . . . . . 44
  Undefined 18:2 . . . . . . . . . . . . . . . . . 1

**References** *INFORM 13:* 151 (2002)

## Kanya Tallow (Fat)
*Pentadesma butyracea*
Specific Gravity (SG)
  15.5/15.5°C
  25/25°C
  Other SG
Refractive Index (RI)
  25°C
  40°C
  Other RI
Iodine Value . . . . . . . . . . . . . . . . . . . 37–47
Saponification Value . . . . . . . . . . . . 188–194
Titer °C
% Unsaponifiable . . . . . . . . . . . . . . . 1.5–1.8
Melting Point °C . . . . . . . . . . . . . . . . 28–40

**Fatty Acid Composition (%)**
  16:0 . . . . . . . . . . . . . . . . . . . . . . . . . 3–8
  9c-16:1 . . . . . . . . . . . . . . . . . . . . . . 0.2
  18:0 . . . . . . . . . . . . . . . . . . . . . . . . . 41–46
  Total 18:1 . . . . . . . . . . . . . . . . . . . . 48–51
  9c,12c-18:2 . . . . . . . . . . . . . . . . . . . 0–1.6

**References** *J. Sci. Food Agric. 28:* 384 (1977)

## Kapok Seed Oil
*Ceiba pentandra (Bombax spp)*

Specific Gravity (SG)
 15.5/15.5°C
 25/25°C
 Other SG . . . . . . . . . . (15/15) 0.920–0.933
Refractive Index (RI)
 25°C . . . . . . . . . . . . . . . . . . . 1.466–1.472
 40°C . . . . . . . . . . . . . . . . . . . 1.460–1.466
 Other RI
Iodine Value . . . . . . . . . . . . . . . . . . 86–110
Saponification Value . . . . . . . . . . . . 189–197
Titer °C
% Unsaponifiable . . . . . . . . . . . . . . . 0.5–1.8
Melting Point °C . . . . . . . . . . . . . . . . . . . 30

**Fatty Acid Composition (%)**
 16:0 . . . . . . . . . . . . . . . . . . . . . . . . 10–28
 18:0 . . . . . . . . . . . . . . . . . . . . . . . . . 2–9
 Total 18:1 . . . . . . . . . . . . . . . . . . . 45–65
 9c,12c-18:2 . . . . . . . . . . . . . . . . . . . 7–35
 20:0 . . . . . . . . . . . . . . . . . . . . . . . . . . . 1
 Other: Cyclopropenoid fatty acids, 0–15

Sterol Composition, %
 Cholesterol
 Brassicasterol
 Campesterol . . . . . . . . . . . . . . . . . . . . . 9
 Stigmasterol . . . . . . . . . . . . . . . . . . . . . 2
 Stigmasta-8,22-dien-3β-ol
 5α-Stigmasta-7,22-dien-3β-ol
 Δ7,25-Stigmastadienol
 β-Sitosterol . . . . . . . . . . . . . . . . . . . . . 86
 Δ5-Avenasterol . . . . . . . . . . . . . . . . . . . 2
 Δ7-Stigmasterol . . . . . . . . . . . . . . . . . . 1
 Δ7-Avenasterol
 Δ7-Campesterol
 Δ7-Ergosterol
 Δ7,25-Stigmasterol
 Sitostanol
 Spinasterol
 Squalene
 24-Methylene Cholesterol
 Other
 % sterols in oil
 Total Sterols, mg/kg

**References** K.A. Williams, *Oils, Fats and Fatty Foods*, 4th edn., Elsevier, NY, 1966, pp. 288
*Prog. Lipid Res. 22:* 161 (1983)

## Karaka Seed Oil
*Corynocapus laevigatus*

Specific Gravity (SG)
 15.5/15.5°C
 25/25°C
 Other SG
Refractive Index (RI)
 25°C
 40°C
 Other RI
Iodine Value
Saponification Value
Titer °C
% Unsaponifiable
Melting Point °C

**Fatty Acid Composition (%)**
 14:0 . . . . . . . . . . . . . . . . . . . . . . . . . 0.1
 16:0 . . . . . . . . . . . . . . . . . . . . . . . . . . 13
 18:0 . . . . . . . . . . . . . . . . . . . . . . . . . 7.0
 Total 18:1 . . . . . . . . . . . . . . . . . . . . . 27
 9c,12c-18:2 . . . . . . . . . . . . . . . . . . . . . 45
 Undefined 18:3 . . . . . . . . . . . . . . . . . . 1
 20:0 . . . . . . . . . . . . . . . . . . . . . . . . . . . 4
 22:0 . . . . . . . . . . . . . . . . . . . . . . . . . 1.4
 24:0 . . . . . . . . . . . . . . . . . . . . . . . . . 0.4

**References** *J. Am. Oil Chem. Soc. 60:* 1894 (1983)

## Karanja Oil, Pongam Oil
*Pongamia glabra*

Specific Gravity (SG)
 15.5/15.5°C
 25/25°C
 Other SG
Refractive Index (RI)

25°C
40°C
Other RI
Iodine Value . . . . . . . . . . . . . . . . . . . . 81–96
Saponification Value . . . . . . . . . . . 177–193
Titer °C
% Unsaponifiable . . . . . . . . . . . . . . . 0.3–9.2
Melting Point °C

**Fatty Acid Composition (%)**
16:0 . . . . . . . . . . . . . . . . . . . . . . . . 3.7–7.9
18:0 . . . . . . . . . . . . . . . . . . . . . . . . 2.4–8.9
Total 18:1 . . . . . . . . . . . . . . . . . 44.5–71.3
Undefined 18:2 . . . . . . . . . . . . . 1.8–18.3
20:0 . . . . . . . . . . . . . . . . . . . . . . . . 2.2–4.7

**References** *INFORM 13:* 151 (2002)

## Katio Fat
*Madhuca mottleyana*

Specific Gravity (SG)
 15.5/15.5°C
 25/25°C
 Other SG . . . . . . . . . . . . . . .(100/15) 0.864
Refractive Index (RI)
 25°C
 40°C . . . . . . . . . . . . . . . . . 1.4609–1.4616
 Other RI
Iodine Value . . . . . . . . . . . . . . . . . . . . 53–67
Saponification Value . . . . . . . . . . . 189–193
Titer °C . . . . . . . . . . . . . . . . . . . . . . . 36–37
% Unsaponifiable
Melting Point °C

**Fatty Acid Composition (%)**
16:0 . . . . . . . . . . . . . . . . . . . . . . . . . . . . . 10
18:0 . . . . . . . . . . . . . . . . . . . . . . . . . . . . . 19
Total 18:1 . . . . . . . . . . . . . . . . . . . . . . . . 69
Undefined 18:2 . . . . . . . . . . . . . . . . . . . 2.5

**References**

## Khakan Fat, Pelu Fat
*Salvadora oleoides and S. persica*

Specific Gravity (SG)
 15.5/15.5°C
 25/25°C
 Other SG
Refractive Index (RI)
 25°C
 40°C
 Other RI
Iodine Value . . . . . . . . . . . . . . . . . . . . . . 5–8
Saponification Value
Titer °C
% Unsaponifiable
Melting Point °C

**Fatty Acid Composition (%)**
8:0 . . . . . . . . . . . . . . . . . . . . . . . . . . . 2.5–4
10:0 . . . . . . . . . . . . . . . . . . . . . . . . . 0.7–5.1
12:0 . . . . . . . . . . . . . . . . . . . . . . . . . . 20–50
14:0 . . . . . . . . . . . . . . . . . . . . . . . . . . 26–55
16:0 . . . . . . . . . . . . . . . . . . . . . . . . . . . 8–14
18:0 . . . . . . . . . . . . . . . . . . . . . . . . . . 10–16

**References** *INFORM 13:*151 (2002)

## Kokum Butter
*Garcinia indica*

Specific Gravity (SG)
 15.5/15.5°C
 25/25°C
 Other SG
Refractive Index (RI)
 25°C
 40°C . . . . . . . . . . . . . . . . . . . . . . . . 1.456
 Other RI
Iodine Value . . . . . . . . . . . . . . . . . . . . 33–37
Saponification Value . . . . . . . . . . . . . . . 192
Titer °C

% Unsaponifiable
Melting Point °C . . . . . . . . . . . . . . . . 39–43

**Fatty Acid Composition (%)**
  14:0. . . . . . . . . . . . . . . . . . . . . . . . . . 0–1
  16:0. . . . . . . . . . . . . . . . . . . . . . . . . . 2–5
  18:0. . . . . . . . . . . . . . . . . . . . . . . . . 49–56
  Total 18:1 . . . . . . . . . . . . . . . . . . . 39–49
  9c,12c-18:2. . . . . . . . . . . . . . . . . . . . 1–2

**References** *JAOCS 76:* 1431–1436 (1999)

## Kombo Butter
*Pycnanthus kombo*
Specific Gravity (SG)
  15.5/15.5°C
  25/25°C
  Other SG
Refractive Index (RI)
  25°C
  40°C
  Other RI
Iodine Value . . . . . . . . . . . . . . . . . . . . 65–67
Saponification Value . . . . . . . . . . . . 224–255
Titer °C
% Unsaponifiable
Melting Point °C

**Fatty Acid Composition (%)**
  12:0. . . . . . . . . . . . . . . . . . . . . . . . . . . . 6
  14:0. . . . . . . . . . . . . . . . . . . . . . . . . . . 62
  14:1. . . . . . . . . . . . . . . . . . . . . . . . . . . 24
  16:0. . . . . . . . . . . . . . . . . . . . . . . . . . . . 4
  Total 18:1 . . . . . . . . . . . . . . . . . . . . . . . 6

**References**

## Korean Pine Seed Oil
*Pinus koraiensis*
Specific Gravity (SG)
  15.5/15.5°C
  25/25°C
  Other SG
Refractive Index (RI)
  25°C
  40°C
  Other RI
Iodine Value
Saponification Value
Titer °C
% Unsaponifiable
Melting Point °C

**Fatty Acid Composition (%)**
  16:0. . . . . . . . . . . . . . . . . . . . . . . . . . . . 5
  18:0. . . . . . . . . . . . . . . . . . . . . . . . . . . . 2
  Total 18:1 . . . . . . . . . . . . . . . . . . . . 26–30
  9c,12c-18:2. . . . . . . . . . . . . . . . . . . 43–45
  Undefined 18:3 . . . . . . . . . . . . . . . . . . 0.1
  5c,9c,12c-18:3 . . . . . . . . . . . . . . . . 15–18
  20:0. . . . . . . . . . . . . . . . . . . . . . . . . . . 0.4
  Total 20:1 . . . . . . . . . . . . . . . . . . . . . . . 1
  20:2 . 5c,11c–20:2, 0.1; 11c,14c–20:2, 0.5
  Unidentified 20:3 . . . . 5c,11c,14c–20:3, 1

**References** *Brit. J. Nutr. 72:* 775 (1994)
  *INFORM 8:* 116 (1997)
  *JAOCS 75:* 865–870 (1998)

## Kusum Oil (Macassar Oil)
*Macassar schleicheratrijuga*
Specific Gravity (SG)
  15.5/15.5°C
  25/25°C
  Other SG
Refractive Index (RI)
  25°C
  40°C . . . . . . . . . . . . . . . . . . 1.459–1.462
  Other RI
Iodine Value . . . . . . . . . . . . . . . . . . . . 48–58
Saponification Value . . . . . . . . . . . . 220–230
Titer °C
% Unsaponifiable
Melting Point °C

**Fatty Acid Composition (%)**
  14:0. . . . . . . . . . . . . . . . . . . . . . . . . . . . 1
  16:0. . . . . . . . . . . . . . . . . . . . . . . . . . . 5–8
  18:0. . . . . . . . . . . . . . . . . . . . . . . . . . . 2–6
  Total 18:1 . . . . . . . . . . . . . . . . . . . . 57–62
  9c,12c-18:2. . . . . . . . . . . . . . . . . . . . . 2–5
  20:0. . . . . . . . . . . . . . . . . . . . . . . . . 20–25
  24:0. . . . . . . . . . . . . . . . . . . . . . . . . . . 2–4

## Lallemantia Oil
*Lallemantia iberica*

Specific Gravity (SG)
  15.5/15.5°C
  25/25°C
  Other SG
Refractive Index (RI)
  25°C
  40°C . . . . . . . . . . . . . . . . . . . . . . . 1.4758
  Other RI . . . . . . . . . . . . (20) 1.424–1.434
Iodine Value . . . . . . . . . . . . . . . . . . 190–209
Saponification Value . . . . . . . . . . . . . . . 194
Titer °C
% Unsaponifiable . . . . . . . . . . . . . . . . . . . 0.5
Melting Point °C

**Fatty Acid Composition (%)**
  16:0 . . . . . . . . . . . . . . . . . . . . . . . . . . . . . 7
  18:0 . . . . . . . . . . . . . . . . . . . . . . . . . . . . . 2
  Total 18:1 . . . . . . . . . . . . . . . . . . . . 7–14
  9c,12c-18:2 . . . . . . . . . . . . . . . . . . 36–38
  Undefined 18:3 . . . . . . . . . . . . . . . 47–53

**References** *Lipids 2:* 371 (1967)

## Larix Sibirica Seed Oil
*Larix sibirica*

Specific Gravity (SG)
  15.5/15.5°C
  25/25°C
  Other SG
Refractive Index (RI)
  25°C
  40°C
  Other RI
Iodine Value
Saponification Value
Titer °C
% Unsaponifiable
Melting Point °C

**Fatty Acid Composition (%)**
  16:0 . . . . . . . . . . . . . . . . . . . . . . . . . . . . . 3
  9c-16:1 . . . . . . . . . . . . . . . . . . . . . . . . 0.1
  18:0 . . . . . . . . . . . . . . . . . . . . . . . . . . . 1–2
  Total 18:1 . . . . . . . . . . . . . . . . . . . . . . 17
  9c,12c-18:2 . . . . . . . . . . . . . . . . . . . . . 43
  Undefined 18:3 . . . . . . . . . . . . . . . . . . 0.3
  5c,9c,12c-18:3 . . . . . . . . . . . . . . . . . . . 31
  20:0 . . . . . . . . . . . . . . . . . . . . . . . . . . . 0.2
  Total 20:1 . . . . . . . . . . . . . . . . . . . . . . 0.5
  5c,11c–20:2, 0.1; 11c,14c–20:2, 0.5
  5c,11c,14c–20:3, 0.7
  Other: 5c,9c–18:2, 2; 5c,9c,12c,15c–18:4, 0.2

**References** *INFORM 8:* 116 (1997)

## Laurel Berry (Bay Berry) Oil
*Laurus nobilis*

Specific Gravity (SG)
  15.5/15.5°C
  25/25°C . . . . . . . . . . . . . . . . . . 0.921–0.941
  Other SG
Refractive Index (RI)
  25°C
  40°C . . . . . . . . . . . . . . . . . . . 1.460–1.465
  Other RI
Iodine Value . . . . . . . . . . . . . . . . . . . . 75–99
Saponification Value
Titer °C
% Unsaponifiable . . . . . . . . . . . . . . . . . 1–6
Melting Point °C

**Fatty Acid Composition (%)**
  12:0 . . . . . . . . . . . . . . . . . . . . . . . . 11–35
  14:0 . . . . . . . . . . . . . . . . . . . . . . . . . . 0.1
  16:0 . . . . . . . . . . . . . . . . . . . . . . . . 10–18
  9c-16:1 . . . . . . . . . . . . . . . . . . . . . . . 1–2
  18:0 . . . . . . . . . . . . . . . . . . . . . . . . . . . . 2
  Total 18:1 . . . . . . . . . . . . . . . . . . . 33–41
  9c,12c-18:2 . . . . . . . . . . . . . . . . . . 18–32
  Undefined 18:3 . . . . . . . . . . . . . . . . . . . 2

**References** *Fette Seifen Anstrichm. 85:* 23 (1983)

**References** K.A. Williams, *Oils, Fats and Fatty Foods,* 4th edn., Elsevier, NY, 1966, pp. 288

## Lawrencia Viridigrisea
*Lawrencia viridigrisea*
Specific Gravity (SG)
  15.5/15.5°C
  25/25°C
  Other SG
Refractive Index (RI)
  25°C
  40°C
  Other RI
Iodine Value
Saponification Value
Titer °C
% Unsaponifiable
Melting Point °C

**Fatty Acid Composition (%)**
  14:0 . . . . . . . . . . . . . . . . . . . . . . . . 0.3
  16:0 . . . . . . . . . . . . . . . . . . . . . . . . 9.9
  16:1 . . . . . . . . . . . . . . . . . . . . . . . . tr
  18:0 . . . . . . . . . . . . . . . . . . . . . . . . 6.7
  Total 18:1 . . . . . . . . . . . . . . . . . . . . 7.8
  Undefined 18:2 . . . . . . . . . . . . . . . 67.8
  Undefined 18:3 . . . . . . . . . . . . . . . . 0.9
  20:0 . . . . . . . . . . . . . . . . . . . . . . . . 0.8

**References** *JAOCS 68:* 518–519 (1991)


## Lemon Seed Oil
*Citrus spp*
Specific Gravity (SG)
  15.5/15.5°C . . . . . . . . . . . . . 0.921–0.923
  25/25°C . . . . . . . . . . . . . . . . 0.916–0.919
  Other SG
Refractive Index (RI)
  25°C
  40°C . . . . . . . . . . . . . . . . . . . 1.463–1.466
  Other RI
Iodine Value . . . . . . . . . . . . . . . . . . 103–109
Saponification Value . . . . . . . . . . . . 188–196
Titer °C
% Unsaponifiable . . . . . . . . . . . . . . . 0.4–0.8
Melting Point °C

**Fatty Acid Composition (%)**
  12:0 . . . . . . . . . . . . . . . . . . . . . . . . 1.8
  14:0 . . . . . . . . . . . . . . . . . . . . . . . . 0.5
  16:0 . . . . . . . . . . . . . . . . . . . . . . . . 41
  9c-16:1 . . . . . . . . . . . . . . . . . . . . . . 5
  18:0 . . . . . . . . . . . . . . . . . . . . . . . . 7
  Total 18:1 . . . . . . . . . . . . . . . . . . . . 34
  9c,12c-18:2 . . . . . . . . . . . . . . . . . . . 5
  Undefined 18:3 . . . . . . . . . . . . . . . . . 1

**References** *Pakistan J. Sci. Ind. Res. 30:* 710 (1987)


## Lesquerella Fendleri Seed Oil
*Lesquerella fendleri*
Specific Gravity (SG)
  15.5/15.5°C
  25/25°C
  Other SG
Refractive Index (RI)
  25°C
  40°C . . . . . . . . . . . . . . . . . . . . . . 1.4710
  Other RI
Iodine Value . . . . . . . . . . . . . . . . . . 104–106
Saponification Value . . . . . . . . . . . . . . . 174
Titer °C
% Unsaponifiable . . . . . . . . . . . . . . . . . 1.82
Melting Point °C

**Fatty Acid Composition (%)**
  16:0 . . . . . . . . . . . . . . . . . . . . . . 1–1.3
  16:1 . . . . . . . . . . . . . . . . . . . . . . . . 0.7
  18:0 . . . . . . . . . . . . . . . . . . . . . . 2–2.1
  Total 18:1 . . . . . . . . . . . . . . . . . 12–18.1
  Undefined 18:2 . . . . . . . . . . . . . . . 5–9.3
  Undefined 18:3 . . . . . . . . . . . . . . . 11–14
  20:0 . . . . . . . . . . . . . . . . . . . . . . . . 0.2
  Total 20:1 . . . . . . . . . . . . . . . . . . 0.6–1.2
  Other: Lesquerolic C20:1(OH), 51.4

**References** *JAOCS 72:* 559 (1995)
*JAOCS 67:* 438–442 (1990)


## Lesquerella Perforata Seed Oil
*Lesquerella perforata*
Specific Gravity (SG)
  15.5/15.5°C
  25/25°C
  Other SG

Refractive Index (RI)
   25°C
   40°C ........................ 1.4753
   Other RI
Iodine Value ..................... 138
Saponification Value
Titer °C
% Unsaponifiable
Melting Point °C

**Fatty Acid Composition (%)**
   16:0 ........................... 5.6
   18:0 ........................... 3–4
   Total 18:1 ................... 18–21
   Undefined 18:2 .................. 2
   Undefined 18:3 ............... 10–12
   Total 20:1 .................... 0.2

**References** *JAOCS 42:* 817 (1965)
   *JAOCS 72:* 559 (1995)

## Lesquerella Recurvata Seed Oil

*Lesquerella recurvata*

Specific Gravity (SG)
   15.5/15.5°C
   25/25°C
   Other SG
Refractive Index (RI)
   25°C
   40°C
   Other RI
Iodine Value
Saponification Value
Titer °C
% Unsaponifiable
Melting Point °C

**Fatty Acid Composition (%)**
   16:0 ............................. 1
   18:0 ............................. 2
   Total 18:1 .................... 11–13
   Undefined 18:2 ................. 5–8
   Undefined 18:3 ................. 3–5
   Total 20:1 ...................... 1

**References** *JAOCS 42:* 817 (1965)
   *JAOCS 72:* 559 (1995)

## Lime Seed Oil

*Citrus aurantifolia*

Specific Gravity (SG)
   15.5/15.5°C
   25/25°C ................ 0.917–0.919
   Other SG
Refractive Index (RI)
   25°C ................... 1.467–1.475
   40°C ................... 1.462–1.469
   Other RI
Iodine Value .................... 93–111
Saponification Value ........... 191–198
Titer °C
% Unsaponifiable
Melting Point °C

**Fatty Acid Composition (%)**
   16:0 ............................ 25
   18:0 ............................. 5
   Total 18:1 ...................... 21
   9c,12c-18:2 ..................... 35
   Undefined 18:3 .................. 12
   20:0 ............................. 1
   22:0 ............................. 1

**References** *Rev. Franc. Corp Gras 40:* 237 (1993)

## Lindera Umbellata Seed Oil

*Lindera umbellata*

Specific Gravity (SG)
   15.5/15.5°C
   25/25°C
   Other SG
Refractive Index (RI)
   25°C ........................ 1.4620
   40°C
   Other RI
Iodine Value ..................... 71
Saponification Value
Titer °C
% Unsaponifiable
Melting Point °C

**Fatty Acid Composition (%)**
   10:0 ............................. 3
   12:0 ............................ 29

14:0 . . . . . . . . . . . . . . . . . . . . . . . . . . 3
Total 18:1 . . . . . . . . . . . . . . . . . . . . . . 6
9c,12c-18:2 . . . . . . . . . . . . . . . . . . . . . . 3
Other 4c–10:1, 4; 4c–12:1, 47; 4c–14:1, 5

**References** *Lipids 1:* 118 (1966)

# Linseed Oil (Flax)
*Linum usitatissimum*
Specific Gravity (SG)
  15.5/15.5°C . . . . . . . . . . . . . 0.930–0.936
  25/25°C . . . . . . . . . . . . . . . . 0.924–0.930
  Other SG
Refractive Index (RI)
  25°C . . . . . . . . . . . . . . . . . . 1.477–1.482
  40°C . . . . . . . . . . . . . . . . . . 1.472–1.475
  Other RI
Iodine Value . . . . . . . . . . . . . . . . . . 170–203
Saponification Value . . . . . . . . . . . . 188–196
Titer °C . . . . . . . . . . . . . . . . . . . . . . . 19–21
% Unsaponifiable . . . . . . . . . . . . . . . . . 0.1–2
Melting Point °C

**Fatty Acid Composition (%)**
  16:0 . . . . . . . . . . . . . . . . . . . . . . . 5.6–7
  18:0 . . . . . . . . . . . . . . . . . . . . . . . . . 3–4
  Total 18:1 . . . . . . . . . . . . . . . . 17.7–20.3
  Undefined 18:2 . . . . . . . . . . . . . . . . 15.7
  9c,12c-18:2 . . . . . . . . . . . . . . . . 17–17.3
  Undefined 18:3 . . . . . . . . . . . . . . 52–57.8
  20:0 . . . . . . . . . . . . . . . . . . . . . . . . 0–0.1

Sterol Composition, %
  Cholesterol . . . . . . . . . . . . . . . . . . . 0–0.9
  Brassicasterol . . . . . . . . . . . . . . . . . 0.1–0.7
  Campesterol . . . . . . . . . . . . . . . . . . 25–31
  Stigmasterol . . . . . . . . . . . . . . . . . . . . 6–9
  Stigmasta-8,22-dien-3β-ol
  5α-Stigmasta-7,22-dien-3β-ol
  D7,25-Stigmastadienol
  β-Sitosterol . . . . . . . . . . . . . . . . . . . 45–53
  Δ5-Avenasterol . . . . . . . . . . . . . . . . . 8–12
  Δ7-Stigmasterol . . . . . . . . . . . . . . . . . . 0–3
  Δ7-Avenasterol . . . . . . . . . . . . . . . . . 0–0.6
  Δ7-Campesterol
  Δ7-Ergosterol
  Δ7,25-Stigmasterol
  Sitostanol

Spinasterol
Squalene
24-Methylene Cholesterol
Other
% sterols in oil
Total Sterols, mg/kg

Tocopherol Composition, mg/kg
  α-Tocopherol . . . . . . . . . . . . . . . . . . 5–10
  β-Tocopherol
  γ-Tocopherol . . . . . . . . . . . . . . . . 430–588
  δ-Tocopherol . . . . . . . . . . . . . . . . . . . 4–8
  Total, mg/kg . . . . . . . . . . . . . . . . 440–588

**References** *INFORM 1:* 937 (1990)
*J. Am. Oil Chem. Soc. 63:* 328 (1986)
*Prog. Lipid Res. 22:* 161 (1983)
*Fat Sci. Technol. 93:* 519 (1991)
*JAOCS 74:* 375–381(1997)
*J. Sci. Food Agric. 72:* 403 (1996)

# Linseed Oil (Low linolenic flax)
Specific Gravity (SG)
  15.5/15.5°C
  25/25°C . . . . . . . . . . . . . . . . . . . . . . 0.917
  Other SG
Refractive Index (RI)
  25°C . . . . . . . . . . . . . . . . . . . . . . . 1.470
  40°C
  Other RI . . . . . . . . . . . . . . . (46) 1.4665
Iodine Value . . . . . . . . . . . . . . . . . . . . 144
Saponification Value . . . . . . . . . . . . . . . 185
Titer °C
% Unsaponifiable
Melting Point °C

**Fatty Acid Composition (%)**
  16:0 . . . . . . . . . . . . . . . . . . . . . . . . . . . 6
  18:0 . . . . . . . . . . . . . . . . . . . . . . . . . . . 4
  Total 18:1 . . . . . . . . . . . . . . . . . . . . . 16
  9c,12c-18:2 . . . . . . . . . . . . . . . . . . . . . 72
  Undefined 18:3 . . . . . . . . . . . . . . . . . . . 2

Sterol Composition, %
  Cholesterol
  Brassicasterol . . . . . . . . . . . . . . . . . . . . . 1
  Campesterol . . . . . . . . . . . . . . . . . . . . 23
  Stigmasterol . . . . . . . . . . . . . . . . . . . . . 4
  Stigmasta-8,22-dien-3β-ol

5α-Stigmasta-7,22-dien-3β-ol
Δ7,25-Stigmastadienol
β-Sitosterol ...................... 54
Δ5-Avenasterol .................. 18
Δ7-Stigmasterol
Δ7-Avenasterol
Δ7-Campesterol
Δ7-Ergosterol
Δ7,25-Stigmasterol
Sitostanol
Spinasterol
Squalene
24-Methylene Cholesterol
Other
% sterols in oil
Total Sterols, mg/kg ............. 2330

Tocopherol Composition, mg/kg
α-Tocopherol ..................... tr
β-Tocopherol
γ-Tocopherol .................... 170
δ-Tocopherol
Total, mg/kg

**References** *INFORM 1:* 937 (1990)
*Lipid Technol. 6:* 29 (1994)
*DSIR Plant Breeding Symp.* N.Z. Agronomy
Soc. Special Publ. #5, p. 266 (1986)

## Litchi Chinensis Seed Oil
*Litchi chinensis*
Specific Gravity (SG)
 15.5/15.5°C
 25/25°C
 Other SG
Refractive Index (RI)
 25°C
 40°C
 Other RI
Iodine Value
Saponification Value
Titer °C
% Unsaponifiable
Melting Point °C

**Fatty Acid Composition (%)**
 14:0 ........................ 0.19

16:0 ........................ 8.36
9c-16:1 ..................... 0.09
18:0 ........................ 3.7
9c-18:1 ..................... 23.80
7c-18:1 ..................... 0.69
9c,12c-18:2 ................. 6.6
9c,12c,15c-18:3 ............. 4.31
20:0 ........................ 0.61
11c-20:1 .................... 0.77
22:0 ........................ 0.26

Tocopherol Composition, mg/kg
 α-Tocopherol ................... 345
 β-Tocopherol .................... 64
 γ-Tocopherol ................... 105
 δ-Tocopherol ................... 121
 Total, mg/kg

Tocotrienols Composition, mg/kg
 α-Tocotrienol .................. 925
 β-Tocotrienol
 γ-Tocotrienol .................. 127
 δ-Tocotrienol ................. 7675
 Total Tocotrienols, mg/kg

**References** *JAOCS 80:* 1013–1020 (2003)

## Longan Seed Oil
*Euphoria longana*
Specific Gravity (SG)
 15.5/15.5°C
 25/25°C
 Other SG
Refractive Index (RI)
 25°C
 40°C
 Other RI
Iodine Value ....................... 64
Saponification Value
Titer °C
% Unsaponifiable
Melting Point °C

**Fatty Acid Composition (%)**
 16:0 ........................... 16
 17:0 ........................... 0.3
 18:0 ............................ 9

Total 18:1 ..................... 0.3
Undefined 18:2 ................. 11
Undefined 18:3 .................. 4
20:0 ............................ 6
Total 20:1 ..................... 0.5
22:0 ............................ 5
24:0 ............................ 2

**References** *Oleagineux Corps Gras Lipides 4:* 459 (1997)

# Louchocarpus Sericens Seed Oil
*Louchocarpus sericens*

Specific Gravity (SG)
  15.5/15.5°C
  25/25°C
  Other SG
Refractive Index (RI)
  25°C
  40°C
  Other RI
Iodine Value
Saponification Value
Titer °C
% Unsaponifiable
Melting Point °C

**Fatty Acid Composition (%)**
  16:0 .......................... 6.6
  9c-16:1 ....................... 0.2
  16:2 .......................... 2.0
  18:0 .......................... 2.3
  9c-18:1 ...................... 18.0
  11c-18:1 ...................... 2.8
  9c,12c-18:2 ................... 6.8
  9c,12c,15c-18:3 .............. 26.5
  20:0 ............................ 1
  Total 20:1 .................... 1.2
  22:0 .......................... 8.5
  24:0 .......................... 3.2

**References** *J. Am. Oil Chem. Soc. 75:* 1031 (1998)

# Luffa Cylindrica Seed Oil
*Luffa cylindrica*

Specific Gravity (SG)
  15.5/15.5°C
  25/25°C
  Other SG
Refractive Index (RI)
  25°C
  40°C
  Other RI
Iodine Value
Saponification Value
Titer °C
% Unsaponifiable
Melting Point °C

**Fatty Acid Composition (%)**
  14:0 .......................... 0.07
  16:0 ......................... 14.02
  9c-16:1 ....................... 0.1
  18:0 .......................... 7.18
  9c-18:1 ...................... 33.07
  7c-18:1 ....................... 0.61
  9c,12c-18:2 .................. 42.98
  9c,12c,15c-18:3 ............... 0.19
  20:0 ............................ .44
  11c-20:1 ...................... 0.09
  22:0 .......................... 0.10
  22:2 .......................... 0.11
  24:0 .......................... 0.09

Tocopherol Composition, mg/kg
  α-Tocopherol .................... 9
  β-Tocopherol .................... 3
  γ-Tocopherol .................. 320
  δ-Tocopherol .................... 2
  Total, mg/kg

**References** *JAOCS 80:* 1013–1020 (2003)

# Lupin (Lupine) Seed Oil
*Lupinus albus*

Specific Gravity (SG)
  15.5/15.5°C ................. 0.923

25/25°C
Other SG
Refractive Index (RI)
  25°C
  40°C
  Other RI . . . . . . . . . . .(20) 1.4725–1.4758
Iodine Value . . . . . . . . . . . . . . . . . . 107–111
Saponification Value . . . . . . . . . . . . 179–183
Titer °C
% Unsaponifiable . . . . . . . . . . . . . . . . . . . 2–3
Melting Point °C

**Fatty Acid Composition (%)**
  16:0. . . . . . . . . . . . . . . . . . . . . . . . . . . . . 8
  18:0. . . . . . . . . . . . . . . . . . . . . . . . . . . . . 2
  Total 18:1 . . . . . . . . . . . . . . . . . . . . . 52–61
  Undefined 18:2 . . . . . . . . . . . . . . . . . 16–23
  Undefined 18:3 . . . . . . . . . . . . . . . . . . 2–8
  20:0. . . . . . . . . . . . . . . . . . . . . . . . . . . . . 1
  22:0. . . . . . . . . . . . . . . . . . . . . . . . . . . . . 3
  Unidentified 22:1 . . . . . . . . . . . . . . . . 2–7

Sterol Composition, %
  Cholesterol
  Brassicasterol
  Campesterol . . . . . . . . . . . . . . . . . . . . . 27
  Stigmasterol . . . . . . . . . . . . . . . . . . . . . 10
  Stigmasta-8,22-dien-3β-ol
  5α-Stigmasta-7,22-dien-3β-ol
  Δ7,25-Stigmastadienol
  β-Sitosterol . . . . . . . . . . . . . . . . . . . . . . 63
  Δ5-Avenasterol
  Δ7-Stigmasterol
  Δ7-Avenasterol
  Δ7-Campesterol
  Δ7-Ergosterol
  Δ7,25-Stigmasterol
  Sitostanol
  Spinasterol
  Squalene
  24-Methylene Cholesterol
  Other
  % sterols in oil
  Total Sterols, mg/kg

**References** *Riv. Ital. Sost. Grasse 57:* 27 (1980)

## Lupin (Lupine) Seed Oil
*Lupinus angustifolius*

Specific Gravity (SG)
  15.5/15.5°C. . . . . . . . . . . . . . . . . . 0.9193
  25/25°C
  Other SG
Refractive Index (RI)
  25°C
  40°C
  Other RI . . . . . . . . . . . . . . . . (19) 1.4790
Iodine Value . . . . . . . . . . . . . . . . . . . . . 105
Saponification Value . . . . . . . . . . . . . . . 183
Titer °C
% Unsaponifiable . . . . . . . . . . . . . . . . . . 4–5
Melting Point °C

**Fatty Acid Composition (%)**
  14:0. . . . . . . . . . . . . . . . . . . . . . . . . . . 0.2
  16:0. . . . . . . . . . . . . . . . . . . . . . . . . . . . . 8
  17:0. . . . . . . . . . . . . . . . . . . . . . . . . . . 0.1
  18:0. . . . . . . . . . . . . . . . . . . . . . . . . . . . . 5
  Total 18:1 . . . . . . . . . . . . . . . . . . . . . . . 32
  9c,12c-18:2 . . . . . . . . . . . . . . . . . . . . . . 48
  Undefined 18:3 . . . . . . . . . . . . . . . . . . . . 5
  20:0. . . . . . . . . . . . . . . . . . . . . . . . . . . . . 1
  22:0. . . . . . . . . . . . . . . . . . . . . . . . . . . . . 2

**References** *J. Sci. Food Agric. 25:* 409 (1974)

## Lupine Seed Oil
*Lupinus luteus*

Specific Gravity (SG)
  15.5/15.5°C
  25/25°C
  Other SG . . . . . . . . . . . . . .(15/15) 0.9193
Refractive Index (RI)
  25°C
  40°C
  Other RI . . . . . . . . . . . . . . . (20) 1.4770
Iodine Value . . . . . . . . . . . . . . . . . . 116–124
Saponification Value . . . . . . . . . . . . 177–185
Titer °C
% Unsaponifiable . . . . . . . . . . . . . . . . . . 4–5
Melting Point °C

**Fatty Acid Composition (%)**
  16:0 . . . . . . . . . . . . . . . . . . . . . . . . . . . 5
  18:0 . . . . . . . . . . . . . . . . . . . . . . . . . . . 2
  Total 18:1 . . . . . . . . . . . . . . . . . . . 24–39
  Undefined 18:2 . . . . . . . . . . . . . . . 45–49
  Undefined 18:3 . . . . . . . . . . . . . . . . . 1–8
  20:0 . . . . . . . . . . . . . . . . . . . . . . . . . . . 2
  20:4 . . . . . . . . . . . . . . . . . . . . . . . . . . . 2
  22:0 . . . . . . . . . . . . . . . . . . . . . . . . . . . 7
  Unidentified 22:1 . . . . . . . . . . . . . . . . 1–6

**References** *Riv. Ital. Sost. Grasse* 57: 27 (1980)

## Lupu Fat
*Theobroma bicolor+B260*

Specific Gravity (SG)
  15.5/15.5°C
  25/25°C
  Other SG
Refractive Index (RI)
  25°C
  40°C . . . . . . . . . . . . . . . . . 1.4565–1.4576
  Other RI
Iodine Value . . . . . . . . . . . . . . . . . . . 38–44
Saponification Value . . . . . . . . . . . . 188–189
Titer °C
% Unsaponifiable . . . . . . . . . . . . . . . 0.4–0.9
Melting Point °C . . . . . . . . . . . . . . . . . . 42

**Fatty Acid Composition (%)**
  16:0 . . . . . . . . . . . . . . . . . . . . . . . . . 5–10
  18:0 . . . . . . . . . . . . . . . . . . . . . . . . 34–50
  Total 18:1 . . . . . . . . . . . . . . . . . . . 39–51
  9c,12c-18:2 . . . . . . . . . . . . . . . . . . . . 3–5
  20:0 . . . . . . . . . . . . . . . . . . . . . . . 1.9–2.1

Sterol Composition, %
  Cholesterol
  Brassicasterol
  Campesterol . . . . . . . . . . . . . . . . . . . . . 3
  Stigmasterol . . . . . . . . . . . . . . . . . . . . . 9
  Stigmasta-8,22-dien-3β-ol
  5α-Stigmasta-7,22-dien-3β-ol
  Δ7,25-Stigmastadienol
  β-Sitosterol . . . . . . . . . . . . . . . . . . . . . 83
  Δ5-Avenasterol
  Δ7-Stigmasterol
  Δ7-Avenasterol
  Δ7-Campesterol
  Δ7-Ergosterol
  Δ7,25-Stigmasterol
  Sitostanol
  Spinasterol
  Squalene
  24-Methylene Cholesterol
  Other . . . . . . . . . . . . . . . . . . . . . . . . . . 5
  % sterols in oil
  Total Sterols, mg/kg . . . . . . . . . . . . . 250

Tocopherol Composition, mg/kg
  α-Tocopherol
  β-Tocopherol
  γ-Tocopherol . . . . . . . . . . . . . . . . . . . . 78
  δ-Tocopherol . . . . . . . . . . . . . . . . . . . . . 8
  Total, mg/kg . . . . . . . . . . . . . . . . . . . . 86

**References** *J. Am. Oil Chem. Soc.* 71: 845 (1994)

## Macadamia Nut Oil
*Macadamia tetraphylla/ ternifolia*

Specific Gravity (SG)
  15.5/15.5°C
  25/25°C
  Other SG
Refractive Index (RI)
  25°C
  40°C
  Other RI
Iodine Value
Saponification Value
Titer °C
% Unsaponifiable
Melting Point °C

**Fatty Acid Composition (%)**
  14:0 . . . . . . . . . . . . . . . . . . . . . . . . . . 0.6
  16:0 . . . . . . . . . . . . . . . . . . . . . . . . . . 8–9
  9c-16:1 . . . . . . . . . . . . . . . . . . . . . 21–22
  18:0 . . . . . . . . . . . . . . . . . . . . . . . . . . 2–4
  Total 18:1 . . . . . . . . . . . . . . . . . . . 56–59
  9c,12c-18:2 . . . . . . . . . . . . . . . . . . . . 2–3
  20:0 . . . . . . . . . . . . . . . . . . . . . . . . . . 2–3

Total 20:1 .................... 1.5–3
22:0......................... 0.8
Unidentified 22:1 ............... 0.3
24:0......................... 0.5
Other 17:1, 0.1

**References**  *J. Am. Soc. Horticultural Sci.*
98: 453 (1973)
*J. Food Technol. 13:* 355 (1978)

## Mahua Oil
*Madhuca indica*
Specific Gravity (SG)
 15.5/15.5°C
 25/25°C
 Other SG
Refractive Index (RI)
 25°C
 40°C
 Other RI
Iodine Value
Saponification Value
Titer °C
% Unsaponifiable
Melting Point °C

**Fatty Acid Composition (%)**
 16:0.........................22–37
 18:0........................18–24
 Total 18:1 ....................32–38
 Undefined 18:2 ...............14–18
 Undefined 18:3 ................... 1
 20:0......................... 1

**References** *INFORM 13:* 151 (2002)

## Mahua Fat
*Madhuca latifolia*
Specific Gravity (SG)
 15.5/15.5°C
 25/25°C
 Other SG
Refractive Index (RI)
 25°C
 40°C
 Other RI

Iodine Value
Saponification Value
Titer °C
% Unsaponifiable
Melting Point °C

**Fatty Acid Composition (%)**
 16:0..........................24
 18:0..........................20
 Total 18:1 .....................39
 Undefined 18:2 ..................17

**References** *JAOCS 76:* 1431 (1999)

## Mammy Apple Seed Oil
*Calocarpum mammosum*
Specific Gravity (SG)
 15.5/15.5°C
 25/25°C................. 0.910–0.913
 Other SG
Refractive Index (RI)
 25°C ................... 1.465–1.469
 40°C
 Other RI
Iodine Value .................... 60–74
Saponification Value ........... 188–199
Titer °C
% Unsaponifiable .................... 1.4
Melting Point °C

**Fatty Acid Composition (%)**
 16:0........................12–18
 18:0........................12–18
 Total 18:1 ....................38–54
 9c,12c-18:2...................13–24

**References** G.S. Jamieson, *Veg. Fats and Oils Chemical Catalog Co.*, 1932, p. 80

## Mango Pulp Oil
*Mangifera indica*
Specific Gravity (SG)
 15.5/15.5°C
 25/25°C
 Other SG
Refractive Index (RI)

25°C
40°C
Other RI
Iodine Value
Saponification Value
Titer °C
% Unsaponifiable
Melting Point °C

**Fatty Acid Composition (%)**
12:0 . . . . . . . . . . . . . . . . . . . . . . . . 0.3–3
14:0 . . . . . . . . . . . . . . . . . . . . . . . . . 1–12
16:0 . . . . . . . . . . . . . . . . . . . . . . . . 22–30
9c-16:1 . . . . . . . . . . . . . . . . . . . . . 16–30
18:0 . . . . . . . . . . . . . . . . . . . . . . . . . . 1–2
Total 18:1 . . . . . . . . . . . . . . . . . . . 24–40
9c,12c-18:2 . . . . . . . . . . . . . . . . . . . 3–10
Undefined 18:3 . . . . . . . . . . . . . . . . . 5–9

**References** *J. Am. Oil Chem. Soc. 52:* 514 (1975)

## Mango Seed Oil
*Mangifera indica*
Specific Gravity (SG)
15.5/15.5°C . . . . . . . . . . . . 0.9133–0.9135
25/25°C
Other SG . . . . . . . . . . . . . . .(30/30) 0.9139
Refractive Index (RI)
25°C . . . . . . . . . . . . . . . . . 1.4609–1.4610
40°C . . . . . . . . . . . . . . . . . 1.4598–1.4600
Other RI
Iodine Value . . . . . . . . . . . . . . . . . . . 39–48
Saponification Value . . . . . . . . . . . 188–195
Titer °C . . . . . . . . . . . . . . . . . . . . 49.1–49.3
% Unsaponifiable . . . . . . . . . . . . . . . . 1.3–3
Melting Point °C

**Fatty Acid Composition (%)**
12:0 . . . . . . . . . . . . . . . . . . . . . . . .0.3–0.4
14:0 . . . . . . . . . . . . . . . . . . . . . . .0.11–0.5
16:0 . . . . . . . . . . . . . . . . . . . . . . . . .4–12
9c-16:1 . . . . . . . . . . . . . . . . . . . . .0.05–0.2
18:0 . . . . . . . . . . . . . . . . . . . . . . 31–48.88
Total 18:1 . . . . . . . . . . . . . . . . . . . . 38–50
7c-18:1 . . . . . . . . . . . . . . . . . . . . . . . 0.15
9c,12c-18:2 . . . . . . . . . . . . . . . . . . 3–6.34
Undefined 18:3 . . . . . . . . . . . . . . . 0.7–1.0

9c,12c,15c-18:3 . . . . . . . . . . . . . . . . 1.25
20:0 . . . . . . . . . . . . . . . . . . . . . . . . . . 2–6
11c-20:1 . . . . . . . . . . . . . . . . . . . . . . 0.29
22:0 . . . . . . . . . . . . . . . . . . . . . . . . . 0.75
13c-22:1 . . . . . . . . . . . . . . . . . . . . . . 0.01
22:2 . . . . . . . . . . . . . . . . . . . . . . . . . 0.22
24:0 . . . . . . . . . . . . . . . . . . . . . .0.14–0.89

Tocopherol Composition, mg/kg
α-Tocopherol . . . . . . . . . . . . . . . . . . . . 103
β-Tocopherol
γ-Tocopherol
δ-Tocopherol
Total, mg/kg

Tocotrienols Composition, mg/kg
α-Tocotrienol . . . . . . . . . . . . . . . . . . . . 179
β-Tocotrienol
γ-Tocotrienol
δ-Tocotrienol
Total Tocotrienols, mg/kg

**References** *J. Am. Oil Chem. Soc. 54:* 494 (1977)
*Fat Sci. Technol. 89:* 306 (1987)
*JAOCS 80:* 1013–1020 (2003)

## Marigold Seed Oil
*Calendula officinalis*
Specific Gravity (SG)
15.5/15.5°C
25/25°C . . . . . . . . . . . . . . . . . . . . . . 0.940
Other SG
Refractive Index (RI)
25°C . . . . . . . . . . . . . . . . . . . . . . . 1.5080
40°C . . . . . . . . . . . . . . . . . . . . . . . 1.5025
Other RI
Iodine Value . . . . . . . . . . . . . . . . . . . . 242
Saponification Value
Titer °C
% Unsaponifiable
Melting Point °C

**Fatty Acid Composition (%)**
16:0 . . . . . . . . . . . . . . . . . . . . . . . . . . . . 4
18:0 . . . . . . . . . . . . . . . . . . . . . . . . . . . 1.5
Total 18:1 . . . . . . . . . . . . . . . . . . . . . . . . 4
9c,12c–18:2 . . . . . . . . . . . . . . . . . . . . . 30

Undefined 18:3 .................. 0.6
Other: 8t,10t,12c–18:3 (calendic) 59

**References** *Ind. Crops Prod. 1:* 57 (1992)

## Marine Microalga Fatty Acid Extract
*Isochrysis galbana*
Specific Gravity (SG)
  15.5/15.5°C
  25/25°C
  Other SG
Refractive Index (RI)
  25°C
  40°C
  Other RI
Iodine Value
Saponification Value
Titer °C
% Unsaponifiable
Melting Point °C

**Fatty Acid Composition (%)**
  14:0 ........................ 10–11
  16:0 ........................... 19
  16:1 ........................... 23
  18:0 .......................... 0.5
  9c-18:1 ....................... 1.7
  11c-18:1 ...................... 3.2
  Undefined 18:2 ................ 0.9
  6c,9c,12c-18:3 ................ 0.2
  9c,12c,15c-18:3 ............... 1.3
  6c,9c,12c,15c-18:4 .............. 7
  8c,11c,14c-20:3 ............... 0.2
  5c,8c,11c,14c-20:4 ............ 0.7
  5c,8c,11c,14c,17c-20:5 ......... 22
  7c, 10c, 13c, 16c-22:4 ........ 1.3
  4c,7c,10c,13c,16c,19c-22:6 ..... 23

**References** *JAOCS 72:* 575 (1995)

## Meadowfoam Seed Oil (Alba)
*Limnanthes alba*
Specific Gravity (SG)
  15.5/15.5°C
  25/25°C .................. 0.905–0.907
  Other SG
Refractive Index (RI)
  25°C ......................... 1.4701
  40°C ................. 1.4644–1.4650
  Other RI
Iodine Value .................. 94–114
Saponification Value
Titer °C
% Unsaponifiable .................. 0.2
Melting Point °C

**Fatty Acid Composition (%)**
  Total 18:1 ..................... 1–2
  9c,12c-18:2 .................. 0–0.5
  20:0 ......................... 0–0.5
  5c-20:1 ...................... 62–63
  Unidentified 22:1 ........... 2.5–4 (5c)
  13c-22:1 ..................... 10–12
  5c,13c-22:2 ..................... 18

**References** *J. Am. Oil Chem. Soc. 64:* 1493 (1987)
*J. Am. Oil Chem. Soc. 41:* 167 (1964)
*New Sources of Fats and Oils,* AOCS Press, Champaign 1981
*Ind. Crops Prod. 1:* 57 (1992)

## Meadowfoam Seed Oil (Douglas)
*Limnanthes douglasii*
Specific Gravity (SG)
  15.5/15.5°C
  25/25°C
  Other SG
Refractive Index (RI)
  25°C
  40°C ................. 1.4628–1.4652
  Other RI
Iodine Value .................... 86–91
Saponification Value ............... 168
Titer °C
% Unsaponifiable
Melting Point °C

**Fatty Acid Composition (%)**
  12:0 ......................... 0–0.1

14:0 . . . . . . . . . . . . . . . . . . . . . . . . 0–0.1
16:0 . . . . . . . . . . . . . . . . . . . . . . . . 0.1–0.4
9c-16:1 . . . . . . . . . . . . . . . . . . . 0.2–0.3
18:0 . . . . . . . . . . . . . . . . . . . . . . . . 0–0.3
Total 18:1 . . . . . . . . . . . . . . . . . . . . . . 1–3
9c,12c-18:2 . . . . . . . . . . . . . . . . . . 0.2–1.0
Undefined 18:3 . . . . . . . . . . . . . . . 0–0.6
5c-20:1 . . . . . . . . . . . . . . . . . . . . . . 58–77
Unidentified 22:1 . . . . . . . . . . . . . . 8–24
5c,13c-22:2 . . . . . . . . . . . . . . . . . . . 7–15

**References** *J. Am. Oil Chem. Soc. 41:* 167 (1964)

## Milletia Thonningii Seed Oil
*Milletia thonningii*
Specific Gravity (SG)
  15.5/15.5°C
  25/25°C
  Other SG
Refractive Index (RI)
  25°C
  40°C
  Other RI
Iodine Value
Saponification Value
Titer °C
% Unsaponifiable
Melting Point °C

**Fatty Acid Composition (%)**
16:0 . . . . . . . . . . . . . . . . . . . . . . . . . . . 4.8
16:2 . . . . . . . . . . . . . . . . . . . . . . . . . . . 1.7
18:0 . . . . . . . . . . . . . . . . . . . . . . . . . . . 2.7
9c-18:1 . . . . . . . . . . . . . . . . . . . . . . . 17.9
11c-18:1 . . . . . . . . . . . . . . . . . . . . . . . 0.3
9c,12c-18:2 . . . . . . . . . . . . . . . . . . . . 7.7
9c,12c,15c-18:3 . . . . . . . . . . . . . . . 23.1
20:0 . . . . . . . . . . . . . . . . . . . . . . . . . . . 1.1
Total 20:1 . . . . . . . . . . . . . . . . . . . . . . 1.7
22:0 . . . . . . . . . . . . . . . . . . . . . . . . . . . 8.9
24:0 . . . . . . . . . . . . . . . . . . . . . . . . . . . 2.5

**References** *J. Am. Oil Chem. Soc. 75:* 1031 (1998)

## Mock Orange
*Cucurbita palmata*
Specific Gravity (SG)
  15.5/15.5°C . . . . . . . . . . . . . . . . . 0.9289
  25/25°C
  Other SG
Refractive Index (RI)
  25°C . . . . . . . . . . . . . . . . . . . . . . . 1.4862
  40°C
  Other RI
Iodine Value . . . . . . . . . . . . . . . . . 131–139
Saponification Value . . . . . . . . . . . 191–193
Titer °C
% Unsaponifiable
Melting Point °C

**Fatty Acid Composition (%)**
16:0 . . . . . . . . . . . . . . . . . . . . . . . . . . . 1–2
18:0 . . . . . . . . . . . . . . . . . . . . . . . . . . . 8–9
Total 18:1 . . . . . . . . . . . . . . . . . . . . . . . 34
Undefined 18:2 . . . . . . . . . . . . . . . . . . 44
Undefined 18:3 . . . . . . . . . . . . . . . . . . 12

**References**

## Momordica Cochinchinensis Seed Oil
*Momordica cochinchinensis*
Specific Gravity (SG)
  15.5/15.5°C
  25/25°C
  Other SG
Refractive Index (RI)
  25°C
  40°C
  Other RI
Iodine Value
Saponification Value
Titer °C
% Unsaponifiable
Melting Point °C

**Fatty Acid Composition (%)**
14:0 . . . . . . . . . . . . . . . . . . . . . . . . . . 0.03

16:0 . . . . . . . . . . . . . . . . . . . . . . . . . 2.05
18:0 . . . . . . . . . . . . . . . . . . . . . . . . 17.99
9c-18:1 . . . . . . . . . . . . . . . . . . . . . . 7.92
7c-18:1 . . . . . . . . . . . . . . . . . . . . . . . 0.1
9c,12c-18:2 . . . . . . . . . . . . . . . . . . 10.49
Undefined 18:3 . . . . . . . . . . . . . . . 58.61
9c,12c,15c-18:3 . . . . . . . . . . . . . . . 0.24
20:0 . . . . . . . . . . . . . . . . . . . . . . . . . 0.25
11c-20:1 . . . . . . . . . . . . . . . . . . . . . 0.25
22:0 . . . . . . . . . . . . . . . . . . . . . . . . . 0.22

Tocopherol Composition, mg/kg
 α-Tocopherol . . . . . . . . . . . . . . . . . . . 176
 β-Tocopherol . . . . . . . . . . . . . . . . . . . . . 3
 γ-Tocopherol . . . . . . . . . . . . . . . . . . . . 93
 δ-Tocopherol . . . . . . . . . . . . . . . . . . . . . 2
 Total, mg/kg

**References** *JAOCS 80:* 1013–1020 (2003)

## Monkey Pod Seed Oil
*Samanea saman*

Specific Gravity (SG)
 15.5/15.5°C
 25/25°C . . . . . . . . . . . . . . . . . . . . . . 0.954
 Other SG
Refractive Index (RI)
 25°C
 40°C
 Other RI
Iodine Value . . . . . . . . . . . . . . . . . . . 90–95
Saponification Value . . . . . . . . . . . . . . . 193
Titer °C
% Unsaponifiable . . . . . . . . . . . . . . . . . . . 4
Melting Point °C

**Fatty Acid Composition (%)**
 14:0 . . . . . . . . . . . . . . . . . . . . . . . . . . 1–2
 16:0 . . . . . . . . . . . . . . . . . . . . . . . . . . . . 2
 9c-16:1 . . . . . . . . . . . . . . . . . . . . . . . 1–2
 Total 18:1 . . . . . . . . . . . . . . . . . . . . . . 61
 9c,12c-18:2 . . . . . . . . . . . . . . . . . . . . . 20
 Undefined 18:3 . . . . . . . . . . . . . . . . . . . 6
 20:0 . . . . . . . . . . . . . . . . . . . . . . . . . . . . 2
 Total 20:1 . . . . . . . . . . . . . . . . . . . . . . . 4

**References** *Riv. Ital. Sost. Grasse 73:* 165 (1996)

## Moringa Oleifera Seed Oil
*Moringa oleifera*

Specific Gravity (SG)
 15.5/15.5°C
 25/25°C
 Other SG
Refractive Index (RI)
 25°C . . . . . . . . . . . . . . . . . . . . . . . 1.4650
 40°C . . . . . . . . . . . . . . . . . . . . . . . 1.4559
 Other RI
Iodine Value . . . . . . . . . . . . . . . . . 65.74–68
Saponification Value . . . . . . . . . . 184.16–188
Titer °C
% Unsaponifiable . . . . . . . . . . . . . . . . . 1–2
Melting Point °C

**Fatty Acid Composition (%)**
 8:0 . . . . . . . . . . . . . . . . . . . . . . . 0.02–0.03
 14:0 . . . . . . . . . . . . . . . . . . . . . . . . . . . 0.1
 16:0 . . . . . . . . . . . . . . . . . . . . . . . . . . 5–6
 9c-16:1 . . . . . . . . . . . . . . . . . . . . . 0.11–1
 7c-16:1 . . . . . . . . . . . . . . . . . . . . . . . 1.10
 17:0 . . . . . . . . . . . . . . . . . . . . . . . . . . 0.04
 18:0 . . . . . . . . . . . . . . . . . . . . . . . . . 5.8–8
 Total 18:1 . . . . . . . . . . . . . . . . . . . . 66–76
 Undefined 18:2 . . . . . . . . . . . . . . . . . 0.71
 9c,12c-18:2 . . . . . . . . . . . . . . . . . . 0.6–3.5
 Undefined 18:3 . . . . . . . . . . . . . . 0.1–0.21
 20:0 . . . . . . . . . . . . . . . . . . . . . . . . . 3–6.8
 Total 20:1 . . . . . . . . . . . . . . . . . . . . . 2–3
 22:0 . . . . . . . . . . . . . . . . . . . . . . . . . . 5–8
 Unidentified 22:1 . . . . . . . . . . . . . . . 0.11
 24:0 . . . . . . . . . . . . . . . . . . . . . . . . . . 0–5
 26:0 . . . . . . . . . . . . . . . . . . . . . . . . 0.98–1

Sterol Composition, %
 Cholesterol . . . . . . . . . . . . . . . . . . . . 0.2
 Brassicasterol . . . . . . . . . . . . . . . . . . 0.1
 Campesterol . . . . . . . . . . . . . . 23.83–24
 Stigmasterol . . . . . . . . . . . . . . 17–17.03
 Stigmasta-8,22-dien-3β-ol
 5α-Stigmasta-7,22-dien-3β-ol . . . . . 1.23

Δ7,25-Stigmastadienol . . . . . . . . . . . 0.39
β-Sitosterol . . . . . . . . . . . . . . . 47–47.07
Δ5-Avenasterol . . . . . . . . . . . . . . . . . . 2.9
Δ7-Stigmasterol . . . . . . . . . . . . . . . . 0.8
Δ7-Avenasterol . . . . . . . . . . . . . . 0.19–0.5
Δ7-Campesterol
Δ7-Ergosterol
Δ7,25-Stigmasterol
Sitostanol
Spinasterol
Squalene
24-Methylene Cholesterol . . . . . . 0.9–1.0
Other: D5,25-Ergostadienol, 0.1–0.30; D5,24 ergostadienol, 0.5; clerosterol, 0.7; stigmastanol, 0.77–0.9; D7, 14-stigmastadienol, 0.5; D5,23-stigmastadienol, 1.2; campestanol,0.4–0.5; 28-Isoavenasterol, 0.25
% sterols in oil
Total Sterols, mg/kg

Tocopherol Composition, mg/kg
  α-Tocopherol . . . . . . . . . . . . . . . . 93–227
  β-Tocopherol
  γ-Tocopherol . . . . . . . . . . . . . . . . . . 26–71
  δ-Tocopherol . . . . . . . . . . . . . 53.98–216
  Total, mg/kg

**References** *Riv. Ital. Sost. Grasse 75:* 21 (1998)
*Riv. Ital. Sost. Grasse 75:* 181 (1998)
*INFORM 13:* 151 (2002)

## Moringa Peregrina Seed Oil
*Moringa peregrina*
Specific Gravity (SG)
  15.5/15.5°C
  25/25°C
  Other SG . . . . . . . . . . . . . . . . (24/24) 0.906
Refractive Index (RI)
  25°C
  40°C . . . . . . . . . . . . . . . . . . . . . . . . 1.460
  Other RI
Iodine Value . . . . . . . . . . . . . . . . . . . . . . 70
Saponification Value . . . . . . . . . . . . . . . 185
Titer °C

% Unsaponifiable
Melting Point °C

**Fatty Acid Composition (%)**
  10:0 . . . . . . . . . . . . . . . . . . . . . . . . . . . 0.1
  14:0 . . . . . . . . . . . . . . . . . . . . . . . . . . . 0.1
  16:0 . . . . . . . . . . . . . . . . . . . . . . . . . . . 8.9
  18:0 . . . . . . . . . . . . . . . . . . . . . . . . . . . 3.8
  Total 18:1 . . . . . . . . . . . . . . . . . . . . . 70.5
  9c,12c-18:2 . . . . . . . . . . . . . . . . . . . . . 0.6
  20:0 . . . . . . . . . . . . . . . . . . . . . . . . . . . 1.9
  Total 20:1 . . . . . . . . . . . . . . . . . . . . . . 1.5
  22:0 . . . . . . . . . . . . . . . . . . . . . . . . . . . 2.4
  Unidentified 22:1 . . . . . . . . . . . . . . . . 0.5

Sterol Composition, %
  Cholesterol . . . . . . . . . . . . . . . . 0.09–0.1
  Brassicasterol . . . . . . . . . . . . . . 0.08–0.4
  Campesterol . . . . . . . . . . . . . . . . . . . . 25
  Stigmasterol . . . . . . . . . . . . . . . . . . . . 27
  Stigmasta-8,22-dien-3β-ol
  5α-Stigmasta-7,22-dien-3β-ol
  Δ7,25-Stigmastadienol
  β-Sitosterol . . . . . . . . . . . . . . . . . . . . . 27
  Δ5-Avenasterol . . . . . . . . . . . . . . . . . . 10
  Δ7-Stigmasterol
  Δ7-Avenasterol . . . . . . . . . . . . . . . . . . . 1
  Δ7-Campesterol
  Δ7-Ergosterol . . . . . . . . . . . . . . . . . 0.09
  Δ7,25-Stigmasterol
  Sitostanol
  Spinasterol
  Squalene
  24-Methylene Cholesterol . . . . . . . . . . . 3
  Other: Clerosterol, 0.8; stigmastanol, 0.8; campestanol, 0.5; D7-campestanol 0.5; D5,23-stigmastadienol, 0.2; D5,24-stigmastadienol, 2.4
% sterols in oil
Total Sterols, mg/kg

Tocopherol Composition, mg/kg
  α-Tocopherol . . . . . . . . . . . . . . . . . . 145
  β-Tocopherol
  γ-Tocopherol . . . . . . . . . . . . . . . . . . . . 58
  δ-Tocopherol . . . . . . . . . . . . . . . . . . . . 66
  Total, mg/kg

**References** *Grasas y Acietes 49:* 170 (1998)

## Mowrah Butter

*Madhuca latifolia* and
*M. longifolia*

Specific Gravity (SG)
 15.5/15.5°C
 25/25°C
 Other SG ......... (100/15) 0.856–0.870
Refractive Index (RI)
 25°C
 40°C ................... 1.458–1.462
 Other RI
Iodine Value .................... 53–70
Saponification Value ........... 188–200
Titer °C ....................... 36–45
% Unsaponifiable ................. 1–3
Melting Point °C

**Fatty Acid Composition (%)**
 16:0 ........................... 24
 18:0 ........................... 19
 9c-18:1 ........................ 43
 Undefined 18:2 ................. 14

**References**

## Munch Seed Oil

*Dimorphotheca pluvialis*

Specific Gravity (SG)
 15.5/15.5°C
 25/25°C
 Other SG
Refractive Index (RI)
 25°C
 40°C
 Other RI
Iodine Value
Saponification Value
Titer °C
% Unsaponifiable
Melting Point °C

**Fatty Acid Composition (%)**
 16:0 .......................... 1.9
 18:0 .......................... 1.5
 Total 18:1 ................... 17.5

 9c,12c-18:2 ................... 12.3
 Undefined 18:3 ................. 1.7
 22:0 ............................. 2
 Unidentified 22:1 ................ 2
 Other: 9-hydroxy10t,12t-18:2 (Dimorphecolic), 61.7; dehydrodimorphecolic, 1.0

**References** *J. Am. Oil Chem. Soc.* 71: 313 (1994)

## Mustard Seed Oil

Black - *Brassica juncea*,
white/yellow - *Sinapsis alba*

Specific Gravity (SG)
 15.5/15.5°C
 25/25°C
 Other SG .......... (20/20) 0.910–0.921
Refractive Index (RI)
 25°C
 40°C ................... 1.461–1.469
 Other RI
Iodine Value .................. 92–125
Saponification Value ........... 170–184
Titer °C
% Unsaponifiable ................ 0–1.5
Melting Point °C

**Fatty Acid Composition (%)**
 14:0 ......................... 0–1.0
 16:0 ......................... 0.5–4.5
 9c-16:1 ...................... 0–0.5
 18:0 ......................... 0.5–2
 Total 18:1 ................... 8–23
 7c-18:1 ...................... 1.22
 11c-18:1 ..................... 7.77
 9c,12c-18:2 .................. 10–24
 Undefined 18:3 ............... 6–18
 9c,12c,15c-18:3 .............. 11.84
 20:0 ......................... 0–1.5
 Total 20:1 ................... 5–13
 11c-20:1 ..................... 5.71
 20:2 ......................... 0–1
 22:0 ......................... 0.2–2.5
 Unidentified 22:1 ............ 22–50
 13c-22:1 ..................... 43.27
 22:2 ......................... 0–1.46

24:0 . . . . . . . . . . . . . . . . . . . . . . . 0–0.67
15c-24:1 . . . . . . . . . . . . . . . . . . . 0.5–2.5

Sterol Composition, %
  Cholesterol
  Brassicasterol . . . . . . . . . . . . . . . . . . . . . 6
  Campesterol . . . . . . . . . . . . . . . . . . . . 33
  Stigmasterol
  Stigmasta-8,22-dien-3β-ol
  5α-Stigmasta-7,22-dien-3β-ol
  Δ7,25-Stigmastadienol
  β-Sitosterol . . . . . . . . . . . . . . . . . . . . . 58
  Δ5-Avenasterol . . . . . . . . . . . . . . . . . . . 2
  Δ7-Stigmasterol
  Δ7-Avenasterol
  Δ7-Campesterol
  Δ7-Ergosterol
  Δ7,25-Stigmasterol
  Sitostanol
  Spinasterol
  Squalene
  24-Methylene Cholesterol
  Other
  % sterols in oil
  Total Sterols, mg/kg

Tocopherol Composition, mg/kg
  α-Tocopherol . . . . . . . . . . . . . . . . 75–138
  β-Tocopherol
  γ-Tocopherol . . . . . . . . . . . . . . . . 308–494
  δ-Tocopherol . . . . . . . . . . . . . . . . . . . 0–31
  Total, mg/kg . . . . . . . . . . . . . . . 446–663

Tocotrienols Composition, mg/kg
  α-Tocotrienol . . . . . . . . . . . . . . . . . . . . . 2
  β-Tocotrienol
  γ-Tocotrienol
  δ-Tocotrienol
  Total Tocotrienols, mg/kg

**References** *Codex* 1993/16
  *Riv. Ital. Sost. Grasse 52:* 79 (1975)
  *J. Nutr. 81:* 335 (1963)
  *J. Am. Oil Chem. Soc. 53:* 732 (1976)
  *JAOCS 80:* 1013–1020 (2003)

## Mustard Seed Oil Oriental

Specific Gravity (SG)
  15.5/15.5°C
  25/25°C
  Other SG
Refractive Index (RI)
  25°C
  40°C
  Other RI
Iodine Value
Saponification Value
Titer °C
% Unsaponifiable
Melting Point °C

**Fatty Acid Composition (%)**
  16:0 . . . . . . . . . . . . . . . . . . . . . . . . . . 2.8
  16:1 . . . . . . . . . . . . . . . . . . . . . . . . . . 0.2
  18:0 . . . . . . . . . . . . . . . . . . . . . . . . . . 1.5
  Total 18:1 . . . . . . . . . . . . . . . . . . . 21.4
  Undefined 18:2 . . . . . . . . . . . . . . . 19.9
  Undefined 18:3 . . . . . . . . . . . . . . . 12.1
  20:0 . . . . . . . . . . . . . . . . . . . . . . . . . . 1.0
  Total 20:1 . . . . . . . . . . . . . . . . . . . 13.5
  20:2 . . . . . . . . . . . . . . . . . . . . . . . . . . 1.1
  22:0 . . . . . . . . . . . . . . . . . . . . . . . . . . 0.5
  Unidentified 22:1 . . . . . . . . . . . . . . 23.1
  22:2 . . . . . . . . . . . . . . . . . . . . . . . . . . 0.4
  24:0 . . . . . . . . . . . . . . . . . . . . . . . . . . 0.3
  15c-24:1 . . . . . . . . . . . . . . . . . . . . . . 1.4

**References** Canadian Grain Commission

## Nectarine Seed Oil

*Prunus persica var nectarina*

Specific Gravity (SG)
  15.5/15.5°C . . . . . . . . . . . . . . . . . . . . . .
  25/25°C
  Other SG
Refractive Index (RI)
  25°C
  40°C
  Other RI
Iodine Value . . . . . . . . . . . . . . . . . . . . . 108
Saponification Value . . . . . . . . . . . . . . . 192
Titer °C
% Unsaponifiable . . . . . . . . . . . . . . . . . . 0.8
Melting Point °C

**Fatty Acid Composition (%)**
  16:0 . . . . . . . . . . . . . . . . . . . . . . . . . . 6.1

16:1 .......................... 0.5
Total 18:1 ..................... 66.3
Undefined 18:2 ................. 26.8
20:0 .......................... 0.3

**References** *JAOCS 69:* 492–494 (1992)

## Neem (Margosa) Oil
*Melia/Azadirachta indica*
Specific Gravity (SG)
 15.5/15.5°C
 25/25°C
 Other SG .......... (30/30) 0.913–0.918
Refractive Index (RI)
 25°C
 40°C ................. 1.4617–1.4627
 Other RI
Iodine Value .................... 68–71
Saponification Value ........... 195–204
Titer °C
% Unsaponifiable ................. 1–2
Melting Point °C

**Fatty Acid Composition (%)**
 16:0 ......................... 13–18
 18:0 ......................... 14–24
 Total 18:1 .................... 49–62
 9c,12c-18:2 .................... 7–15
 20:0 .......................... 1–4

**References** *Food Chem. 26:* 119 (1987)

## Neetsfoot Oil
Specific Gravity (SG)
 15.5/15.5°C
 25/25°C ................. 0.906–0.912
 Other SG
Refractive Index (RI)
 25°C
 40°C ................... 1.458–1.461
 Other RI
Iodine Value .................... 69–75
Saponification Value ........... 192–198
Titer °C
% Unsaponifiable .................. 0–8
Melting Point °C

**Fatty Acid Composition (%)**
 Other ................. unsaturate, 79

**References**

## Nephelium lappaceum
*Nephelium lappaceum*
Specific Gravity (SG)
 15.5/15.5°C
 25/25°C
 Other SG
Refractive Index (RI)
 25°C
 40°C
 Other RI
Iodine Value
Saponification Value
Titer °C
% Unsaponifiable
Melting Point °C

**Fatty Acid Composition (%)**
 14:0 ......................... 0.02
 16:0 ......................... 4.12
 9c-16:1 ....................... 0.34
 18:0 ......................... 5.16
 9c-18:1 ...................... 36.22
 7c-18:1 ....................... 1.18
 9c,12c-18:2 ................... 2.99
 9c,12c,15c-18:3 ................ 0.2
 20:0 ........................ 33.24
 11c-20:1 ...................... 8.24
 22:0 ......................... 3.92
 13c-22:1 ...................... 1.06
 22:2 .......................... 0.6
 24:0 ......................... 0.79

Tocopherol Composition, mg/kg
 α-Tocopherol .................... 7
 β-Tocopherol
 γ-Tocopherol .................... 4
 δ-Tocopherol .................... 4
 Total, mg/kg

Tocotrienols Composition, mg/kg
 α-Tocotrienol ................... 7
 β-Tocotrienol
 γ-Tocotrienol ................... 4

δ-Tocotrienol
Total Tocotrienols, mg/kg

**References**

## Neou Seed Oil
*Parinarium mocrophyllum*
Specific Gravity (SG)
  15.5/15.5°C
  25/25°C
  Other SG . . . . . . . . . . . . . . . (78/4) 0.8901
Refractive Index (RI)
  25°C
  40°C . . . . . . . . . . . . . . . . . . 1.483–1.485
  Other RI
Iodine Value . . . . . . . . . . . . . . . . . . 130–140
Saponification Value . . . . . . . . . . . 184–190
Titer °C
% Unsaponifiable
Melting Point °C

**Fatty Acid Composition (%)**
  16:0 . . . . . . . . . . . . . . . . . . . . . . . . . . 6–12
  18:0 . . . . . . . . . . . . . . . . . . . . . . . . . . . 2–7
  Total 18:1 . . . . . . . . . . . . . . . . . . . . . 33–40
  9c,12c-18:2 . . . . . . . . . . . . . . . . . . . . 15–20
  α-Eleostearic 9c,11t,13t-18:3 . . . . . 31–32
  Total 20:1 . . . . . . . . . . . . . . . . . . . . . . 0.5
  22:0 . . . . . . . . . . . . . . . . . . . . . . . . . . . 0.4

**References** *Fat Sci. Technol. 96:* 64 (1994)

## Nigella Seed Oil
*Nigella sativa*
Specific Gravity (SG)
  15.5/15.5°C
  25/25°C
  Other SG
Refractive Index (RI)
  25°C
  40°C
  Other RI
Iodine Value
Saponification Value
Titer °C

% Unsaponifiable
Melting Point °C

**Fatty Acid Composition (%)**
  16:0 . . . . . . . . . . . . . . . . . . . . . . . . . . 11.4
  16:1 . . . . . . . . . . . . . . . . . . . . . . . . . . . . tr
  18:0 . . . . . . . . . . . . . . . . . . . . . . . . . . . 2.9
  Total 18:1 . . . . . . . . . . . . . . . . . . . . . 21.9
  Undefined 18:2 . . . . . . . . . . . . . . . . 60.8
  Undefined 18:3 . . . . . . . . . . . . . . . . . . tr

Tocopherol Composition, mg/kg
  α-Tocopherol . . . . . . . . . . . . . . . . . . . . 40
  β-Tocopherol . . . . . . . . . . . . . . . . . . . . 50
  γ-Tocopherol . . . . . . . . . . . . . . . . . . . 250
  δ-Tocopherol
  Total, mg/kg

**References** *JAOCS 74:* 375–380 (1997)

## Niger Seed Oil
*Guizotia abyssinica*
Specific Gravity (SG)
  15.5/15.5°C . . . . . . . . . . . . . . 0.923–0.927
  25/25°C
  Other SG
Refractive Index (RI)
  25°C
  40°C . . . . . . . . . . . . . . . . . . 1.467–1.469
  Other RI
Iodine Value . . . . . . . . . . . . . . . . . . 126–135
Saponification Value . . . . . . . . . . . 188–193
Titer °C
% Unsaponifiable . . . . . . . . . . . . . . . 0.5–3.7
Melting Point °C

**Fatty Acid Composition (%)**
  14:0 . . . . . . . . . . . . . . . . . . . . . . . . . . . 1–3
  16:0 . . . . . . . . . . . . . . . . . . . . . . . . . . 5–12
  16:1 . . . . . . . . . . . . . . . . . . . . . . . . . . . . tr
  9c-16:1 . . . . . . . . . . . . . . . . . . . . . . . . . 0.1
  18:0 . . . . . . . . . . . . . . . . . . . . . . . . . . 2–12
  Total 18:1 . . . . . . . . . . . . . . . . . . . . . . 4–10
  11c-18:1 . . . . . . . . . . . . . . . . . . . . . 0.1–0.2
  Undefined 18:2 . . . . . . . . . . . . . . . . . 76.7
  9c,12c-18:2 . . . . . . . . . . . . . . . . . . . . 52–78
  Undefined 18:3 . . . . . . . . . . . . . . . . . . 0–3

20:0 . . . . . . . . . . . . . . . . . . . . . . . 0.2–0.3
22:0 . . . . . . . . . . . . . . . . . . . . . . . 0.3–0.6
24:0 . . . . . . . . . . . . . . . . . . . . . . . 0.2–0.3

Sterol Composition, %
　Cholesterol . . . . . . . . . . . . . . . . . . 0.2–0.8
　Brassicasterol
　Campesterol . . . . . . . . . . . . . . . . . . 12–13
　Stigmasterol . . . . . . . . . . . . . . . . . . 13–14
　Stigmasta-8,22-dien-3β-ol
　5α-Stigmasta-7,22-dien-3β-ol
　Δ7,25-Stigmastadienol
　β-Sitosterol . . . . . . . . . . . . . . . . . . . . 38–43
　Δ5-Avenasterol . . . . . . . . . . . . . . . . . 6–7
　Δ7-Stigmasterol . . . . . . . . . . . . . . . . 4–5
　Δ7-Avenasterol . . . . . . . . . . . . . . . . . . . 4
　Δ7-Campesterol
　Δ7-Ergosterol
　Δ7,25-Stigmasterol
　Sitostanol
　Spinasterol
　Squalene
　24-Methylene Cholesterol
　Other . . . . . . . . . . . . . . . . . . . . . . . 14–22
　% sterols in oil
　Total Sterols, mg/kg

Tocopherol Composition, mg/kg
　α-Tocopherol . . . . . . . . . . . . . . . 600–800
　β-Tocopherol . . . . . . . . . . . . . . . . . . . 6–8
　γ-Tocopherol . . . . . . . . . . . . . . . . . 24–40
　δ-Tocopherol
　Total, mg/kg . . . . . . . . . . . . . . . . 657–853

**References** *J. Am. Oil Chem. Soc. 71:* 839 (1994)
*JAOCS 74:* 375–380 (1997)


## Nutmeg Butter
*Myristica fragrans*
Specific Gravity (SG)
　15.5/15.5°C
　25/25°C
　Other SG
Refractive Index (RI)
　25°C
　40°C . . . . . . . . . . . . . . . . 1.4659–1.4704
　Other RI

Iodine Value . . . . . . . . . . . . . . . . . . . . 48–85
Saponification Value . . . . . . . . . . . . 170–190
Titer °C
% Unsaponifiable
Melting Point °C . . . . . . . . . . . . . . . . 40–50

**Fatty Acid Composition (%)**
　12:0 . . . . . . . . . . . . . . . . . . . . . . . . . . . 3–6
　14:0 . . . . . . . . . . . . . . . . . . . . . . . . . 76–83
　16:0 . . . . . . . . . . . . . . . . . . . . . . . . . . 4–10
　Total 18:1 . . . . . . . . . . . . . . . . . . . . . . 5–11
　9c,12c-18:2 . . . . . . . . . . . . . . . . . . . . . 0–2

**References** *J. Am. Dietetic Assn. 68:* 224 (1976)


## Oat Bean Oil
*Pentaclethera macrophylla*
Specific Gravity (SG)
　15.5/15.5°C
　25/25°C . . . . . . . . . . . . . . . . . . . . . 0.9073
　Other SG
Refractive Index (RI)
　25°C
　40°C
　Other RI . . . . . . . . . . . . . . . . . (30) 1.4723
Iodine Value . . . . . . . . . . . . . . . . . . . . 86–96
Saponification Value . . . . . . . . . . . . 181–187
Titer °C
% Unsaponifiable
Melting Point °C

**Fatty Acid Composition (%)**
　12:0 . . . . . . . . . . . . . . . . . . . . . . . . . . . . 0.3
　14:0 . . . . . . . . . . . . . . . . . . . . . . . . . . . . 0.7
　16:0 . . . . . . . . . . . . . . . . . . . . . . . . . . . . 3.8
　9c-16:1 . . . . . . . . . . . . . . . . . . . . . . . . . 0.1
　18:0 . . . . . . . . . . . . . . . . . . . . . . . . . . . . 2.4
　Total 18:1 . . . . . . . . . . . . . . . . . . . . . . . 31
　9c,12c-18:2 . . . . . . . . . . . . . . . . . . . . . . 36
　20:0 . . . . . . . . . . . . . . . . . . . . . . . . . . . . 2.4
　Total 20:1 . . . . . . . . . . . . . . . . . . . . . . . 1.7
　22:0 . . . . . . . . . . . . . . . . . . . . . . . . . . . . . 4
　15c-24:1 . . . . . . . . . . . . . . . . . . . . . . . . 17

**References** *Riv. Ital. Sost. Grasse 61:* 569 (1984)

## Oat Oil
*Avena sativa*

Specific Gravity (SG)
  15.5/15.5°C
  25/25°C................ 0.919–0.921
  Other SG
Refractive Index (RI)
  25°C
  40°C .................. 1.464–1.470
  Other RI
Iodine Value .................. 105–116
Saponification Value ........... 190–199
Titer °C
% Unsaponifiable ............... 1.3–2.6
Melting Point °C

**Fatty Acid Composition (%)**
  14:0....................... 0.2–4.9
  16:0....................... 13.2–39.4
  16:1....................... 0.1–0.3
  9c-16:1 .................... 0.1
  18:0....................... 0.5–4
  Total 18:1 ................. 17.9–53
  Undefined 18:2 ............. 24–53
  9c,12c-18:2 ................ 24–48
  Undefined 18:3 ............. 0.7–5
  20:0....................... 0.2
  Total lipids (dry wt basis)....... 2–11.8

Tocopherol Composition, mg/kg
  α-Tocopherol .................... 19
  β-Tocopherol..................... 6
  γ-Tocopherol
  δ-Tocopherol
  Total, mg/kg

Tocotrienols Composition, mg/kg
  α-Tocotrienol ................... 51
  β-Tocotrienol ................... 12
  γ-Tocotrienol.................... 7
  δ-Tocotrienol.................... 5
  Total Tocotrienols, mg/kg......... 175

**References** *J. Am. Oil Chem. Soc. 52:* 358 (1975)
  *J. Am. Oil Chem. Soc. 52:* 491 (1975)
  *J. Am. Oil Chem. Soc. 54:* 305 (1977)
  *Anal. Biochem. 32:* 81 (1969)
  *JAOCS 76:* 159–169 (1999)

## Ochoco Butter (kernel fat)
*Scyphocephalium ochocoa*

Specific Gravity (SG)
  15.5/15.5°C
  25/25°C
  Other SG ............... (60/4) 0.8899
Refractive Index (RI)
  25°C
  40°C
  Other RI
Iodine Value ........................ 1.7
Saponification Value ................ 239
Titer °C
% Unsaponifiable
Melting Point °C................... 45–48

**Fatty Acid Composition (%)**
  12:0......................... 17
  14:0......................... 82
  16:0......................... 1
  Total 18:1 ................... 0.5

**References** *Rev. Franc. Corp Gras 39:* 147 (1992)

## Oil Bean Oil
*Pentaclethra macrophylla*

Specific Gravity (SG)
  15.5/15.5°C
  25/25°C..................... 0.9073
  Other SG
Refractive Index (RI)
  25°C
  40°C
  Other RI ................ (30) 1.4723
Iodine Value .................. 86–96
Saponification Value .......... 181–187
Titer °C
% Unsaponifiable
Melting Point °C

**Fatty Acid Composition (%)**
  12:0......................... 0.3
  14:0......................... 0.7
  16:0......................... 3.8
  16:1......................... 0.1
  18:0......................... 2.4

Total 18:1 . . . . . . . . . . . . . . . . . . . . . . . 31
Undefined 18:2 . . . . . . . . . . . . . . . . . . 36
20:0. . . . . . . . . . . . . . . . . . . . . . . . . . . . 2.4
Total 20:1 . . . . . . . . . . . . . . . . . . . . . . 1.7
22:0. . . . . . . . . . . . . . . . . . . . . . . . . . . . . 4
15c-24:1 . . . . . . . . . . . . . . . . . . . . . . . 17

**References** *Riv. Ital. Sost. Grasse 61:* 569 (1984)

## Oiticica Oil
*Licania rigida*

Specific Gravity (SG)
   15.5/15.5°C
   25/25°C
   Other SG . . . . . . . . (20/20) 0.9752–0.9726
Refractive Index (RI)
   25°C . . . . . . . . . . . . . . . . . 1.5121–1.5161
   40°C
   Other RI
Iodine Value . . . . . . . . . . . . . . . . . . . 140–150
Saponification Value . . . . . . . . . . . . 188–193
Titer °C
% Unsaponifiable
Melting Point °C

**Fatty Acid Composition (%)**
   16:0. . . . . . . . . . . . . . . . . . . . . . . . . . . . . 7
   18:0. . . . . . . . . . . . . . . . . . . . . . . . . . . . . 5
   Total 18:1 . . . . . . . . . . . . . . . . . . . . . . 4–7
   α-Eleostearic 9c,11t,13t-18:3 . . . . . . . . . 5
   Other 4-keto-9,11,13–18:3, 70–80

**References**

## Okra Seed Oil
*Hibiscus esculentus*

Specific Gravity (SG)
   15.5/15.5°C
   25/25°C. . . . . . . . . . . . . . . . . 0.916–0.919
   Other SG
Refractive Index (RI)
   25°C . . . . . . . . . . . . . . . . . . 1.468–1.467
   40°C . . . . . . . . . . . . . . . . . . 1.462–1.467
   Other RI

Iodine Value . . . . . . . . . . . . . . . . . . . 90–100
Saponification Value . . . . . . . . . . . . 192–199
Titer °C
% Unsaponifiable . . . . . . . . . . . . . . . 0.7–1.4
Melting Point °C

**Fatty Acid Composition (%)**
   14:0. . . . . . . . . . . . . . . . . . . . . . . . . . . 0–4
   16:0. . . . . . . . . . . . . . . . . . . . . . . . . . 23–33
   9c-16:1 . . . . . . . . . . . . . . . . . . . . . . . 0–0.6
   18:0. . . . . . . . . . . . . . . . . . . . . . . . . . 0.5–13
   Total 18:1 . . . . . . . . . . . . . . . . . . . . . 26–49
   Undefined 18:2 . . . . . . . . . . . . . . . . . 22–42
   9c,12c-18:2 . . . . . . . . . . . . . . . . . . . . 22–42
   20:0. . . . . . . . . . . . . . . . . . . . . . . . . . . . 0–1

Tocopherol Composition, mg/kg
   α-Tocopherol . . . . . . . . . . . . . . . 280–450
   β-Tocopherol
   γ-Tocopherol . . . . . . . . . . . . . . . . 420–660
   δ-Tocopherol
   Total, mg/kg . . . . . . . . . . . . . . . 700–1130

**References** *J. Am. Oil Chem. Soc. 27:* 414 (1950)
*J. Food Sci. Agric. 25:* 401 (1974)

## Olive (Wild) Oil, Kandarakkara Oil
*Ximenia americana*

Specific Gravity (SG)
   15.5/15.5°C. . . . . . . . . . . . . . . . . . 0.9362
   25/25°C
   Other SG
Refractive Index (RI)
   25°C . . . . . . . . . . . . . . . . . 1.4691–1.4731
   40°C
   Other RI
Iodine Value . . . . . . . . . . . . . . . . . . . . 82–95
Saponification Value . . . . . . . . . . . . 167–170
Titer °C
% Unsaponifiable . . . . . . . . . . . . . . . . . 0.5–5
Melting Point °C

**Fatty Acid Composition (%)**
   18:0. . . . . . . . . . . . . . . . . . . . . . . . . . . . . 1
   Total 18:1 . . . . . . . . . . . . . . . . . . . . . . . 61

Undefined 18:2 .................... 7
26:0 ............................. 15
26:1 ............................. 15

**References** *INFORM 13:* 151 (2002)

## Olive Oil (for quality grade reference values see IOOC documentation)
*Olea europaea*

Specific Gravity (SG)
 15.5/15.5°C
 25/25°C
 Other SG .......... (20/20) 0.910–0.916
Refractive Index (RI)
 25°C
 40°C
 Other RI .......... (20) 1.4677–1.4705
Iodine Value ..................... 75–94
Saponification Value ........... 184–196
Titer ° C
% Unsaponifiable .................. 1.5
Melting Point °C .................. -3 - 0

**Fatty Acid Composition (%)**
 14:0 ......................... 0–0.1
 16:0 ......................... 7.5–20
 16:1 ........................... 1.4
 9c-16:1 ..................... 0.3–3.5
 18:0 ......................... 0.5–5.0
 Total 18:1 .................... 55–83
 Undefined 18:2 .................... 9
 9c,12c-18:2 .................. 3.5–21
 Undefined 18:3 ............... 0–1.5
 20:0 ......................... 0–0.8
 22:0 ......................... 0–0.2
 24:0 ......................... 0–1.0

Sterol Composition, %
 Cholesterol ................... 0–0.5
 Brassicasterol ................ 0–0.1
 Campesterol ................... 0–4.0
 Stigmasterol .................. 0–4.0
 Stigmasta-8,22-dien-3β-ol
 5α-Stigmasta-7,22-dien-3β-ol
 Δ7,25-Stigmastadienol
 β-Sitosterol ................... 75–80
 Δ5-Avenasterol ................ 4–14
 Δ7-Stigmasterol ............... 0–0.5
 Δ7-Avenasterol
 Δ7-Campesterol
 Δ7-Ergosterol
 Δ7,25-Stigmasterol
 Sitostanol
 Spinasterol
 Squalene
 24-Methylene Cholesterol
 Other
 % sterols in oil
 Total Sterols, mg/kg .............. 100

Tocopherol Composition, mg/kg
 α-Tocopherol ................. 63–135
 β-Tocopherol. ..................... 6
 γ-Tocopherol ................... 7–15
 δ-Tocopherol
 Total, mg/kg .................. 70–150

**References** *J. Am. Oil Chem. Soc. 63:* 328 (1986)
*J. Nutr. 81:* 335 (1963)
*J. Chromatog. 630:* 213 (1993)
See EU, IOOC and Codex recommendations
*JAOCS. 74:* 375–381 (1997)

## Onion Seed Oil
*Allium cepa*

Specific Gravity (SG)
 15.5/15.5°C ................... 0.9289
 25/25°C
 Other SG
Refractive Index (RI)
 25°C ......................... 1.4730
 40°C
 Other RI
Iodine Value ...................... 112
Saponification Value .............. 197.5
Titer °C
% Unsaponifiable .................. 1.2
Melting Point °C

**Fatty Acid Composition (%)**
- 16:0 .............................. 3
- 18:0 .............................. 1.5
- Total 18:1 ...................... 58
- 9c,12c-18:2 .................... 38

**References**

## Onosmodium Hispidissimum
*Onosmodium hispidissimum*

Specific Gravity (SG)
- 15.5/15.5°C
- 25/25°C
- Other SG

Refractive Index (RI)
- 25°C
- 40°C
- Other RI

Iodine Value
Saponification Value
Titer °C
% Unsaponifiable
Melting Point °C

**Fatty Acid Composition (%)**
- 16:0 ............................ 6.5
- 18:0 ............................ 2.5
- Total 18:1 .................... 13.5
- Undefined 18:2 ............. 18.2
- Undefined 18:3 ............. 46.9
- 18:4 ............................ 8.1
- Total 20:1 .................... 1.8
- Unidentified 22:1 .......... 0.2

**References** *JAOCS 70:* 629 (1993)

## Orange Seed Oil
*Citrus sinensis*

Specific Gravity (SG)
- 15.5/15.5°C
- 25/25°C ................. 0.916–0.920
- Other SG

Refractive Index (RI)
- 25°C ..................... 1.468–1.470
- 40°C ..................... 1.460–1.465
- Other RI

Iodine Value ..................... 98–104
Saponification Value ........... 186–197
Titer °C
% Unsaponifiable ............... 0.4–1.0
Melting Point °C

**Fatty Acid Composition (%)**
- 14:0 ............................... 1
- 16:0 ............................ 21–29
- 9c-16:1 .......................... 1
- 18:0 ............................ 4–8
- Total 18:1 ..................... 20–37
- 9c,12c-18:2 ................... 36–38
- Undefined 18:3 .............. 1–7
- 20:0 ............................. 0.2

Sterol Composition, %
- Cholesterol .................... 0.2
- Brassicasterol
- Campesterol .................... 8
- Stigmasterol .................... 3
- Stigmasta-8,22-dien-3β-ol
- 5α-Stigmasta-7,22-dien-3β-ol
- Δ7,25-Stigmastadienol
- β-Sitosterol .................... 88
- Δ5-Avenasterol ................ 0.2
- Δ7-Stigmasterol
- Δ7-Avenasterol
- Δ7-Campesterol
- Δ7-Ergosterol
- Δ7,25-Stigmasterol
- Sitostanol
- Spinasterol
- Squalene
- 24-Methylene Cholesterol
- Other
- % sterols in oil
- Total Sterols, mg/kg

**References** *Riv. Ital Sost. Grasse 66:* 99 (1989)
*Food Chem. 47:* 77 (1993)
*Grasas y Acietes 39:* 232 (1988)

## Otoba Butter, American Nutmeg Butter
*Virola otoba*

Specific Gravity (SG)
  15.5/15.5°C
  25/25°C
  Other SG
Refractive Index (RI)
  25°C
  40°C . . . . . . . . . . . . . . . . . . . . . . . 1.471
  Other RI
Iodine Value . . . . . . . . . . . . . . . . . . . . . . 54
Saponification Value . . . . . . . . . . . . . . . 185
Titer °C
% Unsaponifiable . . . . . . . . . . . . . . . . . . 20
Melting Point °C . . . . . . . . . . . . . . . . . . 34

**Fatty Acid Composition (%)**
  12:0 . . . . . . . . . . . . . . . . . . . . . . . . . . 21
  14:0 . . . . . . . . . . . . . . . . . . . . . . . . . . 73
  16:0 . . . . . . . . . . . . . . . . . . . . . . . . . 0.3
  Total 18:1 . . . . . . . . . . . . . . . . . . . . . . 6

**References**

## Ouricuri Tallow
*Syagrus coronata/Orbignya cohune*

Specific Gravity (SG)
  15.5/15.5°C
  25/25°C . . . . . . . . . . . . . . . . . . . . . 0.9221
  Other SG
Refractive Index (RI)
  25°C . . . . . . . . . . . . . . . . . . . . . . . 1.4543
  40°C
  Other RI
Iodine Value . . . . . . . . . . . . . . . . . . . . . . 15
Saponification Value . . . . . . . . . . . . . . . 257
Titer °C
% Unsaponifiable
Melting Point °C

**Fatty Acid Composition (%)**
  8:0 . . . . . . . . . . . . . . . . . . . . . . . . . . . 10
  10:0 . . . . . . . . . . . . . . . . . . . . . . . . . . . 9
  12:0 . . . . . . . . . . . . . . . . . . . . . . . . . . 46
  14:0 . . . . . . . . . . . . . . . . . . . . . . . . . . . 9
  16:0 . . . . . . . . . . . . . . . . . . . . . . . . . . . 8
  18:0 . . . . . . . . . . . . . . . . . . . . . . . . . . . 2
  Total 18:1 . . . . . . . . . . . . . . . . . . . . . 13
  9c,12c-18:2 . . . . . . . . . . . . . . . . . . . . . 3

**References**

## Palas Oil
*Butea frondosa/monosperma*

Specific Gravity (SG)
  15.5/15.5°C
  25/25°C
  Other SG . . . . . . . . . . . . . (30/30) 0.9076
Refractive Index (RI)
  25°C
  40°C
  Other RI . . . . . . . . . . . . . . . (30) 1.4791
Iodine Value . . . . . . . . . . . . . . . . . . . . . . 85
Saponification Value . . . . . . . . . . . . . . . 185
Titer °C
% Unsaponifiable . . . . . . . . . . . . . . . . . 1–2
Melting Point °C

**Fatty Acid Composition (%)**
  16:0 . . . . . . . . . . . . . . . . . . . . . . . 20–28
  18:0 . . . . . . . . . . . . . . . . . . . . . . . . . 7–9
  Total 18:1 . . . . . . . . . . . . . . . . . . . 28–31
  Undefined 18:2 . . . . . . . . . . . . . . . 16–26
  9c,12c-18:2 . . . . . . . . . . . . . . . . . . . . . 26
  20:0 . . . . . . . . . . . . . . . . . . . . . . . . . 3–6
  Total 20:1 . . . . . . . . . . . . . . . . . . . . 2–3
  22:0 . . . . . . . . . . . . . . . . . . . . . . . . 5–12
  24:0 . . . . . . . . . . . . . . . . . . . . . . . 2.5–10

**References** *INFORM 13:* 151 (2002)

## Palm Kernel Oil
*Elaeis guineensis*

Specific Gravity (SG)
  15.5/15.5°C
  25/25°C
  Other SG . . . . . . . . . . (40/20) 0.899–0.914
Refractive Index (RI)
  25°C

40°C . . . . . . . . . . . . . . . . . . . . 1.448–1.452
Other RI
Iodine Value . . . . . . . . . . . . . . . . . . . . . 14–21
Saponification Value . . . . . . . . . . . 230–254
Titer °C
% Unsaponifiable . . . . . . . . . . . . . . . . . . 0–1
Melting Point °C . . . . . . . . . . . . . . . . . 24–26

**Fatty Acid Composition (%)**
6:0 . . . . . . . . . . . . . . . . . . . . . . . . . . . 0–0.8
8:0 . . . . . . . . . . . . . . . . . . . . . . . . . . 1.9–6.2
10:0 . . . . . . . . . . . . . . . . . . . . . . . . . 2.6–5.0
12:0 . . . . . . . . . . . . . . . . . . . . . . . . . . 40–55
14:0 . . . . . . . . . . . . . . . . . . . . . . . . . . 14–18
16:0 . . . . . . . . . . . . . . . . . . . . . . . . . 6.5–10.3
18:0 . . . . . . . . . . . . . . . . . . . . . . . . . . 1.3–3
Total 18:1 . . . . . . . . . . . . . . . . . . . . . 12–21
9c,12c-18:2 . . . . . . . . . . . . . . . . . . . . 1–3.5
Undefined 18:3 . . . . . . . . . . . . . . . . . 0–0.7
20:0 . . . . . . . . . . . . . . . . . . . . . . . . . . 0–0.3
Total 20:1 . . . . . . . . . . . . . . . . . . . . . 0–0.5

Sterol Composition, %
Cholesterol . . . . . . . . . . . . . . . . . . 0.6–3.7
Brassicasterol . . . . . . . . . . . . . . . . . 0–0.8
Campesterol . . . . . . . . . . . . . . . . 8.4–12.7
Stigmasterol . . . . . . . . . . . . . . . 12.0–16.6
Stigmasta-8,22-dien-3β-ol
5α-Stigmasta-7,22-dien-3β-ol
Δ7,25-Stigmastadienol
β-Sitosterol . . . . . . . . . . . . . . . . 62.6–73.1
Δ5-Avenasterol . . . . . . . . . . . . . . . 1.4–9.0
Δ7-Stigmasterol . . . . . . . . . . . . . . . 0–2.1
Δ7-Avenasterol . . . . . . . . . . . . . . . . 0–1.4
Δ7-Campesterol
Δ7-Ergosterol
Δ7,25-Stigmasterol
Sitostanol
Spinasterol
Squalene
24-Methylene Cholesterol
Other . . . . . . . . . . . . . . . . . . . . . . . . . 0–2.7
% sterols in oil
Total Sterols, mg/kg . . . . . . . . . 790–1410

Tocopherol Composition, mg/kg
α-Tocopherol . . . . . . . . . . . . . . . . . . 0–40
β-Tocopherol . . . . . . . . . . . . . . . . . . 0–250
γ-Tocopherol . . . . . . . . . . . . . . . . . . 0–260
δ-Tocopherol
Total, mg/kg . . . . . . . . . . . . . . . . . . 0–260

Tocotrienols Composition, mg/kg
α-Tocotrienol
β-Tocotrienol
γ-Tocotrienol . . . . . . . . . . . . . . . . . . . 0–60
δ-Tocotrienol
Total Tocotrienols, mg/kg

**References** *Codex* 1997/17

# Palm Kernel Oil
*Buttia capitata*

Specific Gravity (SG)
15.5/15.5°C
25/25°C
Other SG
Refractive Index (RI)
25°C
40°C
Other RI
Iodine Value
Saponification Value
Titer °C
% Unsaponifiable
Melting Point °C

**Fatty Acid Composition (%)**
8:0 . . . . . . . . . . . . . . . . . . . . . . . . . . . . . 16
10:0 . . . . . . . . . . . . . . . . . . . . . . . . . . . . 16
12:0 . . . . . . . . . . . . . . . . . . . . . . . . . . . . 40
14:0 . . . . . . . . . . . . . . . . . . . . . . . . . . . . 6–7
16:0 . . . . . . . . . . . . . . . . . . . . . . . . . . . . 4–5
18:0 . . . . . . . . . . . . . . . . . . . . . . . . . . . . . 3
Total 18:1 . . . . . . . . . . . . . . . . . . . . . . . 12
9c,12c-18:2 . . . . . . . . . . . . . . . . . . . . . 3–4

**References** *J. Am. Oil Chem. Soc. 56:* 528 (1979)

# Palm Kernel Oil
*Aiphanes acanthophylla*

Specific Gravity (SG)
15.5/15.5°C

25/25°C
Other SG
Refractive Index (RI)
  25°C
  40°C
  Other RI
Iodine Value
Saponification Value
Titer °C
% Unsaponifiable
Melting Point °C

**Fatty Acid Composition (%)**
  8:0 ............................... tr
  10:0 .............................. tr
  12:0 ............................ 41–42
  14:0 ............................ 20–21
  16:0 ............................ 10–11
  18:0 ............................. 3–4
  Total 18:1 ..................... 15–16
  9c,12c-18:2 ..................... 7–8

**References** *J. Am. Oil Chem. Soc. 56:* 528 (1979)

# Palm Kernel Oil
*Elaeis oleifera*

Specific Gravity (SG)
  15.5/15.5°C
  25/25°C
  Other SG
Refractive Index (RI)
  25°C
  40°C
  Other RI
Iodine Value .................... 28–35
Saponification Value
Titer °C
% Unsaponifiable
Melting Point °C

**Fatty Acid Composition (%)**
  6:0 ............................. 0.1
  8:0 ............................. 0.9
  10:0 ............................ 0.8
  12:0 ............................. 29
  14:0 ............................. 26

16:0 ............................. 10
18:0 .............................. 2
Total 18:1 ........................ 26
9c,12c-18:2 ..................... 4–5
Other ........................... 0.4

**References** *J. Am. Oil Chem. Soc. 74:* 1451 (1997)
*J. Sci. Food Agric. 33:* 204 (1982)
*Palm Oil Developments* 27: (1997)
PORIM

# Palm Oil
*Elaeis oleifera*
*Elaeis melanococca*

Specific Gravity (SG)
  15.5/15.5°C
  25/25°C
  Other SG
Refractive Index (RI)
  25°C
  40°C
  Other RI
Iodine Value .................... 61–64
Saponification Value
Titer °C
% Unsaponifiable
Melting Point °C

**Fatty Acid Composition (%)**
  14:0 ............................ 0.2
  16:0 ............................. 19
  9c-16:1 ......................... 1–2
  18:0 .............................. 1
  Total 18:1 ........................ 56
  9c,12c-18:2 ....................... 21
  Other ............................. 1

Sterol Composition, %
  Cholesterol ....................... 2
  Brassicasterol
  Campesterol ...................... 19
  Stigmasterol ..................... 15
  Stigmasta-8,22-dien-3β-ol
  5α-Stigmasta-7,22-dien-3β-ol
  Δ7,25-Stigmastadienol

β-Sitosterol . . . . . . . . . . . . . . . . . . . . . . 64
Δ5-Avenasterol
Δ7-Stigmasterol
Δ7-Avenasterol
Δ7-Campesterol
Δ7-Ergosterol
Δ7,25-Stigmasterol
Sitostanol
Spinasterol
Squalene
24-Methylene Cholesterol
Other
% sterols in oil
Total Sterols, mg/kg . . . . . . . . 3500–4000

Tocopherol Composition, mg/kg
   α-Tocopherol . . . . . . . . . . . . . . . . . . 150
   β-Tocopherol
   γ-Tocopherol
   δ-Tocopherol
   Total, mg/kg

Tocotrienols Composition, mg/kg
   α-Tocotrienol . . . . . . . . . . . . . . . . . . 270
   β-Tocotrienol
   γ-Tocotrienol. . . . . . . . . . . . . . . . . . . 540
   δ-Tocotrienol. . . . . . . . . . . . . . . . . . . . 40
   Total Tocotrienols, mg/kg . . . . . 700–1500

**References** *J. Am. Oil Chem. Soc. 74:* 1451 (1997)
   *J. Sci. Food Agric. 33:* 204 (1982)
   *Palm Oil Developments 27:* (1997) *PORIM*

# Palm Oil
*Elaeis guineensis Dura*

Specific Gravity (SG)
   15.5/15.5°C
   25/25°C
   Other SG
Refractive Index (RI)
   25°C
   40°C
   Other RI
Iodine Value
Saponification Value
Titer °C

% Unsaponifiable
Melting Point °C

**Fatty Acid Composition (%)**
   14:0 . . . . . . . . . . . . . . . . . . . . . . . . . 1.8
   16:0 . . . . . . . . . . . . . . . . . . . . . . . . . . 55
   18:0 . . . . . . . . . . . . . . . . . . . . . . . . . 2.5
   Total 18:1 . . . . . . . . . . . . . . . . . . . . . 30
   9c,12c-18:2 . . . . . . . . . . . . . . . . . . . . 10
   Undefined 18:3 . . . . . . . . . . . . . . . . . 0.4
   20:0 . . . . . . . . . . . . . . . . . . . . . . . . . 0.1

Sterol Composition, %
   Cholesterol . . . . . . . . . . . . . . . . . . . . . 6
   Brassicasterol
   Campesterol . . . . . . . . . . . . . . . . . . . 25
   Stigmasterol . . . . . . . . . . . . . . . . . . . 14
   Stigmasta-8,22-dien-3β-ol
   5α-Stigmasta-7,22-dien-3β-ol
   Δ7,25-Stigmastadienol
   β-Sitosterol . . . . . . . . . . . . . . . . . . . . 55
   Δ5-Avenasterol
   Δ7-Stigmasterol
   Δ7-Avenasterol
   Δ7-Campesterol
   Δ7-Ergosterol
   Δ7,25-Stigmasterol
   Sitostanol
   Spinasterol
   Squalene
   24-Methylene Cholesterol
   Other
   % sterols in oil
   Total Sterols, mg/kg . . . . . . . . 2000–2500

Tocopherol Composition, mg/kg
   α-Tocopherol . . . . . . . . . . . . . . . . . . 310
   β-Tocopherol
   γ-Tocopherol
   δ-Tocopherol
   Total, mg/kg

Tocotrienols Composition, mg/kg
   α-Tocotrienol . . . . . . . . . . . . . . . . . . 210
   β-Tocotrienol
   γ-Tocotrienol. . . . . . . . . . . . . . . . . . . 400
   δ-Tocotrienol. . . . . . . . . . . . . . . . . . . . 80
   Total Tocotrienols, mg/kg

**References** *Palm Oil Developments* 27: (1997) *PORIM*

## Palm Oil
*Elaeis guineensis*

Specific Gravity (SG)
  15.5/15.5°C
  25/25°C
  Other SG . . . . . . . . . . (50/20) 0.891–0.899
Refractive Index (RI)
  25°C
  40°C . . . . . . . . . . . . . . . . 1.454–1.456
  Other RI . . . . . . . . . . . . (50) 1.449–1.455
Iodine Value . . . . . . . . . . . . . . . . . . . 49–55
Saponification Value . . . . . . . . . . . . 190–209
Titer °C
% Unsaponifiable . . . . . . . . . . . . . . . . . 0–1.2
Melting Point °C . . . . . . . . . . . . . . . . . 33–40

**Fatty Acid Composition (%)**
  12:0 . . . . . . . . . . . . . . . . . . . . . . . . . 0–0.4
  14:0 . . . . . . . . . . . . . . . . . . . . . . . . . 0.5–2.0
  16:0 . . . . . . . . . . . . . . . . . . . . . . . . . 40–48
  9c-16:1 . . . . . . . . . . . . . . . . . . . . . . . 0–0.6
  18:0 . . . . . . . . . . . . . . . . . . . . . . . . . 3.5–6.5
  Total 18:1 . . . . . . . . . . . . . . . . . . . . 36–44
  9c,12c-18:2 . . . . . . . . . . . . . . . . . . . 6.5–12.0
  Undefined 18:3 . . . . . . . . . . . . . . . . 0–0.5
  20:0 . . . . . . . . . . . . . . . . . . . . . . . . . 0–1
  Total 20:1 . . . . . . . . . . . . . . . . . . . . 0–0.2
  22:0 . . . . . . . . . . . . . . . . . . . . . . . . . 0–0.1
  24:0 . . . . . . . . . . . . . . . . . . . . . . . . . 0–0.2

Sterol Composition, %
  Cholesterol . . . . . . . . . . . . . . . . . . 2.6–6.7
  Brassicasterol
  Campesterol . . . . . . . . . . . . . . . 18.7–27.5
  Stigmasterol . . . . . . . . . . . . . . . . 8.5–13.9
  Stigmasta-8,22-dien-3β-ol
  5α-Stigmasta-7,22-dien-3β-ol
  Δ7,25-Stigmastadienol
  β-Sitosterol . . . . . . . . . . . . . . . . . 50.2–62.1
  Δ5-Avenasterol . . . . . . . . . . . . . . . . 0–2.8
  Δ7-Stigmasterol . . . . . . . . . . . . . . 0.2–2.4
  Δ7-Avenasterol . . . . . . . . . . . . . . . . 0–5.1
  Δ7-Campesterol
  Δ7-Ergosterol
  Δ7,25-Stigmasterol
  Sitostanol
  Spinasterol
  Squalene
  24-Methylene Cholesterol
  Other
% sterols in oil
  Total Sterols, mg/kg . . . . . . . . . . 362–627

Tocopherol Composition, mg/kg
  α-Tocopherol . . . . . . . . . . . . . . . . . 4–193
  β-Tocopherol . . . . . . . . . . . . . . . . . . 0–234
  γ-Tocopherol . . . . . . . . . . . . . . . . . . 0–526
  δ-Tocopherol . . . . . . . . . . . . . . . . . . 0–123
  Total, mg/kg

Tocotrienols Composition, mg/kg
  α-Tocotrienol . . . . . . . . . . . . . . . . . 4–336
  β-Tocotrienol
  γ-Tocotrienol . . . . . . . . . . . . . . . . . 14–710
  δ-Tocotrienol . . . . . . . . . . . . . . . . . . 0–377
  Total Tocotrienols, mg/kg . . . . . . 98–1500

**References** *Codex* 97/17
  *J. Am. Oil Chem. Soc. 74:* 1451 (1997)
  *Palm Oil Developments* 27: (1997)
  *PORIM*

## Palm Olein

Specific Gravity (SG)
  15.5/15.5°C
  25/25°C
  Other SG . . . . . . . . . . (40/20) 0.899–0.920
Refractive Index (RI)
  25°C
  40°C . . . . . . . . . . . . . . . . 1.4586–1.4592
  Other RI
Iodine Value . . . . . . . . . . . . greater than 56
Saponification Value . . . . . . . . . . . . 194–202
Titer °C
% Unsaponifiable . . . . . . . . . . . . . . . . . 0–1.3
Melting Point °C

**Fatty Acid Composition (%)**
  12:0 . . . . . . . . . . . . . . . . . . . . . . . . . 0.1–0.5
  14:0 . . . . . . . . . . . . . . . . . . . . . . . . . 0.9–1.4
  16:0 . . . . . . . . . . . . . . . . . . . . . . . . . 38.2–42.9
  9c-16:1 . . . . . . . . . . . . . . . . . . . . . . . 0.1–0.3
  18:0 . . . . . . . . . . . . . . . . . . . . . . . . . 3.7–4.8
  Total 18:1 . . . . . . . . . . . . . . . . . . . . 39.8–43.9

9c,12c-18:2 . . . . . . . . . . . . . . . . 10.4–13.4
Undefined 18:3 . . . . . . . . . . . . . . 0.1–0.6
20:0 . . . . . . . . . . . . . . . . . . . . . . . 0.2–0.6

**References** *Codex* 97/17

## Palm Stearin

Specific Gravity (SG)
  15.5/15.5°C
  25/25°C
  Other SG . . . . . . . . . . (60/20) 0.881–0.891
Refractive Index (RI)
  25°C
  40°C . . . . . . . . . . . . . . . . . 1.4472–1.4511
  Other RI
Iodine Value . . . . . . . . . . . . . . . . less than 48
Saponification Value . . . . . . . . . . . . 193–205
Titer °C
% Unsaponifiable . . . . . . . . . . . . . . . . . 0–0.9
Melting Point °C

**Fatty Acid Composition (%)**
  12:0 . . . . . . . . . . . . . . . . . . . . . . . 0.1–0.4
  14:0 . . . . . . . . . . . . . . . . . . . . . . . 1.1–1.8
  16:0 . . . . . . . . . . . . . . . . . . . . . . 48.4–73.8
  9c-16:1 . . . . . . . . . . . . . . . . . . . . . 0.05–0.2
  18:0 . . . . . . . . . . . . . . . . . . . . . . . 3.9–5.6
  Total 18:1 . . . . . . . . . . . . . . . . . . 15.6–36.0
  9c,12c-18:2 . . . . . . . . . . . . . . . . . . . 3.2–9.8
  Undefined 18:3 . . . . . . . . . . . . . . . . 0.1–0.6
  20:0 . . . . . . . . . . . . . . . . . . . . . . . 0.3–0.6

**References** *Codex* 97/17

## Papaya Seed Oil

*Carica papaya*

Specific Gravity (SG)
  15.5/15.5°C
  25/25°C
  Other SG
Refractive Index (RI)
  25°C
  40°C
  Other RI . . . . . . . . . . . . (20) 1.466–1.4679
Iodine Value . . . . . . . . . . . . . . . . . . . . 65–73
Saponification Value . . . . . . . . . . . . 185–199
Titer °C

% Unsaponifiable . . . . . . . . . . . . . . . . . . 1–3
Melting Point °C

**Fatty Acid Composition (%)**
  12:0 . . . . . . . . . . . . . . . . . . . . . . . . . . 0.1
  14:0 . . . . . . . . . . . . . . . . . . . . . . . . 0.4–1.0
  16:0 . . . . . . . . . . . . . . . . . . . . . . . . . 16–18
  9c-16:1 . . . . . . . . . . . . . . . . . . . . . . 0.7–1.3
  18:0 . . . . . . . . . . . . . . . . . . . . . . . . . . 3–6
  Total 18:1 . . . . . . . . . . . . . . . . . . . . . 63–77
  9c,12c-18:2 . . . . . . . . . . . . . . . . . . . 0.4–10
  20:0 . . . . . . . . . . . . . . . . . . . . . . . . 0.1–0.6
  22:0 . . . . . . . . . . . . . . . . . . . . . . . . 0.1–2.0

Sterol Composition, %
  Cholesterol . . . . . . . . . . . . . . . . . . . . . . 3
  Brassicasterol
  Campesterol . . . . . . . . . . . . . . . . . . . 9–11
  Stigmasterol . . . . . . . . . . . . . . . . . . . . 6–7
  Stigmasta-8,22-dien-3β-ol
  5α-Stigmasta-7,22-dien-3β-ol
  Δ7,25-Stigmastadienol
  β-Sitosterol . . . . . . . . . . . . . . . . . . . 69–72
  Δ5-Avenasterol . . . . . . . . . . . . . . . . . . 6–7
  Δ7-Stigmasterol . . . . . . . . . . . . . . . . . . 1–2
  Δ7-Avenasterol . . . . . . . . . . . . . . . . . 0.5–1
  Δ7-Campesterol
  Δ7-Ergosterol
  Δ7,25-Stigmasterol
  Sitostanol
  Spinasterol
  Squalene
  24-Methylene Cholesterol . . . . . . . . . 1–2
  Other: Fucosterol, 0.8
  % sterols in oil
  Total Sterols, mg/kg

Tocopherol Composition, mg/kg
  α-Tocopherol . . . . . . . . . . . . . . . . . . . 85.5
  β-Tocopherol . . . . . . . . . . . . . . . . . . . . 7.5
  γ-Tocopherol . . . . . . . . . . . . . . . . . . . . 7.5
  δ-Tocopherol . . . . . . . . . . . . . . . . . . . . 7.0
Total, mg/kg

**References** *Riv. Ital. Sost. Grasse* 67: 257 (1990)
*Riv. Ital. Sost. Grasse* 54: 429 (1990)
*Riv. Ital. Sost. Grasse* 58: 324 (1981)
*J. Food Sci.* 43: 255 (1978)
*Pakistan J. Sci. Ind. Res.* 35: 43 (1992)

## Parkia Biglandulosa Seed Fat
*Parkia biglandulosa*
Specific Gravity (SG)
  15.5/15.5°C . . . . . . . . . . . . . . . . . . 0.9208
  25/25°C
  Other SG
Refractive Index (RI)
  25°C
  40°C
  Other RI . . . . . . . . . . . . . . . . . (21) 1.4701
Iodine Value . . . . . . . . . . . . . . . . . . . . . . . 81
Saponification Value . . . . . . . . . . . . . . . 190
Titer °C
% Unsaponifiable . . . . . . . . . . . . . . . . . . . 1
Melting Point °C

**Fatty Acid Composition (%)**
  14:0 . . . . . . . . . . . . . . . . . . . . . . . . . . . 0.2
  16:0 . . . . . . . . . . . . . . . . . . . . . . . . . . . 26
  9c-16:1 . . . . . . . . . . . . . . . . . . . . . . . . . 7
  18:0 . . . . . . . . . . . . . . . . . . . . . . . . . . . 33
  Total 18:1 . . . . . . . . . . . . . . . . . . . . . . 27
  9c,12c-18:2 . . . . . . . . . . . . . . . . . . . . . 3
  20:0 . . . . . . . . . . . . . . . . . . . . . . . . . . . . 4

**References** *J. Am. Oil Chem. Soc. 61:* 1023 (1984)

## Parsley Seed Oil
*Petroselinum sativum*
Specific Gravity (SG)
  15.5/15.5°C
  25/25°C
  Other SG
Refractive Index (RI)
  25°C
  40°C . . . . . . . . . . . . . . . . . . . . . . . 1.4800
  Other RI
Iodine Value . . . . . . . . . . . . . . . . . 110–120
Saponification Value
Titer °C
% Unsaponifiable . . . . . . . . . . . . . . . . . . . 2
Melting Point °C

**Fatty Acid Composition (%)**
  16:0 . . . . . . . . . . . . . . . . . . . . . . . . . . . . 2
  18:0 . . . . . . . . . . . . . . . . . . . . . . . . . . . . 1
  Total 18:1 . . . . . . . . . . . . . . . . . . . . 12–15
  6c-18:1 . . . . . . . . . . . . . . . . . . . . . . 69–76
  9c,12c-18:2 . . . . . . . . . . . . . . . . . . . 6–14

**References**

## Pataua Palm Oil (Pulp)
*Jessenia bataua M*
Specific Gravity (SG)
  15.5/15.5°C
  25/25°C . . . . . . . . . . . . . . . . . 0.911–0.918
  Other SG
Refractive Index (RI)
  25°C . . . . . . . . . . . . . . . . . . . . 1.468–1.470
  40°C
  Other RI
Iodine Value . . . . . . . . . . . . . . . . . . . 75–80
Saponification Value . . . . . . . . . . . 190–196
Titer °C
% Unsaponifiable
Melting Point °C

**Fatty Acid Composition (%)**
  16:0 . . . . . . . . . . . . . . . . . . . . . . . . . . . . 9
  18:0 . . . . . . . . . . . . . . . . . . . . . . . . . . . . 6
  Total 18:1 . . . . . . . . . . . . . . . . . . . . . . 81
  9c,12c-18:2 . . . . . . . . . . . . . . . . . . . . . 4

**References**

## Paullinia Elegans Seed Oil
*Sapindaceae (Soapberry) Family*
Specific Gravity (SG)
  15.5/15.5°C
  25/25°C
  Other SG
Refractive Index (RI)
  25°C
  40°C
  Other RI
Iodine Value
Saponification Value
Titer °C

% Unsaponifiable
Melting Point °C

**Fatty Acid Composition (%)**
16:0 .......................... 2.2
9c-16:1 ....................... 3.4
11c-16:1 ...................... 0.2
18:0 .......................... 1.7
9c-18:1 ....................... 12.2
11c-18:1 ...................... 19.8
Undefined 18:2 ................. 3.1
Undefined 18:3 ................. 1.8
20:0 .......................... 5
Total 20:1 .................... 4
11c-20:1 ...................... 4
13c-20:1 ...................... 44
15c-20:1 ...................... 0.7
22:0 .......................... 0.3
Unidentified 22:1 .............. 0.4
13c-22:1 ...................... 0.4
15c-22:1 ...................... 0.8

**References** *J. High Resol. Chromatogr. 18:* 443 (1995)


# Peach Kernel Oil
*Prunus persica*

Specific Gravity (SG)
15.5/15.5°C
25/25°C ....................... 0.913
Other SG
Refractive Index (RI)
25°C .................. 1.468–1.470
40°C .................. 1.459–1.462
Other RI
Iodine Value ............... 94–110
Saponification Value ....... 189–194
Titer °C
% Unsaponifiable ........... 0.7–0.71
% Sterols in crude oil .......... 0.2
Melting Point °C

**Fatty Acid Composition (%)**
14:0 ......................... 0.2–1
16:0 ......................... 8.1

16:1 ......................... 0.4
9c-16:1 ..................... 0.2–0.4
18:0 ......................... 2–6
Total 18:1 ................. 58.5–70
Undefined 18:2 ............ 15–32.8
9c,12c-18:2 ................. 15–29
Undefined 18:3 .............. 0–0.2
20:0 ........................ 0.3–0.6

Sterol Composition, %
Cholesterol .................... 0.2
Brassicasterol
Campesterol .................... 4–8
Stigmasterol ................... 1–6
Stigmasta-8,22-dien-3β-ol
5α-Stigmasta-7,22-dien-3β-ol
Δ7,25-Stigmastadienol
β-Sitosterol ................. 64–90
Δ5-Avenasterol ............... 2–10
Δ7-Stigmasterol ............... 1–2
Δ7-Avenasterol ................. 2
Δ7-Campesterol ................. 2
Δ7-Ergosterol
Δ7,25-Stigmasterol
Sitostanol
Spinasterol
Squalene
24-Methylene Cholesterol
Other
% sterols in oil
Total Sterols, mg/kg

Tocopherol Composition, mg/kg
α-Tocopherol ................... 74
β-Tocopherol ................... 15
γ-Tocopherol ................... 11
δ-Tocopherol .................... 0
Total, mg/kg ................... 88

**References** *J. Am. Oil Chem. Soc. 48:* 902 (1965)
*Riv. Ital. Sost. Grasse 52:* 82 (1975)
*Lebensmittelchem. gerichtl. Chem. 36:* 53 (1982)
*Food Chem. 28:* 31 (1988)
*Riv. Ital. Sostanze Grasse 75:* 405 (1998)
*JAOCS 69:* 492–494 (1992)
*Rev. Franc. Corps Gras 33:* 115 (1986)

## Peanut/Groundnut Oil
*Arachis hypogaea*

Specific Gravity (SG)
  15.5/15.5°C
  25/25°C
  Other SG . . . . . . . . . (20/20) 0.914–0.917
Refractive Index (RI)
  25°C
  40°C . . . . . . . . . . . . . . . . . 1.460–1.465
  Other RI
Iodine Value . . . . . . . . . . . . . . . . . . . 86–107
Saponification Value . . . . . . . . . . . . 187–196
Titer °C
% Unsaponifiable . . . . . . . . . . . . . . . . . . <10
Melting Point °C . . . . . . . . . . . . . . . . . . . . -2

**Fatty Acid Composition (%)**
  12:0 . . . . . . . . . . . . . . . . . . . . . . . . 0–0.1
  14:0 . . . . . . . . . . . . . . . . . . . . . . . . 0–0.1
  16:0 . . . . . . . . . . . . . . . . . . . . . . 8.3–14.0
  16:1 . . . . . . . . . . . . . . . . . . . . . . . 0.0–0.2
  9c-16:1 . . . . . . . . . . . . . . . . . . . . . 0–0.2
  18:0 . . . . . . . . . . . . . . . . . . . . . . . 1.9–4.4
  Total 18:1 . . . . . . . . . . . . . . . . . 36.4–67.1
  Undefined 18:2 . . . . . . . . . . . . . . . 14.0–43
  9c,12c-18:2 . . . . . . . . . . . . . . . . 14.0–43.0
  Undefined 18:3 . . . . . . . . . . . . . . . 0.0–0.1
  20:0 . . . . . . . . . . . . . . . . . . . . . . . 1.1–1.7
  Total 20:1 . . . . . . . . . . . . . . . . . . . 0.7–1.7
  22:0 . . . . . . . . . . . . . . . . . . . . . . . 2.1–4.4
  Unidentified 22:1 . . . . . . . . . . . . . . . 0–0.3
  24:0 . . . . . . . . . . . . . . . . . . . . . . . 1.1–2.2
  15c-24:1 . . . . . . . . . . . . . . . . . . . . . 0–0.3

Sterol Composition, %
  Cholesterol . . . . . . . . . . . . . . . . . . . 0–3.8
  Brassicasterol . . . . . . . . . . . . . . . . . 0–0.2
  Campesterol . . . . . . . . . . . . . . . 12.0–19.8
  Stigmasterol . . . . . . . . . . . . . . . . 5.4–13.2
  Stigmasta-8,22-dien-3β-ol
  5α-Stigmasta-7,22-dien-3β-ol
  Δ7,25-Stigmastadienol
  β-Sitosterol . . . . . . . . . . . . . . . . 47.4–67.7
  Δ5-Avenasterol . . . . . . . . . . . . . . 8.3–18.8
  Δ7-Stigmasterol . . . . . . . . . . . . . . . . 0–5.1
  Δ7-Avenasterol . . . . . . . . . . . . . . . 0.0–5.5
  Δ7-Campesterol
  Δ7-Ergosterol
  Δ7,25-Stigmasterol
  Sitostanol
  Spinasterol
  Squalene
  24-Methylene Cholesterol
  Other . . . . . . . . . . . . . . . . . . . . . . . 0.0–1.4
% sterols in oil
Total Sterols, mg/kg . . . . . . . . . 901–2854

Tocopherol Composition, mg/kg
  α-Tocopherol . . . . . . . . . . . . . . . . 49–373
  β-Tocopherol . . . . . . . . . . . . . . . . . . . 0–41
  γ-Tocopherol . . . . . . . . . . . . . . . . . 88–389
  δ-Tocopherol . . . . . . . . . . . . . . . . . . . 0–22
  Total, mg/kg . . . . . . . . . . . . . . . 176–1291

**References** *Codex* 1993/16
*J. Am. Oil Chem. Soc. 64:* 534 (1987)

## Peanut Oil (high oleic)

Specific Gravity (SG)
  15.5/15.5°C
  25/25°C
  Other SG
Refractive Index (RI)
  25°C
  40°C
  Other RI
Iodine Value
Saponification Value
Titer °C
% Unsaponifiable
Melting Point °C

**Fatty Acid Composition (%)**
  16:0 . . . . . . . . . . . . . . . . . . . . . . . . . . . . 7
  18:0 . . . . . . . . . . . . . . . . . . . . . . . . . . . . 3
  Total 18:1 . . . . . . . . . . . . . . . . . . . . . . 76
  9c,12c-18:2 . . . . . . . . . . . . . . . . . . . . . . 4
  19:0 . . . . . . . . . . . . . . . . . . . . . . . . . . . . 1
  20:0 . . . . . . . . . . . . . . . . . . . . . . . . . . . . 1
  Total 20:1 . . . . . . . . . . . . . . . . . . . . . . . 2
  22:0 . . . . . . . . . . . . . . . . . . . . . . . . . . . . 4
  24:0 . . . . . . . . . . . . . . . . . . . . . . . . . . . . 2

**References** *J. Anim. Sci. 70:* 3734 (1992)

## Pear Seed Oil
*Pyrus domestica*

Specific Gravity (SG)
  15.5/15.5°C
  25/25°C
  Other SG . . . . . . . . . . . . . . .(20/20) 0.912
Refractive Index (RI)
  25°C
  40°C . . . . . . . . . . . . . . . . . . 1.465–1.468
  Other RI
Iodine Value . . . . . . . . . . . . . . . . . . 121–127
Saponification Value . . . . . . . . . . . 189–197
Titer °C
% Unsaponifiable . . . . . . . . . . . . . . . 0.5–1.1
% Sterols in crude oil . . . . . . . . . . . . . . . 0.4
Melting Point °C

**Fatty Acid Composition (%)**
  14:0. . . . . . . . . . . . . . . . . . . . . . . . . . 0.2
  16:0. . . . . . . . . . . . . . . . . . . . . . . . . . . 10
  16:1. . . . . . . . . . . . . . . . . . . . . . . . . . . . 1
  18:0. . . . . . . . . . . . . . . . . . . . . . . . . . . . 1
  Total 18:1 . . . . . . . . . . . . . . . . . . . . . . 19
  Undefined 18:2 . . . . . . . . . . . . . . . . . . 69
  Undefined 18:3 . . . . . . . . . . . . . . . . . 0.4

Sterol Composition, %
  Cholesterol . . . . . . . . . . . . . . . . . . . . . 0.9
  Brassicasterol . . . . . . . . . . . . . . . . . . . 0.1
  Campesterol . . . . . . . . . . . . . . . . . . . . 2.5
  Stigmasterol . . . . . . . . . . . . . . . . . . . . 5.6
  Stigmasta-8,22-dien-3β-ol
  5α-Stigmasta-7,22-dien-3β-ol
  Δ7,25-Stigmastadienol
  β-Sitosterol . . . . . . . . . . . . . . . . . . . . . . 86
  Δ5-Avenasterol . . . . . . . . . . . . . . . . . . 0.4
  Δ7-Stigmasterol . . . . . . . . . . . . . . . . . . . 1
  Δ7-Avenasterol
  Δ7-Campesterol . . . . . . . . . . . . . . . . . 0.4
  Δ7-Ergosterol
  Δ7,25-Stigmasterol . . . . . . . . . . . . . . . . . 3
  Sitostanol
  Spinasterol
  Squalene
  24-Methylene Cholesterol
  Other
  % sterols in oil
  Total Sterols, mg/kg

**References** *Riv. Ital. Sost. Grasse 75:* 405 (1998)

## Pecan Nut Oil
*Carya illinoensis*

Specific Gravity (SG)
  15.5/15.5°C
  25/25°C . . . . . . . . . . . . . . . . . 0.912–0.915
  Other SG
Refractive Index (RI)
  25°C . . . . . . . . . . . . . . . . . . . 1.469–1.470
  40°C
  Other RI
Iodine Value . . . . . . . . . . . . . . . . . . 100–106
Saponification Value . . . . . . . . . . . 190–198
Titer °C
% Unsaponifiable . . . . . . . . . . . . . . . 0.4–1.5
Melting Point °C

**Fatty Acid Composition (%)**
  16:0. . . . . . . . . . . . . . . . . . . . . . . . . . 5–11
  9c-16:1 . . . . . . . . . . . . . . . . . . . . . 0.1–0.2
  18:0. . . . . . . . . . . . . . . . . . . . . . . . . . . 1–6
  Total 18:1 . . . . . . . . . . . . . . . . . . . . 49–69
  9c-18:1 . . . . . . . . . . . . . . . . . 60.55–67.47
  9c,12c-18:2 . . . . . . . . . . . . . . . . . . 19–40
  Undefined 18:3 . . . . . . . . . . . . . . . . . . 0–3
  20:0. . . . . . . . . . . . . . . . . . . . . . . . 0.1–0.2
  Total 20:1 . . . . . . . . . . . . . . . . . . . 0.2–0.3
  22:0. . . . . . . . . . . . . . . . . . . . . . . . 0.1–0.2

Sterol Composition, %
  Cholesterol
  Brassicasterol
  Campesterol . . . . . . . . . . . . . . . . . 3.5–4.5
  Stigmasterol . . . . . . . . . . . . . . . . . . . . 1–2
  Stigmasta-8,22-dien-3β-ol
  5α-Stigmasta-7,22-dien-3β-ol
  Δ7,25-Stigmastadienol
  β-Sitosterol . . . . . . . . . . . . . . . . . . . 81–93
  Δ5-Avenasterol . . . . . . . . . . . . . . . . . . 1–8
  Δ7-Stigmasterol . . . . . . . . . . . . . . . 0.2–0.6
  Δ7-Avenasterol . . . . . . . . . . . . . . . . 0.2–0.4
  Δ7-Campesterol
  Δ7-Ergosterol
  Δ7,25-Stigmasterol
  Sitostanol

Spinasterol
Squalene
24-Methylene Cholesterol
Other
% sterols in oil
Total Sterols, mg/kg ........ 1000–2900

Tocopherol Composition, mg/kg
   α-Tocopherol ................ 50–370
   β-Tocopherol
   γ-Tocopherol .............. 20–182.81
   δ-Tocopherol ................... 0–79
   Total, mg/kg ................. 88–420

**References** *J. Am. Oil Chem. Soc. 45:* 437 (1968)
*J. Am. Oil Chem. Soc. 27:* 414 (1950)
*J. Am. Dietetic Assn. 73:* 39 (1978)
*Riv. Ital. Sost. Grasse 73:* 29 (1996)
*JAOCS 76:* 957–965 (1999)

## Perilla Oil
*Perilla frutescens*

Specific Gravity (SG)
   15.5/15.5°C ............. 0.930–0.937
   25/25°C ................ 0.923–0.930
   Other SG
Refractive Index (RI)
   25°C .................. 1.476–1.478
   40°C .................. 1.470–1.477
   Other RI
Iodine Value .................. 192–208
Saponification Value ........... 188–197
Titer °C
% Unsaponifiable ............... 1.3–1.5
Melting Point °C

**Fatty Acid Composition (%)**
   14:0 .......................... 0.03
   16:0 ......................... 6–8.92
   16:1 ........................... tr
   9c-16:1 ....................... 0.11
   18:0 ......................... 1–2.15
   Total 18:1 ................... 13–15
   9c-18:1 .................. 11.41–12.92
   7c-18:1 ....................... 0.97
   Undefined 18:2 ................ 15.5
   9c,12c-18:2 ............... 14–17.93
   Undefined 18:3 ................ 44–64
   9c,12c,15c-18:3 .............. 59.37
   20:0 .......................... 0.17
   11c-20:1 ...................... 0.13
   22:0 .......................... 0.04
   22:2 .......................... 0.04

Tocopherol Composition, mg/kg
   α-Tocopherol ................. 10–57
   β-Tocopherol ..................... 37
   γ-Tocopherol ............... 526–538
   δ-Tocopherol .................. 31–40
   Total, mg/kg

**References** *J. Am. Oil Chem. Soc. 36:* 477 (1959)
*J. Am. Oil Chem. Soc. 71:* 619 (1994)
*JAOCS 74:* 375–380 (1997)
*JAOCS 80:* 1013–1020 (2003)
*JAOCS 68:* 781–783 (1991)

## Phulwara Butter
*Madhuca butyraceae*

Specific Gravity (SG)
   15.5/15.5°C
   25/25°C
   Other SG ......... (100/15) 0.856–0.870
Refractive Index (RI)
   25°C
   40°C .................. 1.455–1.461
   Other RI
Iodine Value ................... 40–51
Saponification Value ............ 188–200
Titer °C ........................ 48–52
% Unsaponifiable ................ 1.4–5
Melting Point °C ................ 39–47

**Fatty Acid Composition (%)**
   16:0 ......................... 57–61
   18:0 .......................... 3–4
   Total 18:1 ................... 30–36
   Undefined 18:2 ................ 4–5
   9c,12c-18:2 ................... 3–4

**References** *J. Am. Oil Chem. Soc. 55:* 621 (1978)
*INFORM 13:* 151 (2002)

## Physic Nut Oil, Ratanjyor Oil
*Jatropha curcas*
Specific Gravity (SG)
  15.5/15.5°C ............. 0.918–0.923
  25/25°C
  Other SG
Refractive Index (RI)
  25°C
  40°C ................... 1.462–1.465
  Other RI ................ (30) 1.47
Iodine Value ................... 93–107
Saponification Value ........... 188–196
Titer °C
% Unsaponifiable ................ 0.4–1
Melting Point °C

**Fatty Acid Composition (%)**
  14:0 ....................... 0–0.5
  16:0 ....................... 4–17
  18:0 ....................... 5–10
  Total 18:1 .................. 37–63
  Undefined 18:2 .............. 19–41
  Other ....................... 1.4

**References** *INFORM 13:* 151 (2002)

## Pili Nut Oil
*Canarium ovatum*
Specific Gravity (SG)
  15.5/15.5°C
  25/25°C
  Other SG
Refractive Index (RI)
  25°C
  40°C
  Other RI
Iodine Value
Saponification Value
Titer °C
% Unsaponifiable
Melting Point °C

**Fatty Acid Composition (%)**
  14:0 ....................... 0.05
  16:0 ..................... 26.46–33.3
  16:1 ....................... 0.30

  18:0 ..................... 6.53–10.9
  Total 18:1 ............... 44.7–56.55
  Undefined 18:2 ........... 10.09–10.1
  Undefined 18:3 ............ 0.37–0.53
  20:0 ........................ tr-0.24

**References** *JAOCS 77:* 991–996 (2000)
*JAOCS 75:* 807–811 (1998)

## Pimpinella Acuminata Seed Oil
*Pimpinella acuminata*
Specific Gravity (SG)
  15.5/15.5°C
  25/25°C
  Other SG
Refractive Index (RI)
  25°C
  40°C
  Other RI
Iodine Value
Saponification Value
Titer °C
% Unsaponifiable
Melting Point °C

**Fatty Acid Composition (%)**
  14:0 ............................ 6
  16:0 ........................... 17
  18:0 ............................ 6
  Total 18:1 ...................... 47
  9c,12c-18:2 ..................... 14
  Undefined 18:3 .................. 10

**References** *Fat Sci. Technol. 97:* 455 (1995)

## Pindo Palm Kernel Oil
*Anecastrum romanozoffianum*
Specific Gravity (SG)
  15.5/15.5°C
  25/25°C
  Other SG
Refractive Index (RI)
  25°C
  40°C

Other RI
Iodine Value ........................ 14
Saponification Value ................ 237
Titer °C
% Unsaponifiable
Melting Point °C

**Fatty Acid Composition (%)**
    8:0 ........................... 1–2
    10:0 ............................. 2
    12:0 ......................... 56–58
    14:0 ......................... 20–22
    16:0 ........................... 6–7
    18:0 ........................... 2–3
    Total 18:1 .................... 8–10
    9c,12c-18:2 .................... 1–2

**References** *J. Am. Oil Chem. Soc. 56:* 528 (1979)
*Food Chem. 28:* 177 (1988)

# Pine Nut Oil
*Pinus monophylla*

Specific Gravity (SG)
    15.5/15.5°C
    25/25°C
    Other SG
Refractive Index (RI)
    25°C ....................... 1.4698
    40°C
    Other RI
Iodine Value ...................... 102
Saponification Value ........... 184–189
Titer °C
% Unsaponifiable ................... 2
Melting Point °C

**Fatty Acid Composition (%)**
    14:0 ............................. 5
    16:0 ............................. 3
    18:0 ........................... 0.4
    Total 18:1 ................... 58–62
    9c,12c-18:2 .................. 30–33

**References**

# Pine Nut Oil
*Pinus pinea*

Specific Gravity (SG)
    15.5/15.5°C .................. 0.9199
    25/25°C
    Other SG
Refractive Index (RI)
    25°C
    40°C
    Other RI
Iodine Value ................. 118–121
Saponification Value ........... 193–197
Titer °C
% Unsaponifiable ............. 0.6–2.0
Melting Point °C

**Fatty Acid Composition (%)**
    16:0 ........................... 6–8
    9c-16:1 ..................... 0.1–0.2
    18:0 ........................... 2–3
    Total 18:1 ................... 36–39
    9c,12c-18:2 .................. 47–51
    Undefined 18:3 ................. 0.6
    9c,12c,15c-18:3 ................ 0.6
    5c,9c,12c-18:3 ................. 0.4
    20:0 ........................... 0.5
    Total 20:1 ..................... 0.7
    20:2 ........................... 0.5
    Other 5c,11c-20:2, 0.1; 11c,14c-20:2, 0.5; 5c,11c,14c-20:3, 2.5

**References** *J. Am. Oil Chem. Soc. 72:* 1043 (1995)
*INFORM 8:* 116 (1997)

# Pine Nut Oil
*Pinus halepensis spp*

Specific Gravity (SG)
    15.5/15.5°C
    25/25°C
    Other SG
Refractive Index (RI)
    25°C
    40°C

Other RI
Iodine Value
Saponification Value
Titer °C
% Unsaponifiable
Melting Point °C

**Fatty Acid Composition (%)**
16:0 . . . . . . . . . . . . . . . . . . . . . . . . 4.0–4.7
9c-16:1 . . . . . . . . . . . . . . . . . . . . . . . . 0.1
18:0 . . . . . . . . . . . . . . . . . . . . . . . . . 3.3–3.7
9c-18:1 . . . . . . . . . . . . . . . . . . . . 18.8–23.7
11c-18:1 . . . . . . . . . . . . . . . . . . . . . 0.3–0.4
9c,12c-18:2 . . . . . . . . . . . . . . . . . 55.5–60.5
9c,12c,15c-18:3 . . . . . . . . . . . . . . . . . 0.7
5c,9c,12c-18:3 . . . . . . . . . . . . . . . . 3.1–4.4
20:0 . . . . . . . . . . . . . . . . . . . . . . . . . . . 0.5
Total 20:1 . . . . . . . . . . . . . . . . . . . . 0.5–0.9
20:2 . . . . . . . . . . . . . . . . . . . . . . . . 0.5–1.1
8c,11c,14c-20:3 . . . . . . . . . . . . . . . tr-0.04
22:0 . . . . . . . . . . . . . . . . . . . . . . . . . . tr-0.1
Other: 5c,9c-18:2, 0.5–1.0; 5c,9c,12c,
15c-18:4, 0–0.02; 5c,11c-20:2,
0.4–0.5; 5c,11c,14c-20:3, 3.6–5.4

**References** *J. Am. Oil Chem. Soc. 75:* 45 (1998)

# Pine Nut Oil
*Pinus ponderosa spp*

Specific Gravity (SG)
  15.5/15.5°C
  25/25°C
  Other SG
Refractive Index (RI)
  25°C
  40°C
  Other RI
Iodine Value
Saponification Value
Titer °C
% Unsaponifiable
Melting Point °C

**Fatty Acid Composition (%)**
16:0 . . . . . . . . . . . . . . . . . . . . . . . . 3.4–5.0
9c-16:1 . . . . . . . . . . . . . . . . . . . . . . . . 0.1
18:0 . . . . . . . . . . . . . . . . . . . . . . . . 1.5–2.2
9c-18:1 . . . . . . . . . . . . . . . . . . . . 16.7–30.9
11c-18:1 . . . . . . . . . . . . . . . . . . . . . 0.4–0.5
9c,12c-18:2 . . . . . . . . . . . . . . . . . 42.6–48.6
9c,12c,15c-18:3 . . . . . . . . . . . . . . . 0.5–0.6
5c,9c,12c-18:3 . . . . . . . . . . . . . . 11.3–18.4
20:0 . . . . . . . . . . . . . . . . . . . . . . . . . 0.4–0.5
Total 20:1 . . . . . . . . . . . . . . . . . . . . 0.9–1.2
20:2 . . . . . . . . . . . . . . . . . . . . . . . . 0.3–0.8
8c,11c,14c-20:3 . . . . . . . . . . . . . . . . tr-0.2
22:0 . . . . . . . . . . . . . . . . . . . . . . . . . . . . tr
Other: 5c,9c-18:2, 2.3–3.9; 5c,9c,12c,
15c-18:4, 0.1; 5c,11c-20:2, 0.3–0.4;
5c,11c,14c-20:3, 1.4–3.8

**References** *J. Am. Oil Chem. Soc. 75:* 45 (1998)

# Pine Nut Oil
*Pinus Banksiana spp*

Specific Gravity (SG)
  15.5/15.5°C
  25/25°C
  Other SG
Refractive Index (RI)
  25°C
  40°C
  Other RI
Iodine Value
Saponification Value
Titer °C
% Unsaponifiable
Melting Point °C

**Fatty Acid Composition (%)**
16:0 . . . . . . . . . . . . . . . . . . . . . . . . 3.2–5.8
9c-16:1 . . . . . . . . . . . . . . . . . . . . . 0.1–0.2
18:0 . . . . . . . . . . . . . . . . . . . . . . . . 1.5–2.7
9c-18:1 . . . . . . . . . . . . . . . . . . . . 15.4–20.3
11c-18:1 . . . . . . . . . . . . . . . . . . . . . 0.3–1.7
9c,12c-18:2 . . . . . . . . . . . . . . . . . 43.4–52.2
9c,12c,15c-18:3 . . . . . . . . . . . . . . . 0.3–1.4
5c,9c,12c-18:3 . . . . . . . . . . . . . . . . 7.9–22.9
20:0 . . . . . . . . . . . . . . . . . . . . . . . . . 0.2–0.6
Total 20:1 . . . . . . . . . . . . . . . . . . . . 0.7–1.2
20:2 . . . . . . . . . . . . . . . . . . . . . . . . 0.6–0.9

8c,11c,14c-20:3 .............. 0.2–0.7
22:0 .......................... tr-0.2
Other: 5c,9c-18:2, 0.9–3.2; 5c,9c,12c,15c-18:4, tr-0.1; 5c,11c-20:2, 0.2–0.9; 5c,11c,14c-20:3, 1.8–7.0

**References** *J. Am. Oil Chem. Soc. 75:* 45 (1998)

## Pine Nut Oil
*Pinus cembroides edulis*
Specific Gravity (SG)
  15.5/15.5°C
  25/25°C
  Other SG
Refractive Index (RI)
  25°C
  40°C
  Other RI
Iodine Value
Saponification Value
Titer °C
% Unsaponifiable
Melting Point °C

**Fatty Acid Composition (%)**
  14:0 ........................... 0.1
  16:0 ........................... 7.1
  9c-16:1 ........................ 0.2
  17:0 ........................... 0.03
  18:0 ........................... 2.3
  9c-18:1 ....................... 46.9
  11c-18:1 ....................... 0.6
  9c,12c-18:2 ................... 40.7
  9c,12c,15c-18:3 ................ 0.2
  5c,9c,12c-18:3 ................. 0.4
  20:0 ........................... 0.5
  Total 20:1 ..................... 0.5
  20:2 ........................... 0.2
  Other: 5c,9c-18:2, 0.1; 5c,9c,12c-20:3, 0.3

**References** *J. Am. Oil Chem. Soc. 74:* 613 (1997)

## Pine Nut Oil
*Pinus pinaster*
Specific Gravity (SG)
  15.5/15.5°C
  25/25°C
  Other SG
Refractive Index (RI)
  25°C
  40°C
  Other RI
Iodine Value
Saponification Value
Titer °C
% Unsaponifiable
Melting Point °C

**Fatty Acid Composition (%)**
  16:0 ........................ 3.6–4.0
  9c-16:1 ..................... 0.1–0.2
  18:0 ........................ 2–2.4
  Total 18:1 ..................... 18
  9c,12c-18:2 .................... 56
  Undefined 18:3 ............... 1–1.3
  5c,9c,12c-18:3 ................. 7.1
  20:0 ........................... 0.3
  Total 20:1 ..................... 1.0
  20:2 ............................. 1
  Other: 5c,9c-18:2, 0.7; 11c,14c-20:2, 0.8; 5c,11c,14c-20:3, 7.1

**References** *INFORM 8:* 116 (1997)

## Piper Nigrum Seed Oil
*Piper nigrum*
Specific Gravity (SG)
  15.5/15.5°C
  25/25°C
  Other SG
Refractive Index (RI)
  25°C
  40°C
  Other RI
Iodine Value ....................... 66

Saponification Value . . . . . . . . . . . . . . . . 203
Titer °C
% Unsaponifiable . . . . . . . . . . . . . . . . . . 1.8
Melting Point °C

**Fatty Acid Composition (%)**
   10:0 . . . . . . . . . . . . . . . . . . . . . . . . . . . . 4
   12:0 . . . . . . . . . . . . . . . . . . . . . . . . . . 2.5
   14:0 . . . . . . . . . . . . . . . . . . . . . . . . . . . . 3
   16:0 . . . . . . . . . . . . . . . . . . . . . . . . . . . 27
   18:0 . . . . . . . . . . . . . . . . . . . . . . . . . . . . 7
   Total 18:1 . . . . . . . . . . . . . . . . . . . . . . 30
   9c,12c-18:2 . . . . . . . . . . . . . . . . . . . . 7–8
   Other: Malvalic, 6; Sterculic, 4; Vernolic, 8

**References** *Fat Sci. Technol. 97:* 453 (1995)

## Pisa Oil
*Actinodaphne hookeri*

Specific Gravity (SG)
   15.5/15.5°C
   25/25°C
   Other SG . . . . . . . . . . . . . . . . (25/4) 0.925
Refractive Index (RI)
   25°C
   40°C
   Other RI . . . . . . . . . . . . . . . . (30) 1.4490
Iodine Value . . . . . . . . . . . . . . . . . . . . . . 11
Saponification Value . . . . . . . . . . . . . . . . 256
Titer °C
% Unsaponifiable
Melting Point °C . . . . . . . . . . . . . . . . 43–44

**Fatty Acid Composition (%)**
   12:0 . . . . . . . . . . . . . . . . . . . . . . . . 88–98
   14:0 . . . . . . . . . . . . . . . . . . . . . . . . . 2–3
   16:0 . . . . . . . . . . . . . . . . . . . . . . . 0.5–1.0
   Total 18:1 . . . . . . . . . . . . . . . . . . . . . 2–5

**References**

## Pistachio Nut Oil
*Pistacia vera*

Specific Gravity (SG)
   15.5/15.5°C . . . . . . . . . . . . . 0.915–0.920
   25/25°C
   Other SG
Refractive Index (RI)
   25°C . . . . . . . . . . . . . . . . . . . 1.467–1.470
   40°C . . . . . . . . . . . . . . . . . . . 1.460–1.466
   Other RI
Iodine Value . . . . . . . . . . . . . . . . . . . 84–96
Saponification Value . . . . . . . . . . . 189–195
Titer °C . . . . . . . . . . . . . . . . . . . . . . . 13–16
% Unsaponifiable
Melting Point °C . . . . . . . . . . . . . . . . . . . . 5

**Fatty Acid Composition (%)**
   14:0 . . . . . . . . . . . . . . . . . . . . . . . . 0–0.6
   16:0 . . . . . . . . . . . . . . . . . . . . . . . . . 8–13
   9c-16:1 . . . . . . . . . . . . . . . . . . . . . 0.5–1.0
   18:0 . . . . . . . . . . . . . . . . . . . . . . . 0.5–2.0
   Total 18:1 . . . . . . . . . . . . . . . . . . . . 56–70
   9c,12c-18:2 . . . . . . . . . . . . . . . . . . . 18–31
   Undefined 18:3 . . . . . . . . . . . . . . . 0.1–0.4
   20:0 . . . . . . . . . . . . . . . . . . . . . . . . . . 0.3
   Total 20:1 . . . . . . . . . . . . . . . . . . . . . . 0.6

Sterol Composition, %
   Cholesterol
   Brassicasterol
   Campesterol . . . . . . . . . . . . . . . . . . . . . . 5
   Stigmasterol . . . . . . . . . . . . . . . . . . . . . . 2
   Stigmasta-8,22-dien-3β-ol
   5α-Stigmasta-7,22-dien-3β-ol
   Δ7,25-Stigmastadienol
   β-Sitosterol . . . . . . . . . . . . . . . . . . . . . . 77
   Δ5-Avenasterol . . . . . . . . . . . . . . . . . . . . 7
   Δ7-Stigmasterol . . . . . . . . . . . . . . . . . . . 1
   Δ7-Avenasterol
   Δ7-Campesterol
   Δ7-Ergosterol
   Δ7,25-Stigmasterol
   Sitostanol
   Spinasterol
   Squalene
   24-Methylene Cholesterol
   Other
   % sterols in oil
   Total Sterols, mg/kg . . . . . . . . . . . . 2010

**References** *J. Am. Oil Chem. Soc. 52:* 512 (1975)
   *J. Am. Dietetic Assn. 73:* 39 (1978)
   *J. Food Technol. 13:* 355 (1978)

## Pistacia Atlantica Fruit Oil
*Pistacia atlantica*

Specific Gravity (SG)
  15.5/15.5°C
  25/25°C
  Other SG
Refractive Index (RI)
  25°C
  40°C
  Other RI
Iodine Value ....................... 88
Saponification Value ................ 199
Titer °C
% Unsaponifiable
Melting Point °C

**Fatty Acid Composition (%)**
  16:0 ........................... 24
  16:1 ............................ 1.2
  18:0 ............................ 1.8
  Total 18:1 ...................... 46
  Undefined 18:2 ................. 27.4

Sterol Composition, %
  Cholesterol ..................... 1.5
  Brassicasterol
  Campesterol ..................... 4.3
  Stigmasterol
  Stigmasta-8,22-dien-3β-ol
  5α-Stigmasta-7,22-dien-3β-ol
  Δ7,25-Stigmastadienol
  β-Sitosterol .................... 87
  Δ5-Avenasterol .................... 4
  Δ7-Stigmasterol
  Δ7-Avenasterol ................. 3.2
  Δ7-Campesterol
  Δ7-Ergosterol
  Δ7,25-Stigmasterol
  Sitostanol
  Spinasterol
  Squalene
  24-Methylene Cholesterol
  Other
  % sterols in oil
  Total Sterols, mg/kg

**References** *JAOCS* 79:1049–1050 (2002)

## Plum Kernel Oil
*Prunus domestica*

Specific Gravity (SG)
  15.5/15.5°C
  25/25°C ................. 0.911–0.916
  Other SG
Refractive Index (RI)
  25°C .................. 1.4690–1.4692
  40°C .................... 1.462–1.465
  Other RI
Iodine Value ................. 100–110
Saponification Value ........... 180–195
Titer °C ....................... 5.7–5.8
% Unsaponifiable ............... 0.4–1.1
% Sterols in crude oil ............. 0.3
Melting Point °C

**Fatty Acid Composition (%)**
  12:0 ........................... 0.5
  14:0 ........................... 0.9–1.1
  16:0 ........................... 6–13
  16:1 ........................... 0.1–0.5
  9c-16:1 ........................ 0.1–0.3
  18:0 ............................ 1–8
  Total 18:1 ..................... 62–72
  Undefined 18:2 ................. 13–30
  9c,12c-18:2 .................... 13–30
  Undefined 18:3 ................. 0–0.3
  20:0 ............................ 0.3

Sterol Composition, %
  Cholesterol ..................... 0.6
  Brassicasterol
  Campesterol ....................... 4
  Stigmasterol ...................... 6
  Stigmasta-8,22-dien-3β-ol
  5α-Stigmasta-7,22-dien-3β-ol
  Δ7,25-Stigmastadienol
  β-Sitosterol ..................... 75
  Δ5-Avenasterol .................... 5
  Δ7-Stigmasterol ................... 5
  Δ7-Avenasterol .................... 3
  Δ7-Campesterol .................... 1
  Δ7-Ergosterol
  Δ7,25-Stigmasterol
  Sitostanol
  Spinasterol

Squalene
24-Methylene Cholesterol
Other
% sterols in oil
Total Sterols, mg/kg

**References** *Palm Oil Tech. Bull. 2:* 8 (1996)
*Fat Sci. Technol. 89:* 304 (1987)
*Riv. Ital. Sost. Grasse 75:* 405 (1998)
*Fat Sci. Technol. 89 (8):* 304 (1987)
*JAOCS 69:* 492–494 (1992)

## Poga Oleosa Kernel Oil
*Poga oleosa*

Specific Gravity (SG)
   15.5/15.5°C
   25/25°C
   Other SG
Refractive Index (RI)
   25°C
   40°C
   Other RI
Iodine Value . . . . . . . . . . . . . . . . . . . . 84–94
Saponification Value . . . . . . . . . . . . 184–193
Titer °C . . . . . . . . . . . . . . . . . . . . . . . . 22–25
% Unsaponifiable . . . . . . . . . . . . . . . . . . . 0.4
Melting Point °C

**Fatty Acid Composition (%)**
   16:0 . . . . . . . . . . . . . . . . . . . . . . . . . . . 11
   18:0 . . . . . . . . . . . . . . . . . . . . . . . . . . . . 7
   Total 18:1 . . . . . . . . . . . . . . . . . . . . . . . 70
   9c,12c-18:2 . . . . . . . . . . . . . . . . . . . . . . 13

**References** *Rev. Franc. Corp Gras 39:* 147 (1992)

## Poli Oil, Wild Safflower Seed Oil
*Carthamus oxycanthus*

Specific Gravity (SG)
   15.5/15.5°C
   25/25°C
   Other SG
Refractive Index (RI)
   25°C . . . . . . . . . . . . . . . . . . . . . . 1.4729
   40°C
   Other RI
Iodine Value . . . . . . . . . . . . . . . . . . . . . . 113
Saponification Value . . . . . . . . . . . . . . . . 195
Titer °C
% Unsaponifiable . . . . . . . . . . . . . . . . . . . 0.4
Melting Point °C

**Fatty Acid Composition (%)**
   16:0 . . . . . . . . . . . . . . . . . . . . . . . . . . . . 9
   18:0 . . . . . . . . . . . . . . . . . . . . . . . . . . . . 2
   Total 18:1 . . . . . . . . . . . . . . . . . . . . . . . 17
   Undefined 18:2 . . . . . . . . . . . . . . . . . . . 71
   Undefined 18:3 . . . . . . . . . . . . . . . . . . 6.5

**References** *J. Oil Technol. Assoc. India (JOTAI), 11:* 8–10 (1970)

## Poppyseed Oil
*Papaver somniferum*

Specific Gravity (SG)
   15.5/15.5°C . . . . . . . . . . . . . . 0.924–0.927
   25/25°C . . . . . . . . . . . . . . . . . 0.918–0.920
   Other SG
Refractive Index (RI)
   25°C
   40°C . . . . . . . . . . . . . . . . . . . 1.467–1.470
   Other RI
Iodine Value . . . . . . . . . . . . . . . . . . 132–146
Saponification Value . . . . . . . . . . . . 188–196
Titer °C
% Unsaponifiable . . . . . . . . . . . . . . . 0.4–1.2
Melting Point °C

**Fatty Acid Composition (%)**
   14:0 . . . . . . . . . . . . . . . . . . . . . . . 0.1–0.7
   16:0 . . . . . . . . . . . . . . . . . . . . . . . . . 7–11
   9c-16:1 . . . . . . . . . . . . . . . . . . . . . 0.8–1.6
   18:0 . . . . . . . . . . . . . . . . . . . . . . . . . . 1–4
   Total 18:1 . . . . . . . . . . . . . . . . . . . . 16–30
   9c,12c-18:2 . . . . . . . . . . . . . . . . . . . 62–73

Sterol Composition, %
   Cholesterol
   Brassicasterol

Campesterol . . . . . . . . . . . . . . . . . . . . . 22
Stigmasterol . . . . . . . . . . . . . . . . . . . . . . 3
Stigmasta-8,22-dien-3β-ol
5α-Stigmasta-7,22-dien-3β-ol
Δ7,25-Stigmastadienol
β-Sitosterol . . . . . . . . . . . . . . . . . . . . . . 68
Δ5-Avenasterol . . . . . . . . . . . . . . . . . . . . 2
Δ7-Stigmasterol . . . . . . . . . . . . . . . . . . . 2
Δ7-Avenasterol
Δ7-Campesterol
Δ7-Ergosterol
Δ7,25-Stigmasterol
Sitostanol
Spinasterol
Squalene
24-Methylene Cholesterol
Other
% sterols in oil
Total Sterols, mg/kg

**References** *Lipids 9:* 921 (1974)

## Mediterranean Seagrass
*Posidonia oceanica*
Specific Gravity (SG)
   15.5/15.5°C
   25/25°C
   Other SG
Refractive Index (RI)
   25°C
   40°C
   Other RI
Iodine Value
Saponification Value
Titer °C
% Unsaponifiable
Melting Point °C

**Fatty Acid Composition (%)**
   12:0 . . . . . . . . . . . . . . . . . . . . . . . . . . . 6
   14:0 . . . . . . . . . . . . . . . . . . . . . . . . . . 0.2
   16:0 . . . . . . . . . . . . . . . . . . . . . . . . . . . 21
   16:1 . . . . . . . . . . . . . . . . . . . . . . . . . . 0.5
   16:2 . . . . . . . . . . . . . . . . . . . . . . . . . . 0.3
   18:0 . . . . . . . . . . . . . . . . . . . . . . . . . . . 3
   Total 18:1 . . . . . . . . . . . . . . . . . . . . . 2–3

Undefined 18:2 . . . . . . . . . . . . . . . . . . 28
Undefined 18:3 . . . . . . . . . . . . . . . . . . 37
20:0 . . . . . . . . . . . . . . . . . . . . . . . . . . 0.2
22:0 . . . . . . . . . . . . . . . . . . . . . . . . . . 0.2
Unidentified 22:1 . . . . . . . . . . . . . . . 0.3
24:0 . . . . . . . . . . . . . . . . . . . . . . . . . . 0.3

**References** *Phytochemistry 34:* 381 (1993)

## Proso Millet
*Ponicum miliaceum*
Specific Gravity (SG)
   15.5/15.5°C . . . . . . . . . . . . . . . . . . 0.9383
   25/25°C
   Other SG
Refractive Index (RI)
   25°C
   40°C . . . . . . . . . . . . . . . . . . . . . . . 1.4577
   Other RI
Iodine Value . . . . . . . . . . . . . . . . . . . . . 129
Saponification Value . . . . . . . . . . . . . . . 192
Titer °C
% Unsaponifiable
Melting Point °C

**Fatty Acid Composition (%)**
   16:0 . . . . . . . . . . . . . . . . . . . . . . . . . . . . 6
   18:0 . . . . . . . . . . . . . . . . . . . . . . . . . . . . 1
   Total 18:1 . . . . . . . . . . . . . . . . . . . . . . 25
   9c,12c-18:2 . . . . . . . . . . . . . . . . . . . . . 66
   Undefined 18:3 . . . . . . . . . . . . . . . . . . . 1
   20:0 . . . . . . . . . . . . . . . . . . . . . . . . . . 0.5
   5c,8c,11c,14c-20:4 . . . . . . . . . . . . . . . 0.1
   22:0 . . . . . . . . . . . . . . . . . . . . . . . . . . 0.4
   Unidentified 22:1 . . . . . . . . . . . . . . . 0.1

**References** *Cereal Chem. 71:* 355 (1994)

## Prosopis Africana Seed Oil
*Prosopis africana*
Specific Gravity (SG)
   15.5/15.5°C
   25/25°C
   Other SG

Refractive Index (RI)
  25°C
  40°C
  Other RI
Iodine Value
Saponification Value
Titer °C
% Unsaponifiable
Melting Point °C

**Fatty Acid Composition (%)**
  16:0 .......................... 9.2
  9c-16:1 ....................... 1.2
  16:2 .......................... 1.9
  18:0 .......................... 4.5
  9c-18:1 ...................... 29.4
  11c-18:1 ...................... 1.7
  9c,12c-18:2 .................. 29.8
  6c,9c,12c-18:3 ................ 0.3
  9c,12c,15c-18:3 ............... 2.0
  20:0 .......................... 0.9
  Total 20:1 .................... 0.3
  22:0 .......................... 0.8
  24:0 .......................... 0.6

**References** *J. Am. Oil Chem. Soc. 75:* 1031 (1998)

## Prune Kernel Oil
*Prunus cerasifera*

Specific Gravity (SG)
  15.5/15.5°C
  25/25°C
  Other SG
Refractive Index (RI)
  25°C
  40°C ...................... 1.4712
  Other RI
Iodine Value .................. 96–104
Saponification Value .......... 210
Titer °C
% Unsaponifiable .............. 1.4
Melting Point °C

**Fatty Acid Composition (%)**
  16:0 .......................... 4–8
  9c-16:1 ....................... 0.4
  18:0 .......................... 1.5–2
  Total 18:1 .................... 61–79
  9c,12c-18:2 ................... 15–29

Tocopherol Composition, mg/kg
  α-Tocopherol .................. 85
  β-Tocopherol
  γ-Tocopherol .................. 656
  δ-Tocopherol .................. 36
  Total, mg/kg .................. 777

**References** *Fette Seifen Anstrichm. 86:* 160 (1984)
*Rev. Franc. Corps Gras 33:* 115 (1986)

## Pseudotsuga Menziesii Seed Oil
*Pseudotsuga menziesii*

Specific Gravity (SG)
  15.5/15.5°C
  25/25°C
  Other SG
Refractive Index (RI)
  25°C
  40°C
  Other RI
Iodine Value
Saponification Value
Titer °C
% Unsaponifiable
Melting Point °C

**Fatty Acid Composition (%)**
  14:0 .......................... 0.1
  9c-14:1 ....................... tr
  15:0 .......................... tr
  16:0 .......................... 3.5
  9c-16:1 ....................... 0.2
  7c-16:1 ....................... 0.1
  17:0 .......................... 0.1
  18:0 .......................... 1.8
  9c-18:1 ...................... 18.1
  11c-18:1 ...................... 0.8
  5,9–18:2 ...................... 2.8
  9c,12c-18:2 ................... 44
  9c,12c,15c-18:3 ............... 0.6

5,9c,12c,15c-18:4 .............. 0.1
20:0 ......................... 0.6
11c-20:1 ..................... 0.9
20:2 ......................... 0.4
5,11c,14c-20:3 ............... 1.7
22:0 ......................... 0.5
24:0 ......................... 0.2
26:0 ......................... 0.1

**References** *JAOCS 75:* 1761–1765 (1998)

## Pterocarpus Osun Seed Oil
*Pterocarpus osun*

Specific Gravity (SG)
    15.5/15.5°C
    25/25°C
    Other SG
Refractive Index (RI)
    25°C
    40°C
    Other RI
Iodine Value
Saponification Value
Titer °C
% Unsaponifiable
Melting Point °C

**Fatty Acid Composition (%)**
    16:0 ......................... 12.0
    16:2 ......................... 1.4
    18:0 ......................... 5.2
    9c-18:1 ...................... 18.3
    11c-18:1 ..................... 0.3
    9c,12c-18:2 .................. 27.8
    6c,9c,12c-18:3 ............... 0.4
    9c,12c,15c-18:3 .............. 0.6
    20:0 ......................... 2.4
    Total 20:1 ................... 1.3
    22:0 ......................... 10.4
    24:0 ......................... 2.5

**References** *J. Am. Oil Chem. Soc. 75:* 1031 (1998)

## Pterocarpus Santalinoides Seed Oil
*Pterocarpus santalinoides*

Specific Gravity (SG)
    15.5/15.5°C
    25/25°C
    Other SG
Refractive Index (RI)
    25°C
    40°C
    Other RI
Iodine Value
Saponification Value
Titer °C
% Unsaponifiable
Melting Point °C

**Fatty Acid Composition (%)**
    14:0 ......................... 0.2
    16:0 ......................... 7.0
    9c-16:1 ...................... 0.5
    16:2 ......................... 0.5
    18:0 ......................... 2.9
    9c-18:1 ...................... 5.6
    11c-18:1 ..................... 0.5
    9c,12c-18:2 .................. 11.5
    9c,12c,15c-18:3 .............. 1.3
    20:0 ......................... 0.6
    22:0 ......................... 1.1
    24:0 ......................... 1.3

**References** *J. Am. Oil Chem. Soc. 75:* 1031 (1998)

## Pumpkin Seed Oil
*Cucurbita pepo*

Specific Gravity (SG)
    15.5/15.5°C
    25/25°C
    Other SG .......... (20/20) 0.903–0.926
Refractive Index (RI)
    25°C
    40°C ................ 1.4653–1.4661

Other RI . . . . . . . . . . . . . (20) 1.466–1.474
Iodine Value . . . . . . . . . . . . . . . . . 103–133
Saponification Value . . . . . . . . . . . 174–203
Titer °C
% Unsaponifiable . . . . . . . . . . . . . . . 0.5–1.8
Melting Point °C

**Fatty Acid Composition (%)**
   14:0. . . . . . . . . . . . . . . . . . . . . . . . . . . . 0.1
   16:0. . . . . . . . . . . . . . . . . . . . . . . . . . . . 7–15
   18:0. . . . . . . . . . . . . . . . . . . . . . . . . . . . 3–13
   Total 18:1 . . . . . . . . . . . . . . . . . . . . . . 21–47
   9c-18:1 . . . . . . . . . . . . . . . . . . . . . . . . . . . 38
   Undefined 18:2 . . . . . . . . . . . . . . . . . 42.1
   9c,12c-18:2 . . . . . . . . . . . . . . . . . . . . 36–61
   Undefined 18:3 . . . . . . . . . . . . . . . . . . . 0.2
   20:0. . . . . . . . . . . . . . . . . . . . . . . . . . 0.3–0.5
   Total 20:1 . . . . . . . . . . . . . . . . . . . . . . . . 0.1
   22:0. . . . . . . . . . . . . . . . . . . . . . . . . . 0.1–0.2
   Unidentified 22:1 . . . . . . . . . . . . . . . 0–0.2
   24:0. . . . . . . . . . . . . . . . . . . . . . . . . . 0.2–0.54
   26:0. . . . . . . . . . . . . . . . . . . . . . . . . . . . 0.13

Sterol Composition, %
   Cholesterol
   Brassicasterol
   Campesterol . . . . . . . . . . . . . . . . . . . . 0.9
   Stigmasterol . . . . . . . . . . . . . . . . . 1–3.51
   Stigmasta-8,22-dien-3β-ol
   5α-Stigmasta-7,22-dien-3β-ol
   Δ7,25-Stigmastadienol . . . . . . . . . . . . 22
   β-Sitosterol
   Δ5-Avenasterol
   Δ7-Stigmasterol . . . . . . . . . . . . . . . . . 3–4
   Δ7-Avenasterol . . . . . . . . . . . . . . . . . . . 10
   Δ7-Campesterol
   Δ7-Ergosterol
   Δ7,25-Stigmasterol
   Sitostanol
   Spinasterol . . . . . . . . . . . . . . . . . . . . . . 27
   Squalene
   24-Methylene Cholesterol
   Other: 24-Methyl-cholest-7-enol, 6;
      D7,22,25-stigmastatrienol, 29
   % sterols in oil
   Total Sterols, mg/kg

Tocopherol Composition, mg/kg
   α-Tocopherol . . . . . . . . . . . . . . . . . . . . 12
   β-Tocopherol
   γ-Tocopherol . . . . . . . . . . . . . . . . . . . . 285
   δ-Tocopherol. . . . . . . . . . . . . . . . . . . . . . 4
   Total, mg/kg

Tocotrienols Composition, mg/kg
   α-Tocotrienol
   β-Tocotrienol . . . . . . . . . . . . . . . . . . . . . 9
   γ-Tocotrienol. . . . . . . . . . . . . . . . . . . . . . 5
   δ-Tocotrienol
   Total Tocotrienols, mg/kg

**References** *J. Am. Oil Chem. Soc. 53:* 42
  (1976)
  *J. Am. Oil Chem. Soc. 54:* 525 (1977)
  *J. Am. Dietetic Assn. 73:* 39 (1978)
  *Z. Lebensmittel Unters. Forsch. 203:* 216
  (1996)
  *Grasas y Aceites 48:* 267–272 (1977)

# Quamoclit Seed Oil
*Quamoclit phoenicea*

Specific Gravity (SG)
   15.5/15.5°C
   25/25°C
   Other SG
Refractive Index (RI)
   25°C
   40°C
   Other RI
Iodine Value . . . . . . . . . . . . . . . . . . . . . 90.3
Saponification Value . . . . . . . . . . . . . . 200.6
Titer °C
% Unsaponifiable . . . . . . . . . . . . . . . . . . . 2
Melting Point °C

**Fatty Acid Composition (%)**
   16:0. . . . . . . . . . . . . . . . . . . . . . . . . . . 22.2
   18:0. . . . . . . . . . . . . . . . . . . . . . . . . . . 11.3
   9c-18:1 . . . . . . . . . . . . . . . . . . . . . . . . 13.5
   9c,12c-18:2 . . . . . . . . . . . . . . . . . . . . 40.1
   20:0. . . . . . . . . . . . . . . . . . . . . . . . . . . . 3.5
   22:0. . . . . . . . . . . . . . . . . . . . . . . . . . . . . 3
   Other: vernolic, 6.4

**References** *JAOCS 69:* 190–191 (1992)

## Quamoclit Seed Oil
*Quamoclit coccinea*

Specific Gravity (SG)
　15.5/15.5°C
　25/25°C
　Other SG
Refractive Index (RI)
　25°C
　40°C
　Other RI
Iodine Value . . . . . . . . . . . . . . . . . . . . . . 77.7
Saponification Value . . . . . . . . . . . . . . . 201.8
Titer °C
% Unsaponifiable . . . . . . . . . . . . . . . . . . . 2.2
Melting Point °C

**Fatty Acid Composition (%)**
　16:0 . . . . . . . . . . . . . . . . . . . . . . . . . . 33.3
　18:0 . . . . . . . . . . . . . . . . . . . . . . . . . . . 1.7
　9c-18:1 . . . . . . . . . . . . . . . . . . . . . . . 14.6
　9c,12c-18:2 . . . . . . . . . . . . . . . . . . . . 30.8
　20:0 . . . . . . . . . . . . . . . . . . . . . . . . . . . 6.8
　22:0 . . . . . . . . . . . . . . . . . . . . . . . . . . . 2.6
　Other: vernolic, 10.2

**References** *JAOCS 69:* 190–191 (1992)

## Quince Seed Oil
*Cydonia Mill*

Specific Gravity (SG)
　15.5/15.5°C
　25/25°C
　Other SG . . . . . . . . . (20/20) 0.923–0.926
Refractive Index (RI)
　25°C
　40°C . . . . . . . . . . . . . . . . . 1.4656–1.467
　Other RI
Iodine Value . . . . . . . . . . . . . . . . . . 113–122
Saponification Value . . . . . . . . . . . . 186–194
Titer °C
% Unsaponifiable . . . . . . . . . . . . . . . 0.3–1.6
% Sterols in crude oil . . . . . . . . . . . . 0.1–0.3
Melting Point °C

**Fatty Acid Composition (%)**
　14:0 . . . . . . . . . . . . . . . . . . . . . . . . . . . 0.1
　16:0 . . . . . . . . . . . . . . . . . . . . . . . . . . . . . 5
　16:1 . . . . . . . . . . . . . . . . . . . . . . . . . . . 0.2
　18:0 . . . . . . . . . . . . . . . . . . . . . . . . . . . . . 1
　Total 18:1 . . . . . . . . . . . . . . . . . . . . . . . 41
　Undefined 18:2 . . . . . . . . . . . . . . . . . . . 52

Sterol Composition, %
　Cholesterol . . . . . . . . . . . . . . . . . . . . . . 0.4
　Brassicasterol . . . . . . . . . . . . . . . . . . . . 0.3
　Campesterol . . . . . . . . . . . . . . . . . . . . . . . 4
　Stigmasterol . . . . . . . . . . . . . . . . . . . . . . . 5
　Stigmasta-8,22-dien-3β-ol
　5α-Stigmasta-7,22-dien-3β-ol
　Δ7,25-Stigmastadienol
　β-Sitosterol . . . . . . . . . . . . . . . . . . . . . . . 88
　Δ5-Avenasterol . . . . . . . . . . . . . . . . . . . . . 1
　Δ7-Stigmasterol
　Δ7-Avenasterol . . . . . . . . . . . . . . . . . . . 0.3
　Δ7-Campesterol . . . . . . . . . . . . . . . . . . 0.4
　Δ7-Ergosterol
　Δ7,25-Stigmasterol
　Sitostanol
　Spinasterol
　Squalene
　24-Methylene Cholesterol
　Other
% sterols in oil
Total Sterols, mg/kg

**References** *Riv. Ital. Sost. Grasse 75:* 405 (1998)

## Radyera farragei
*Radyera farragei*

Specific Gravity (SG)
　15.5/15.5°C
　25/25°C
　Other SG
Refractive Index (RI)
　25°C
　40°C
　Other RI
Iodine Value
Saponification Value
Titer °C
% Unsaponifiable
Melting Point °C

**Fatty Acid Composition (%)**
- 14:0 . . . . . . . . . . . . . . . . . . . . . . . . . . . 0.1
- 16:0 . . . . . . . . . . . . . . . . . . . . . . . . . . . 10.5
- 16:1 . . . . . . . . . . . . . . . . . . . . . . . . . . . 0.2
- 18:0 . . . . . . . . . . . . . . . . . . . . . . . . . . . 2.8
- Total 18:1 . . . . . . . . . . . . . . . . . . . . 15.8
- Undefined 18:2 . . . . . . . . . . . . . . . . 68.2
- Undefined 18:3 . . . . . . . . . . . . . . . . . . 1
- 20:0 . . . . . . . . . . . . . . . . . . . . . . . . . . . 0.3

**References** *JAOCS 68:* 518–519 (1991)

## Rambutan Tallow
*Nephelium lappaceum*

Specific Gravity (SG)
   15.5/15.5°C
   25/25°C
   Other SG . . . . . . . . (99/15.5) 0.859–0.863
Refractive Index (RI)
   25°C
   40°C . . . . . . . . . . . . . . . . . . . . 1.458–1.459
   Other RI
Iodine Value . . . . . . . . . . . . . . . . . . . . 39–44
Saponification Value . . . . . . . . . . . 193–195
Titer °C
% Unsaponifiable . . . . . . . . . . . . . . . . . . . 0.5
Melting Point °C . . . . . . . . . . . . . . . . 38–42

**Fatty Acid Composition (%)**
- 16:0 . . . . . . . . . . . . . . . . . . . . . . . . . . . . 2
- 18:0 . . . . . . . . . . . . . . . . . . . . . . . . . . . 14
- Total 18:1 . . . . . . . . . . . . . . . . . . . . . . 45
- 20:0 . . . . . . . . . . . . . . . . . . . . . . . . . . . 35

**References**

## Rapeseed Oil
*Brassica napus*

Specific Gravity (SG)
   15.5/15.5°C
   25/25°C
   Other SG . . . . . . . . . . (20/20) 0.910–0.920
Refractive Index (RI)
   25°C
   40°C . . . . . . . . . . . . . . . . . . . . 1.465–1.469
   Other RI
Iodine Value . . . . . . . . . . . . . . . . . . . 94–120
Saponification Value . . . . . . . . . . . 168–181
Titer °C
% Unsaponifiable . . . . . . . . . . . . . . . . . . 0–2
Melting Point -10°C

**Fatty Acid Composition (%)**
- 12:0 . . . . . . . . . . . . . . . . . . . . . . . . . . . 0.1
- 14:0 . . . . . . . . . . . . . . . . . . . . . . . . . . . 0.2
- 16:0 . . . . . . . . . . . . . . . . . . . . . . . . . . 1.5–6
- 16:1 . . . . . . . . . . . . . . . . . . . . . . . . . . . 0.3
- 9c-16:1 . . . . . . . . . . . . . . . . . . . . . . . . . 0–3
- 18:0 . . . . . . . . . . . . . . . . . . . . . . . . . 0.5–3.1
- Total 18:1 . . . . . . . . . . . . . . . . . . . 8–60.1
- Undefined 18:2 . . . . . . . . . . . . . . . . . 21.4
- 9c,12c-18:2 . . . . . . . . . . . . . . . . . . . 11–23
- Undefined 18:3 . . . . . . . . . . . . . . . . . 5–13
- 20:0 . . . . . . . . . . . . . . . . . . . . . . . . . . . 0–3
- Total 20:1 . . . . . . . . . . . . . . . . . . . . . 3–15
- 20:2 . . . . . . . . . . . . . . . . . . . . . . . . . . . 0–1
- 22:0 . . . . . . . . . . . . . . . . . . . . . . . . . . . 0–2
- Unidentified 22:1 . . . . . . . . . . . . . . . 2–60
- 22:2 . . . . . . . . . . . . . . . . . . . . . . . . . . . 0–2
- 24:0 . . . . . . . . . . . . . . . . . . . . . . . . . . . 0–2
- 15c-24:1 . . . . . . . . . . . . . . . . . . . . . . . 0–3

Sterol Composition, %
   Cholesterol
   Brassicasterol . . . . . . . . . . . . . . . . 12–13
   Campesterol . . . . . . . . . . . . . . . . . . 30–33
   Stigmasterol . . . . . . . . . . . . . . . . . 0.4–0.6
   Stigmasta-8,22-dien-3β-ol
   5α-Stigmasta-7,22-dien-3β-ol
   Δ7,25-Stigmastadienol
   β-Sitosterol . . . . . . . . . . . . . . . . . . . 49–55
   Δ5-Avenasterol . . . . . . . . . . . . . . . . . . 1–2
   Δ7-Stigmasterol
   Δ7-Avenasterol
   Δ7-Campesterol
   Δ7-Ergosterol
   Δ7,25-Stigmasterol
   Sitostanol
   Spinasterol
   Squalene
   24-Methylene Cholesterol
   Other
   % sterols in oil
   Total Sterols, mg/kg . . . . . . . . . . . . . 881

Tocopherol Composition, mg/kg
α-Tocopherol . . . . . . . . . . . . . . 116–180
β-Tocopherol. . . . . . . . . . . . . . . . . . . 34
γ-Tocopherol. . . . . . . . . . . . . . 340–737
δ-Tocopherol. . . . . . . . . . . . . . . . . . 275
Total, mg/kg . . . . . . . . . . . . . . . . . 1165

**References**  Codex 1997/17
Riv. Ital. Sost. Grasse 52: 79 (1975)
R.G. Ackman in Canola and Rapeseed, F. Shahidi, ed., Van Nostrum Reinhold, NY, 1990, p. 88
JAOCS 74: 375–381(1997)

## Rapeseed Oil
## (low linolenic, Canola)

Specific Gravity (SG)
  15.5/15.5°C
  25/25°C
  Other SG
Refractive Index (RI)
  25°C
  40°C
  Other RI
Iodine Value . . . . . . . . . . . . . . . . . . . . 91
Saponification Value
Titer °C
% Unsaponifiable
Melting Point °C

**Fatty Acid Composition (%)**
  16:0. . . . . . . . . . . . . . . . . . . . . . . . . 4–5
  9c-16:1 . . . . . . . . . . . . . . . . . . . . . . 0.2
  18:0. . . . . . . . . . . . . . . . . . . . . . . . . 1–2
  Total 18:1 . . . . . . . . . . . . . . . . . . 59–66
  9c,12c-18:2. . . . . . . . . . . . . . . . . 24–29
  Undefined 18:3 . . . . . . . . . . . . . . . . 2–3
  20:0. . . . . . . . . . . . . . . . . . . . . . . . . 0.7
  Total 20:1 . . . . . . . . . . . . . . . . . . . . 1.2
  22:0. . . . . . . . . . . . . . . . . . . . . . . . 0–0.5
  Unidentified 22:1 . . . . . . . . . . . . 0–0.05

**References**  J. Am. Oil Chem. Soc. 67: 161 (1990)
J. Am. Oil Chem. Soc. 70: 983 (1993)

## Rapeseed Oil
## (low erucic, Canola)

Specific Gravity (SG)
  15.5/15.5°C
  25/25°C
  Other SG . . . . . . . . . (20/20) 0.914–0.920
Refractive Index (RI)
  25°C
  40°C . . . . . . . . . . . . . . . . . . . 1.465–1.467
  Other RI
Iodine Value . . . . . . . . . . . . . . . . 110–126
Saponification Value . . . . . . . . . . 182–193
Titer °C
% Unsaponifiable . . . . . . . . . . . . . . . . 0–2
Melting Point °C

**Fatty Acid Composition (%)**
  14:0. . . . . . . . . . . . . . . . . . . . . . . . . 0–0.2
  16:0. . . . . . . . . . . . . . . . . . . . . . . . 3.3–6.0
  9c-16:1 . . . . . . . . . . . . . . . . . . . . . 0.1–0.6
  17:0. . . . . . . . . . . . . . . . . . . . . . . . . . 0.3
  18:0. . . . . . . . . . . . . . . . . . . . . . . . 1.1–2.5
  Total 18:1 . . . . . . . . . . . . . . . . . . . 52–67
  9c,12c-18:2. . . . . . . . . . . . . . . . . . 16–25
  Undefined 18:3 . . . . . . . . . . . . . . . . 6–14
  20:0. . . . . . . . . . . . . . . . . . . . . . . . 0.2–0.8
  Total 20:1 . . . . . . . . . . . . . . . . . . . 0.1–3.4
  20:2. . . . . . . . . . . . . . . . . . . . . . . . . 0–0.1
  22:0. . . . . . . . . . . . . . . . . . . . . . . . . 0–0.5
  Unidentified 22:1 . . . . . . . . . . . . . . 0–4.7
  22:2. . . . . . . . . . . . . . . . . . . . . . . . . 0–0.1
  24:0. . . . . . . . . . . . . . . . . . . . . . . . . 0–0.2
  15c-24:1 . . . . . . . . . . . . . . . . . . . . . 0–0.4
  Other: 17:1, 0–0.3

Sterol Composition, %
  Cholesterol . . . . . . . . . . . . . . . . . . 0.5–1.3
  Brassicasterol . . . . . . . . . . . . . . . 5.0–13.0
  Campesterol . . . . . . . . . . . . . . . 24.7–38.6
  Stigmasterol . . . . . . . . . . . . . . . . . 0–0.7
  Stigmasta-8,22-dien-3β-ol
  5α-Stigmasta-7,22-dien-3β-ol
  Δ7,25-Stigmastadienol
  β-Sitosterol . . . . . . . . . . . . . . . . . . 45–58
  Δ5-Avenasterol . . . . . . . . . . . . . . 3.1–6.6
  Δ7-Stigmasterol . . . . . . . . . . . . . . 0–1.3

Δ7-Avenasterol .............. 0–0.8
Δ7-Campesterol
Δ7-Ergosterol
Δ7,25-Stigmasterol
Sitostanol
Spinasterol
Squalene
24-Methylene Cholesterol
Other....................... 0–4.2
% sterols in oil
Total Sterols, mg/kg ....... 4820–11280

Tocopherol Composition, mg/kg
  α-Tocopherol .............. 100–386
  β-Tocopherol............... 0–140
  γ-Tocopherol .............. 189–753
  δ-Tocopherol............... 0–22
  Total, mg/kg .............. 424–2680

**References** *Codex* 1997/17

## Raphanus Sativus Seed Oil
*Raphanus sativus*
Specific Gravity (SG)
  15.5/15.5°C
  25/25°C
  Other SG
Refractive Index (RI)
  25°C
  40°C
  Other RI
Iodine Value
Saponification Value
Titer °C
% Unsaponifiable
Melting Point °C

**Fatty Acid Composition (%)**
  14:0........................ 0.05
  16:0........................ 4.51
  9c-16:1 .................... 0.17
  18:0........................ 1.75
  9c-18:1 .................... 16.99
  9c,12a-18:2................. 13.86
  9c,12c,15c-18:3 ............ 8.74
  20:0........................ 1.27
  11c-20:1 ................... 9.41
  22:0........................ 0.98

13c-22:1 ................... 35.69
22:2........................ 0.35
24:0........................ 0.87

Tocopherol Composition, mg/kg
  α-Tocopherol
  β-Tocopherol
  γ-Tocopherol................ 516
  δ-Tocopherol................ 21
  Total, mg/kg

Tocotrienols Composition, mg/kg
  α-Tocotrienol
  β-Tocotrienol
  γ-Tocotrienol............... 3
  δ-Tocotrienol
  Total Tocotrienols, mg/kg

**References** *JAOCS 80:* 1013–1020 (2003)

## Ravison Oil
*Brassica campestris*
Specific Gravity (SG)
  15.5/15.5°C ............. 0.917–0.922
  25/25°C
  Other SG
Refractive Index (RI)
  25°C .................... 1.470–1.473
  40°C
  Other RI
Iodine Value ................ 106–122
Saponification Value ........ 177–183
Titer °C
% Unsaponifiable ............ 0.8–2.2
Melting Point °C

**Fatty Acid Composition (%)**
  14:0........................ 0.06
  16:0........................ 3.1–4
  9c-16:1 .................... 0.25
  18:0........................ 0.77–2
  Total 18:1 ................. 16
  9c-18:1 .................... 6.22
  7c-18:1 .................... 1.12
  11c-18:1 ................... 6.22
  9c,12c-18:2................. 16.68–21
  Undefined 18:3 ............. 10
  9c,12c,15c-18:3 ............ 10.78

20:0 . . . . . . . . . . . . . . . . . . . . . . . 0.69–2
Total 20:1 . . . . . . . . . . . . . . . . . . . . . 4.1
11c-20:1 . . . . . . . . . . . . . . . . . . . . . 4.96
22:0 . . . . . . . . . . . . . . . . . . . . . . . 0.5–0.92
Unidentified 22:1 . . . . . . . . . . . . . . . . 39
13c-22:1 . . . . . . . . . . . . . . . . . . . . . 44.11
22:2 . . . . . . . . . . . . . . . . . . . . . . . . . 2.07
24:0 . . . . . . . . . . . . . . . . . . . . . . . . 0.6–0.69

Tocopherol Composition, mg/kg
  α-Tocopherol . . . . . . . . . . . . . . . . . . . 114
  β-Tocopherol
  γ-Tocopherol . . . . . . . . . . . . . . . . . . . 445
  δ-Tocopherol . . . . . . . . . . . . . . . . . . . . 8
  Total, mg/kg

Tocotrienols Composition, mg/kg
  α-Tocotrienol . . . . . . . . . . . . . . . . . . . . 5
  β-Tocotrienol
  γ-Tocotrienol
  δ-Tocotrienol
  Total Tocotrienols, mg/kg

**References** *JAOCS 80:* 1013–1020 (2003)


# Red Pepper Seed Oil
*Capsicum annuum*

Specific Gravity (SG)
  15.5/15.5°C
  25/25°C
  Other SG
Refractive Index (RI)
  25°C
  40°C
  Other RI
Iodine Value
Saponification Value
Titer °C
% Unsaponifiable
Melting Point °C

**Fatty Acid Composition (%)**
  14:0 . . . . . . . . . . . . . . . . . . . . . . . 0.1–0.3
  16:0 . . . . . . . . . . . . . . . . . . . . . . . . 11.4
  9c-16:1 . . . . . . . . . . . . . . . . . . . . . 0.2–0.3
  17:0 . . . . . . . . . . . . . . . . . . . . . . . . 0.1
  18:0 . . . . . . . . . . . . . . . . . . . . . . 2.5–3.0

Total 18:1 . . . . . . . . . . . . . . . . . . . . . 8–9
Undefined 18:2 . . . . . . . . . . . . . . . . 76–78
9c,12c-18:2 . . . . . . . . . . . . . . . . . . . 75.8
Undefined 18:3 . . . . . . . . . . . . . . . . . 0.3
20:0 . . . . . . . . . . . . . . . . . . . . . . . 0.1–0.2
22:0 . . . . . . . . . . . . . . . . . . . . . . . . . 0.1
24:0 . . . . . . . . . . . . . . . . . . . . . . . . . 0.2

**References** *Riv. Ital. Sost. Grasse 68:* 309 (1991)
*J. Am. Oil Chem. Soc. 76:* 1449 (1999)


# Ribes Alpinum Seed Oil
*Ribes alpinum*

Specific Gravity (SG)
  15.5/15.5°C
  25/25°C
  Other SG
Refractive Index (RI)
  25°C
  40°C
  Other RI
Iodine Value
Saponification Value
Titer °C
% Unsaponifiable
Melting Point °C

**Fatty Acid Composition (%)**
  16:0 . . . . . . . . . . . . . . . . . . . . . . . . . . 6
  16:1 . . . . . . . . . . . . . . . . . . . . . . . . . 0.2
  18:0 . . . . . . . . . . . . . . . . . . . . . . . . . . 1
  epoxy 18:1 . . . . . . . . . . . . . . . . . . . . 18
  Undefined 18:2 . . . . . . . . . . . . . . . . . 39
  Undefined 18:3 . . . . . . . . . . . . . . . . . 22
  6c,9c,12c-18:3 . . . . . . . . . . . . . . . . . . 9
  18:4 . . . . . . . . . . . . . . . . . . . . . . . . . . 4
  Total 20:1 . . . . . . . . . . . . . . . . . . . . 0.1

**References** *JAOCS 60:* 1858 (1983)


# Rice Bran Oil
*Oryza sativa*

Specific Gravity (SG)
  15.5/15.5°C

25/25°C ................ 0.916–0.921
Other SG
Refractive Index (RI)
   25°C .................. 1.470–1.473
   40°C .................. 1.465–1.468
   Other RI
Iodine Value ................... 92–108
Saponification Value ........... 181–189
Titer °C
% Unsaponifiable .................. 3–5
Melting Point °C

**Fatty Acid Composition (%)**
   14:0 ........................ 0.5–0.7
   16:0 ......................... 16–28
   9c-16:1 ....................... 0.5
   18:0 .......................... 2–4
   Total 18:1 ................... 38–48
   9c,12c-18:2 .................. 16–36
   Undefined 18:3 .............. 0.2–2.2
   20:0 ........................ 0.5–0.8
   Total 20:1 .................. 0.3–0.5
   22:0 ........................ 0.1–0.5
   24:0 .......................... 0–0.5

Sterol Composition, %
   Cholesterol
   Brassicasterol
   Campesterol .................. 20–28
   Stigmasterol .................. 8–15
   Stigmasta-8,22-dien-3β-ol
   5α-Stigmasta-7,22-dien-3β-ol
   Δ7,25-Stigmastadienol
   β-Sitosterol .................. 49–54
   Δ5-Avenasterol ................ 5–11
   Δ7-Stigmasterol ................ 1–2
   Δ7-Avenasterol ................. 2–4
   Δ7-Campesterol
   Δ7-Ergosterol
   Δ7,25-Stigmasterol
   Sitostanol
   Spinasterol
   Squalene
   24-Methylene Cholesterol
   Other
   % sterols in oil
   Total Sterols, mg/kg ............ 10550

Tocopherol Composition, mg/kg
   α-Tocopherol .................... 600
   β-Tocopherol
   γ-Tocopherol .................... 300
   δ-Tocopherol
   Total, mg/kg .................... 900

**References** *J. Am. Oil Chem. Soc. 45:* 68 (1968)
*J. Am. Oil Chem. Soc. 50:* 122 (1973)
*J. Am. Oil Chem. Soc. 27:* 414 (1950)
*J. Am. Dietetic Assn. 73:* 39 (1978)

# Ricinodendron Heudelotii Kernel Oil
*Ricinodendron heudelotii*

Specific Gravity (SG)
   15.5/15.5°C
   25/25°C
   Other SG
Refractive Index (RI)
   25°C ................... 1.503–1.506
   40°C
   Other RI
Iodine Value ...................... 151
Saponification Value ............... 193
Titer °C
% Unsaponifiable ................. 0.5–1
Melting Point °C

**Fatty Acid Composition (%)**
   16:0 ........................... 10
   18:0 ............................ 7
   Total 18:1 ..................... 8–9
   9c,12c-18:2 .................... 36
   α-Eleostearic 9c,11t,13t-18:3 ........ 30
   Other: 9t,11t,13t-18:3, 8 (β-eleostearic); 9c,11t,13c-18:3, 2 (catalpic)

**References** *Rev. Franc. Corp Gras 39:* 147 (1992)

# Rubber Seed Oil
*Hevea brasiliensis*

Specific Gravity (SG)
   15.5/15.5°C ............. 0.924–0.930
   25/25°C

Other SG
Refractive Index (RI)
  25°C
  40°C . . . . . . . . . . . . . . . . . 1.466–1.469
  Other RI
Iodine Value . . . . . . . . . . . . . . . . . . 132–141
Saponification Value . . . . . . . . . . . 190–195
Titer °C
% Unsaponifiable . . . . . . . . . . . . . . . . 0.5–1
Melting Point °C

**Fatty Acid Composition (%)**
  16:0 . . . . . . . . . . . . . . . . . . . . . . . . . . 9–11
  18:0 . . . . . . . . . . . . . . . . . . . . . . . . . . 8–12
  Total 18:1 . . . . . . . . . . . . . . . . . . . 17–30
  9c,12c-18:2 . . . . . . . . . . . . . . . . . . 35–41
  Undefined 18:3 . . . . . . . . . . . . . . . 14–26
  20:0 . . . . . . . . . . . . . . . . . . . . . . . . . . . . . 1

**References**

# Rye Germ Oil
*Secale cercale*

Specific Gravity (SG)
  15.5/15.5°C . . . . . . . . . . . 0.9324–0.9412
  25/25°C
  Other SG
Refractive Index (RI)
  25°C . . . . . . . . . . . . . . . . . . 1.472–1.478
  40°C
  Other RI
Iodine Value . . . . . . . . . . . . . . . . . . 111–142
Saponification Value . . . . . . . . . . . 172–192
Titer °C
% Unsaponifiable . . . . . . . . . . . . . . . . 1–11
Melting Point °C

**Fatty Acid Composition (%)**
  14:0 . . . . . . . . . . . . . . . . . . . . . . . . . . . . . 2
  16:0 . . . . . . . . . . . . . . . . . . . . . . . . . . 9–21
  18:0 . . . . . . . . . . . . . . . . . . . . . . . . . . . 0.2
  Total 18:1 . . . . . . . . . . . . . . . . . . . . 7–35
  9c,12c-18:2 . . . . . . . . . . . . . . . . . . 48–72
  Undefined 18:3 . . . . . . . . . . . . . . . . . 3–8

**References**

# Safflower Oil (high linoleic)

Specific Gravity (SG)
  15.5/15.5°C
  25/25°C
  Other SG
Refractive Index (RI)
  25°C . . . . . . . . . . . . . . . . . . 1.472–1.476
  40°C . . . . . . . . . . . . . . . . . . 1.467–1.469
  Other RI
Iodine Value . . . . . . . . . . . . . . . . . . 132–150
Saponification Value . . . . . . . . . . . 186–198
Titer °C
% Unsaponifiable . . . . . . . . . . . . . . . . 0–15
Melting Point °C

**Fatty Acid Composition (%)**
  14:0 . . . . . . . . . . . . . . . . . . . . . . . . . . . 0.1
  16:0 . . . . . . . . . . . . . . . . . . . . . . . . . . . 6.2
  9c-16:1 . . . . . . . . . . . . . . . . . . . . . . . . 0.4
  18:0 . . . . . . . . . . . . . . . . . . . . . . . . . . . 2.2
  Total 18:1 . . . . . . . . . . . . . . . . . . . . . 11.7
  9c,12c-18:2 . . . . . . . . . . . . . . . . . . . . 74.1
  Undefined 18:3 . . . . . . . . . . . . . . . . . 0.4
  20:0 . . . . . . . . . . . . . . . . . . . . . . . . . . . 0.3
  Total 20:1 . . . . . . . . . . . . . . . . . . . . . 0.2
  22:0 . . . . . . . . . . . . . . . . . . . . . . . . . . . 0.5
  Unidentified 22:1 . . . . . . . . . . . . . . . 0.9
  24:0 . . . . . . . . . . . . . . . . . . . . . . . . . . . 0.1
  15c-24:1 . . . . . . . . . . . . . . . . . . . . . . 0.1

**References**

# Safflower Seed Oil
*Carthamus tinctorius*

Specific Gravity (SG)
  15.5/15.5°C . . . . . . . . . . . . . . 0.922–0.927
  25/25°C
  Other SG
Refractive Index (RI)
  25°C . . . . . . . . . . . . . . . . . . 1.472–1.476
  40°C . . . . . . . . . . . . . . . . . . 1.467–1.470
  Other RI
Iodine Value . . . . . . . . . . . . . . . . . . 136–148
Saponification Value . . . . . . . . . . . 186–198
Titer °C

% Unsaponifiable . . . . . . . . . . . . . . . . . 0–1.5
Melting Point °C

**Fatty Acid Composition (%)**
   14:0. . . . . . . . . . . . . . . . . . . . . . . . 0–0.2
   16:0. . . . . . . . . . . . . . . . . . . . . . . . 5.3–8.0
   9c-16:1 . . . . . . . . . . . . . . . . . . . . . 0–0.2
   18:0. . . . . . . . . . . . . . . . . . . . . . . . 1.9–2.9
   Total 18:1 . . . . . . . . . . . . . . . . . 8.4–30.0
   9c,12c-18:2 . . . . . . . . . . . . . . . . 67.8–83.2
   Undefined 18:3 . . . . . . . . . . . . . . . 0–0.1
   20:0. . . . . . . . . . . . . . . . . . . . . . . . 0.2–0.4
   Total 20:1 . . . . . . . . . . . . . . . . . . 0.1–0.3
   22:0. . . . . . . . . . . . . . . . . . . . . . . . 0.2–0.8
   Unidentified 22:1 . . . . . . . . . . . . . . 0–1.8
   24:0. . . . . . . . . . . . . . . . . . . . . . . . 0–0.2
   15c-24:1 . . . . . . . . . . . . . . . . . . . . 0–0.2

Sterol Composition, %
   Cholesterol . . . . . . . . . . . . . . . . . . . 0–0.5
   Brassicasterol
   Campesterol . . . . . . . . . . . . . . . . 9.2–13.0
   Stigmasterol . . . . . . . . . . . . . . . . . 6.5–9.6
   Stigmasta-8,22-dien-3β-ol
   5α-Stigmasta-7,22-dien-3β-ol
   Δ7,25-Stigmastadienol
   β-Sitosterol . . . . . . . . . . . . . . . . . 40.2–49.8
   Δ5-Avenasterol . . . . . . . . . . . . . . . 2.1–4.0
   Δ7-Stigmasterol . . . . . . . . . . . . . 15.7–22.4
   Δ7-Avenasterol . . . . . . . . . . . . . . . 2.9–5.3
   Δ7-Campesterol
   Δ7-Ergosterol
   Δ7,25-Stigmasterol
   Sitostanol
   Spinasterol
   Squalene
   24-Methylene Cholesterol
   Other. . . . . . . . . . . . . . . . . . . . . . . 0.5–2.8
   % sterols in oil
   Total Sterols, mg/kg . . . . . . . . 2095–2650

Tocopherol Composition, mg/kg
   α-Tocopherol . . . . . . . . . . . . . . 230–660
   β-Tocopherol. . . . . . . . . . . . . . . . . . 0–20
   γ-Tocopherol. . . . . . . . . . . . . . . . . . 0–15
   δ-Tocopherol
   Total, mg/kg . . . . . . . . . . . . . . . 245–690

Tocotrienols Composition, mg/kg
   α-Tocotrienol

   β-Tocotrienol
   γ-Tocotrienol. . . . . . . . . . . . . . . . . . 0–15
   δ-Tocotrienol
   Total Tocotrienols, mg/kg

**References** *Codex* 1997/17

## Safflower Seed Oil (high oleic)

Specific Gravity (SG)
   15.5/15.5°C
   25/25°C
   Other SG . . . . . . . . . . (20/20) 0.920–0.925
Refractive Index (RI)
   25°C . . . . . . . . . . . . . . . . . 1.4680–1.4720
   40°C
   Other RI
Iodine Value . . . . . . . . . . . . . . . . . . . . 91–95
Saponification Value
Titer °C
% Unsaponifiable . . . . . . . . . . . . . . . . . 0–1.5
Melting Point °C

**Fatty Acid Composition (%)**
   14:0. . . . . . . . . . . . . . . . . . . . . . . . 0–0.1
   16:0. . . . . . . . . . . . . . . . . . . . . . . . . . 5–6
   9c-16:1 . . . . . . . . . . . . . . . . . . . . . 0–0.2
   18:0. . . . . . . . . . . . . . . . . . . . . . . . 1.5–2.0
   Total 18:1 . . . . . . . . . . . . . . . . . . . 74–80
   9c,12c-18:2 . . . . . . . . . . . . . . . . . . . 13–18
   Undefined 18:3 . . . . . . . . . . . . . . . 0–0.2
   20:0. . . . . . . . . . . . . . . . . . . . . . . . . 0–0.3
   Total 20:1 . . . . . . . . . . . . . . . . . . . 0–0.2
   22:0. . . . . . . . . . . . . . . . . . . . . . . . . 0–0.2

Sterol Composition, %
   Cholesterol . . . . . . . . . . . . . . . . . . . 0–0.2
   Brassicasterol
   Campesterol . . . . . . . . . . . . . . . . . . 10–16
   Stigmasterol . . . . . . . . . . . . . . . . . . . 8–15
   Stigmasta-8,22-dien-3β-ol
   5α-Stigmasta-7,22-dien-3β-ol
   Δ7,25-Stigmastadienol
   β-Sitosterol . . . . . . . . . . . . . . . . . . . 52–60
   Δ5-Avenasterol . . . . . . . . . . . . . . . . . 5–6
   Δ7-Stigmasterol . . . . . . . . . . . . . . . 13–18
   Δ7-Avenasterol . . . . . . . . . . . . . . . . . 5–6
   Δ7-Campesterol
   Δ7-Ergosterol

Δ7,25-Stigmasterol
Sitostanol
Spinasterol
Squalene
24-Methylene Cholesterol
Other
% sterols in oil
Total Sterols, mg/kg

Tocopherol Composition, mg/kg
   α-Tocopherol . . . . . . . . . . . . . . 480–600
   β-Tocopherol
   γ-Tocopherol
   δ-Tocopherol
   Total, mg/kg

**References** *J. Am. Oil Chem. Soc. 60:* 2003 (1983)
*J. Am. Oil Chem. Soc. 53:* 713 (1976)
*Riv. Ital. Sost. Grasse 65:* 49 (1988)

## Safou Oil
*Dacryodes edulis*
Specific Gravity (SG)
   15.5/15.5°C
   25/25°C
   Other SG
Refractive Index (RI)
   25°C
   40°C
   Other RI
Iodine Value
Saponification Value
Titer °C
% Unsaponifiable
Melting Point °C

**Fatty Acid Composition (%)**
   14:0 . . . . . . . . . . . . . . . . . . . . . . . . 0.1
   16:0 . . . . . . . . . . . . . . . . . . . . . . . 45–46
   18:0 . . . . . . . . . . . . . . . . . . . . . . . . 2–3
   Total 18:1 . . . . . . . . . . . . . . . . . . . 30–32
   9c,12c-18:2 . . . . . . . . . . . . . . . . . . 19–20
   Undefined 18:3 . . . . . . . . . . . . . . . . . 0.7
   20:0 . . . . . . . . . . . . . . . . . . . . . . . . 0.8
   Total 20:1 . . . . . . . . . . . . . . . . . . . . 0.1
   22:0 . . . . . . . . . . . . . . . . . . . . . . . . 0.1

**References** *Fruits 46:* 271 (1991)

## Sal Fat
*Shorea robusta*
Specific Gravity (SG)
   15.5/15.5°C
   25/25°C
   Other SG
Refractive Index (RI)
   25°C
   40°C . . . . . . . . . . . . . . . . . . . 1.456–1.457
   Other RI
Iodine Value . . . . . . . . . . . . . . . . . . . . 31–45
Saponification Value . . . . . . . . . . . . 175–192
Titer °C
% Unsaponifiable . . . . . . . . . . . . . . . 0.6–1.3
Melting Point °C . . . . . . . . . . . . . . . . . 30–36

**Fatty Acid Composition (%)**
   16:0 . . . . . . . . . . . . . . . . . . . . . . . . 6–23
   18:0 . . . . . . . . . . . . . . . . . . . . . . . . 33–57
   Total 18:1 . . . . . . . . . . . . . . . . . . . 31–52
   9c,12c-18:2 . . . . . . . . . . . . . . . . . . . 0.3–5
   20:0 . . . . . . . . . . . . . . . . . . . . . . . . 1–8
   Other: Hydroxystearic, 0.8; epoxystearic, 0.4–1.2

**References** *J. Oil Technol. Assn. India 13:* 114 (1981)
*J. Oil Technol. Assn. India 13:* 120 (1981)
*J. Food Sci. Technol. India 21:* 322 (1984)

## Salicornia Seed Oil
*Salicornia bigelovii; glasswort; maroh samphire*
Specific Gravity (SG)
   15.5/15.5°C
   25/25°C
   Other SG
Refractive Index (RI)
   25°C
   40°C
   Other RI
Iodine Value
Saponification Value
Titer °C
% Unsaponifiable
Melting Point °C

**Fatty Acid Composition (%)**
  16:0 .......................... 8
  18:0 .......................... 2
  Total 18:1 .................... 12
  Undefined 18:2 ................ 74
  Undefined 18:3 ............... 2–3

Sterol Composition, %
  Cholesterol
  Brassicasterol
  Campesterol
  Stigmasterol
  Stigmasta-8,22-dien-3β-ol
  5α-Stigmasta-7,22-dien-3β-ol
  Δ7,25-Stigmastadienol
  β-Sitosterol ..................... 23
  Δ5-Avenasterol
  Δ7-Stigmasterol .................. 42
  Δ7-Avenasterol
  Δ7-Campesterol
  Δ7-Ergosterol
  Δ7,25-Stigmasterol
  Sitostanol
  Spinasterol ...................... 17
  Squalene
  24-Methylene Cholesterol
  Other
  % sterols in oil
  Total Sterols, mg/kg

Tocopherol Composition, mg/kg
  α-Tocopherol ..................... 49
  β-Tocopherol
  γ-Tocopherol ..................... 48
  δ-Tocopherol
  Total, mg/kg

**References** *INFORM 11:* 418 (2000)


# Sapindus mukorossi
*Sapindus mukorossi*
Specific Gravity (SG)
  15.5/15.5°C
  25/25°C
  Other SG
Refractive Index (RI)
  25°C
  40°C

Other RI
Iodine Value
Saponification Value
Titer °C
% Unsaponifiable
Melting Point °C

**Fatty Acid Composition (%)**
  14:0 ........................ 0.03
  16:0 ........................ 5.27
  9c-16:1 ..................... 0.22
  18:0 ........................ 1.39
  9c-18:1 ..................... 52.39
  7c-18:1 ..................... 2.43
  9c,12c-18:2 ................. 8.35
  9c,12c,15c-18:3 ............. 1.37
  20:0 ........................ 4.93
  11c-20:1 .................... 20.57
  22:0 ........................ 0.86
  13c-22:1 .................... 0.75
  24:0 ........................ 0.5

Tocopherol Composition, mg/kg
  α-Tocopherol .................... 66
  β-Tocopherol
  γ-Tocopherol ................... 208
  δ-Tocopherol .................... 26
  Total, mg/kg

Tocotrienols Composition, mg/kg
  α-Tocotrienol
  β-Tocotrienol
  γ-Tocotrienol ................... 31
  δ-Tocotrienol
  Total Tocotrienols, mg/kg

**References** *JAOCS 80:* 1013–1020 (2003)


# Samanea Saman Seed Oil (Monkey pod)
*Samanea saman*
Specific Gravity (SG)
  15.5/15.5°C
  25/25°C ....................... 0.954
  Other SG
Refractive Index (RI)
  25°C
  40°C

Other RI
Iodine Value . . . . . . . . . . . . . . . . . . . . . 90–95
Saponification Value . . . . . . . . . . . . . . . . 193
Titer °C
% Unsaponifiable . . . . . . . . . . . . . . . . . . . . 4
Melting Point °C

**Fatty Acid Composition (%)**
    14:0 . . . . . . . . . . . . . . . . . . . . . . . . . . . 1–2
    16:0 . . . . . . . . . . . . . . . . . . . . . . . . . . . . . 2
    16:1 . . . . . . . . . . . . . . . . . . . . . . . . . . . 1–2
    Total 18:1 . . . . . . . . . . . . . . . . . . . . . . . 61
    Undefined 18:2 . . . . . . . . . . . . . . . . . . . 20
    Undefined 18:3 . . . . . . . . . . . . . . . . . . . . 6
    20:0 . . . . . . . . . . . . . . . . . . . . . . . . . . . . . 2
    Total 20:1 . . . . . . . . . . . . . . . . . . . . . . . . 4

**References** *Riv. Ital. Sost. Grasse 73(4):* 165 (1996)

# Schizochytrium aggregatum (ATCC 28209) Fungal Lipids
*Schizochytrium aggregatum*

Specific Gravity (SG)
    15.5/15.5°C
    25/25°C
    Other SG
Refractive Index (RI)
    25°C
    40°C
    Other RI
Iodine Value
Saponification Value
Titer °C
% Unsaponifiable
Melting Point °C

**Fatty Acid Composition (%)**
    14:0 . . . . . . . . . . . . . . . . . . . . . . . . . . . . . 4
    14:1 . . . . . . . . . . . . . . . . . . . . . . . . . . . . . 1
    16:0 . . . . . . . . . . . . . . . . . . . . . . . . . . . . 17
    16:1 . . . . . . . . . . . . . . . . . . . . . . . . . . . . . 6
    16:2 . . . . . . . . . . . . . . . . . . . . . . . . . . . . . 2
    18:0 . . . . . . . . . . . . . . . . . . . . . . . . . . . . . 6
    Total 18:1 . . . . . . . . . . . . . . . . . . . . . . . 41
    Undefined 18:2 . . . . . . . . . . . . . . . . . . . 15
    Undefined 18:3 . . . . . . . . . . . . . . . . . . . . 3
    20:2 . . . . . . . . . . . . . . . . . . . . . . . . . . . . . 1
    20:5 . . . . . . . . . . . . . . . . . . . . . . . . . . . . . 1
    22:6 . . . . . . . . . . . . . . . . . . . . . . . . . . . . . 4

**References** *Lipids 27:* 15 (1992)

# Schizonepeta tenuifolia
*Schizonepeta tenuifolia*

Specific Gravity (SG)
    15.5/15.5°C
    25/25°C
    Other SG
Refractive Index (RI)
    25°C
    40°C
    Other RI
Iodine Value
Saponification Value
Titer °C
% Unsaponifiable
Melting Point °C

**Fatty Acid Composition (%)**
    14:0 . . . . . . . . . . . . . . . . . . . . . . . . . . 0.05
    16:0 . . . . . . . . . . . . . . . . . . . . . . . . . . 9.11
    9c-16:1 . . . . . . . . . . . . . . . . . . . . . . . . 0.12
    18:0 . . . . . . . . . . . . . . . . . . . . . . . . . . 1.65
    9c-18:1 . . . . . . . . . . . . . . . . . . . . . . . 14.26
    7c-18:1 . . . . . . . . . . . . . . . . . . . . . . . . 1.48
    9c,12c-18:2 . . . . . . . . . . . . . . . . . . . . 29.18
    9c,12c,15c-18:3 . . . . . . . . . . . . . . . . . 42.45
    20:0 . . . . . . . . . . . . . . . . . . . . . . . . . . 0.21
    11c-20:1 . . . . . . . . . . . . . . . . . . . . . . . 0.19
    22:0 . . . . . . . . . . . . . . . . . . . . . . . . . . 0.06
    13c-22:1 . . . . . . . . . . . . . . . . . . . . . . . 0.03
    22:2 . . . . . . . . . . . . . . . . . . . . . . . . . . 0.05

Tocopherol Composition, mg/kg
    α-Tocopherol . . . . . . . . . . . . . . . . . . . . . 64
    β-Tocopherol
    γ-Tocopherol . . . . . . . . . . . . . . . . . . . . 546
    δ-Tocopherol . . . . . . . . . . . . . . . . . . . . . 37
    Total, mg/kg

**References** *JAOCS 80:* 1013–1020 (2003)

## Sciadopytis Verticillata Seed Oil
*Sciadopytis verticillata*

Specific Gravity (SG)
 15.5/15.5°C
 25/25°C
 Other SG
Refractive Index (RI)
 25°C
 40°C
 Other RI
Iodine Value
Saponification Value
Titer °C
% Unsaponifiable
Melting Point °C

**Fatty Acid Composition (%)**
 16:0 . . . . . . . . . . . . . . . . . . . . . . . . . . . . 3
 9c-16:1 . . . . . . . . . . . . . . . . . . . . . . . 0.1
 18:0 . . . . . . . . . . . . . . . . . . . . . . . . . . . . 2
 Total 18:1 . . . . . . . . . . . . . . . . . . . . . 22
 9c,12c-18:2 . . . . . . . . . . . . . . . . . . . . 46
 Undefined 18:3 . . . . . . . . . . . . . . . . . . 2
 20:0 . . . . . . . . . . . . . . . . . . . . . . . . . 0.3
 Total 20:1 . . . . . . . . . . . . . . . . . . . . 1–2
 20:2 . . . . . . . . . . . . . . . . . . . . . . . . . 4–5
 Other: 11c,14c,17c-20:3, 0.2; 5c,11c-20:2, 0.8; 5c,11c,14c-20:3, 15; 5c,11c,14c,17c-20:4, 2

**References** *INFORM 8:* 116 (1997)


## Sequa Oil
*Fevillea cordifloria*

Specific Gravity (SG)
 15.5/15.5°C
 25/25°C
 Other SG
Refractive Index (RI)
 25°C . . . . . . . . . . . . . . . . . . . . . . . . . . . .
 40°C . . . . . . . . . . . . . . . . . 1.4751–1.4772
 Other RI
Iodine Value . . . . . . . . . . . . . . . . . . . . 52–75
Saponification Value . . . . . . . . . . . . 192–195

Titer °C
% Unsaponifiable . . . . . . . . . . . . . . . 0.7–0.8
Melting Point °C

**Fatty Acid Composition (%)**
 16:0 . . . . . . . . . . . . . . . . . . . . . . . . . . . . 4
 17:0 . . . . . . . . . . . . . . . . . . . . . . . . . . 0.1
 18:0 . . . . . . . . . . . . . . . . . . . . . . . . . . . 53
 Total 18:1 . . . . . . . . . . . . . . . . . . . . . . 5
 9c,12c-18:2 . . . . . . . . . . . . . . . . . . . . . 4
 α-Eleostearic 9c,11t,13t-18:3 . . . . . . . . 31
 20:0 . . . . . . . . . . . . . . . . . . . . . . . . . . . . 1
 Other: 18:3 conjugated isomers, 2

**References** *Fat Sci. Technol. 94:* 294 (1992)


## Sesame Seed Oil
*Sesamum indicum*

Specific Gravity (SG)
 15.5/15.5°C
 25/25°C
 Other SG . . . . . . . . . (20/20) 0.915–0.923
Refractive Index (RI)
 25°C
 40°C . . . . . . . . . . . . . . . . . 1.465–1.469
 Other RI
Iodine Value . . . . . . . . . . . . . . . . . . 104–120
Saponification Value . . . . . . . . . . . 187–195
Titer °C
% Unsaponifiable . . . . . . . . . . . . . . . . . 0–2
Melting Point °C

**Fatty Acid Composition (%)**
 14:0 . . . . . . . . . . . . . . . . . . . . . . . . . 0–0.1
 16:0 . . . . . . . . . . . . . . . . . . . . . . 7.9–10.2
 16:1 . . . . . . . . . . . . . . . . . . . . . . . . . . 0.2
 9c-16:1 . . . . . . . . . . . . . . . . . . . . . 0.1–0.2
 18:0 . . . . . . . . . . . . . . . . . . . . . . . . 4.4–6.7
 Total 18:1 . . . . . . . . . . . . . . . . . 33.5–44.1
 Undefined 18:2 . . . . . . . . . . . . . . . . . 41.2
 9c,12c-18:2 . . . . . . . . . . . . . . . 40.3–50.8
 Undefined 18:3 . . . . . . . . . . . . . . . 0.3–0.7
 20:0 . . . . . . . . . . . . . . . . . . . . . . . . 0.3–0.7
 Total 20:1 . . . . . . . . . . . . . . . . . . . . 0–0.3
 22:0 . . . . . . . . . . . . . . . . . . . . . . . . . 0–0.3
 Unidentified 22:1 . . . . . . . . . . . . . . . . . tr
 24:0 . . . . . . . . . . . . . . . . . . . . . . . . . 0–0.3

Sterol Composition, %
  Cholesterol . . . . . . . . . . . . . . . . . . 0.1–0.2
  Brassicasterol . . . . . . . . . . . . . . . 0.1–0.2
  Campesterol . . . . . . . . . . . . . . 10.1–20.0
  Stigmasterol . . . . . . . . . . . . . . . . . 3.4–6.4
  Stigmasta-8,22-dien-3β-ol
  5α-Stigmasta-7,22-dien-3β-ol
  Δ7,25-Stigmastadienol
  β-Sitosterol . . . . . . . . . . . . . . . . . 57.7–61.9
  Δ5-Avenasterol . . . . . . . . . . . . . . 6.2–7.8
  Δ7-Stigmasterol . . . . . . . . . . . . . 1.8–7.6
  Δ7-Avenasterol . . . . . . . . . . . . . . 1.2–5.6
  Δ7-Campesterol
  Δ7-Ergosterol
  Δ7,25-Stigmasterol
  Sitostanol
  Spinasterol
  Squalene
  24-Methylene Cholesterol
  Other . . . . . . . . . . . . . . . . . . . . . . . 0.7–9.2
  % sterols in oil
  Total Sterols, mg/kg . . . . . . . 4500–18960

Tocopherol Composition, mg/kg
  α-Tocopherol . . . . . . . . . . . . . . . . . . 0–4
  β-Tocopherol
  γ-Tocopherol . . . . . . . . . . . . . . . 521–983
  δ-Tocopherol . . . . . . . . . . . . . . . . . 4–21
  Total, mg/kg . . . . . . . . . . . . . . 531–1000

Tocotrienols Composition, mg/kg
  α-Tocotrienol
  β-Tocotrienol
  γ-Tocotrienol . . . . . . . . . . . . . . . . . 0–20
  δ-Tocotrienol
  Total Tocotrienols, mg/kg

**References** *Codex* 1993/16
  *J. Sci. Food Agric. 59:* 327 (1992)
  *Fat Sci. Technol. 94:* 254 (1992)
  *J. Sci. Food Agric. 27:* 165 (1976)
  *J. Sci. Food Technol. 59:* 327 (1992)
  *JAOCS 74:* 375–380 (1997)
  *JAOCS 71:* 149 (1994)

# Sesame Seed Oil
*Sesamum radiatum*

Specific Gravity (SG)
  15.5/15.5°C
  25/25°C
  Other SG
Refractive Index (RI)
  25°C
  40°C
  Other RI
Iodine Value
Saponification Value
Titer °C
% Unsaponifiable . . . . . . . . . . . . . . . 2.5–2.7
Melting Point °C

**Fatty Acid Composition (%)**
  16:0 . . . . . . . . . . . . . . . . . . . . . . . . . . . 10
  18:0 . . . . . . . . . . . . . . . . . . . . . . . . . . . 10
  9c-18:1 . . . . . . . . . . . . . . . . . . . . . . . . 38
  11c-18:1 . . . . . . . . . . . . . . . . . . . . . . 0.6
  9c,12c-18:2 . . . . . . . . . . . . . . . . . . . . 41
  Undefined 18:3 . . . . . . . . . . . . . . . . 0.5
  20:0 . . . . . . . . . . . . . . . . . . . . . . . . . . . . 1

Sterol Composition, %
  Cholesterol . . . . . . . . . . . . . . . . . . . . 0.2
  Brassicasterol
  Campesterol . . . . . . . . . . . . . . . . . . . 12
  Stigmasterol . . . . . . . . . . . . . . . . . . . 4–6
  Stigmasta-8,22-dien-3β-ol
  5α-Stigmasta-7,22-dien-3β-ol
  Δ7,25-Stigmastadienol
  β-Sitosterol . . . . . . . . . . . . . . . . . . . . . 60
  Δ5-Avenasterol . . . . . . . . . . . . . . . 12–13
  Δ7-Stigmasterol . . . . . . . . . . . . . . . . 2–3
  Δ7-Avenasterol . . . . . . . . . . . . . . . . . 2–4
  Δ7-Campesterol
  Δ7-Ergosterol
  Δ7,25-Stigmasterol
  Sitostanol
  Spinasterol
  Squalene
  24-Methylene Cholesterol
  Other: 4–5 (monomethyl- and dimethyl-sterols)
  % sterols in oil
  Total Sterols, mg/kg

Tocopherol Composition, mg/kg
  α-Tocopherol . . . . . . . . . . . . . . . . . . 0.8
  β-Tocopherol
  γ-Tocopherol . . . . . . . . . . . . . . . . . 97–99
  δ-Tocopherol . . . . . . . . . . . . . . . . . 0.4–2
  Total, mg/kg

## Sesbania Pachycarpa Seed Oil
*Sesbania pachycarpa*
Specific Gravity (SG)
   15.5/15.5°C
   25/25°C
   Other SG
Refractive Index (RI)
   25°C
   40°C
   Other RI
Iodine Value
Saponification Value
Titer °C
% Unsaponifiable
Melting Point °C

**Fatty Acid Composition (%)**
   16:0 . . . . . . . . . . . . . . . . . . . . . . . . . . . 6.9
   16:2 . . . . . . . . . . . . . . . . . . . . . . . . . . . 1.4
   18:0 . . . . . . . . . . . . . . . . . . . . . . . . . . . 4.3
   9c-18:1 . . . . . . . . . . . . . . . . . . . . . . . . 16.1
   11c-18:1 . . . . . . . . . . . . . . . . . . . . . . . . 0.4
   9c,12c-18:2 . . . . . . . . . . . . . . . . . . . . . 18.7
   9c,12c,15c-18:3 . . . . . . . . . . . . . . . . . 20.8
   20:0 . . . . . . . . . . . . . . . . . . . . . . . . . . . 0.7
   Total 20:1 . . . . . . . . . . . . . . . . . . . . . . 0.3
   22:0 . . . . . . . . . . . . . . . . . . . . . . . . . . . 0.9
   24:0 . . . . . . . . . . . . . . . . . . . . . . . . . . . 0.3

**References** *J. Am. Oil Chem. Soc. 75:* 1031 (1998)

## Sheanut Butter
*Butyrospermum parkii/Vitellaria paradoxa*
Specific Gravity (SG)
   15.5/15.5°C . . . . . . . . . . . . . . 0.916–0.918
   25/25°C
   Other SG . . . . . . . . . (100/15) 0.859–0.869
Refractive Index (RI)
   25°C
   40°C . . . . . . . . . . . . . . . . . . . . 1.463–1.467
Other RI
Iodine Value . . . . . . . . . . . . . . . . . . . . . 52–66
Saponification Value . . . . . . . . . . . . 178–198
Titer °C . . . . . . . . . . . . . . . . . . . . . . . 48–54
% Unsaponifiable . . . . . . . . . . . . . . . . . 2–11
Melting Point °C . . . . . . . . . . . . . . . . . 32–45

**Fatty Acid Composition (%)**
   12:0 . . . . . . . . . . . . . . . . . . . . . . . . . . . 0.4
   14:0 . . . . . . . . . . . . . . . . . . . . . . . . . . . 0.3
   16:0 . . . . . . . . . . . . . . . . . . . . . . . . . . . 4–8
   18:0 . . . . . . . . . . . . . . . . . . . . . . . . . 36–41
   Total 18:1 . . . . . . . . . . . . . . . . . . . . . 45–50
   9c,12c-18:2 . . . . . . . . . . . . . . . . . . . . . 4–8
   Undefined 18:3 . . . . . . . . . . . . . . . . . 0–0.4
   20:0 . . . . . . . . . . . . . . . . . . . . . . . . . . . 1–2

Sterol Composition, %
   Cholesterol
   Brassicasterol
   Campesterol
   Stigmasterol
   Stigmasta-8,22-dien-3β-ol
   5α-Stigmasta-7,22-dien-3β-ol
   Δ7,25-Stigmastadienol
   β-Sitosterol
   Δ5-Avenasterol
   Δ7-Stigmasterol . . . . . . . . . . . . . . . . . . 38
   Δ7-Avenasterol . . . . . . . . . . . . . . . . . . . 11
   Δ7-Campesterol
   Δ7-Ergosterol
   Δ7,25-Stigmasterol
   Sitostanol
   Spinasterol
   Squalene
   24-Methylene Cholesterol
   Other: 24-methyl-Δ7-cholestanol, 6;
      Δ7,22-Stigmastadien-3β-ol, 45
% sterols in oil
Total Sterols, mg/kg . . . . . . . . . . . . . 2470

**References** *J. Am. Dietetics Assn. 73:* 39 (1978)

## Sida Humilis Seed Oil
*Sida humilis*
Specific Gravity (SG)
   15.5/15.5°C

25/25°C
Other SG
Refractive Index (RI)
  25°C
  40°C
  Other RI
Iodine Value .................... 73–107
Saponification Value
Titer °C
% Unsaponifiable .................... 1
Melting Point °C

**Fatty Acid Composition (%)**
  12:0 .......................... 0.4
  14:0 .......................... 0.5
  16:0 ........................... 17
  18:0 ............................ 4
  Total 18:1 ..................... 65
  9c,12c-18:2 ..................... 6
  Undefined 18:3 .................. 2
  20:0 ............................ 3
  22:0 .......................... 2–3

**References** *Fette Seifen Anstrichm. 86:* 167 (1984)

# Simarouba Oil (Paradise Tree)
*Simarouba glauca*
Specific Gravity (SG)
  15.5/15.5°C
  25/25°C
  Other SG
Refractive Index (RI)
  25°C ...................... 1.4556
  40°C ...................... 1.4596
  Other RI
Iodine Value .................... 54–58
Saponification Value ........... 191–192
Titer °C
% Unsaponifiable ................. 0.4
Melting Point °C ................ 25–28

**Fatty Acid Composition (%)**
  16:0 ..................... 10.9–12
  18:0 ....................... 25–33
  Total 18:1 ................. 55–59
  9c,12c-18:2 .................. 3.3
  Undefined 18:3 ............... 0.4

**References** *INFORM 13:* 151 (2002)

# Soap Tree Seed Oil
*Sapindus trifoliatus*
Specific Gravity (SG)
  15.5/15.5°C
  25/25°C
  Other SG
Refractive Index (RI)
  25°C ................. 1.4764–1.4880
  40°C
  Other RI
Iodine Value .................... 58–64
Saponification Value ........... 180–194
Titer °C
% Unsaponifiable ................ 1–1.5
Melting Point °C

**Fatty Acid Composition (%)**
  16:0 .......................... 5–7
  9c-16:1 ......................... 1
  18:0 .......................... 4–8
  Total 18:1 ................... 55–62
  9c,12c-18:2 ................... 2–8
  Undefined 18:3 ................ 0–1
  20:0 ........................ 16–22
  Total 20:1 .................... 0–9
  22:0 .......................... 1–2
  Unidentified 22:1 ............. 0.5
  24:0 .......................... 0.3

**References** *Fat Sci. Technol. 96:* 69 (1994)

# Soapberry (Chinese) Seed Oil
*Sapindus murorossi*
Specific Gravity (SG)
  15.5/15.5°C
  25/25°C ..................... 0.9040
  Other SG
Refractive Index (RI)
  25°C
  40°C ....................... 1.4632
  Other RI ............... (28) 1.4680
Iodine Value .................... 78–80
Saponification Value .............. 197

Titer °C
% Unsaponifiable
Melting Point °C

**Fatty Acid Composition (%)**
    16:0 . . . . . . . . . . . . . . . . . . . . . . . . . . 4–6
    16:1 . . . . . . . . . . . . . . . . . . . . . . . . . . . 0.5
    18:0 . . . . . . . . . . . . . . . . . . . . . . . . . . 0.2–1
    Total 18:1 . . . . . . . . . . . . . . . . . . . . 54–63
    Undefined 18:2 . . . . . . . . . . . . . . . . . 5–14
    Undefined 18:3 . . . . . . . . . . . . . . . . . . 1–6
    20:0 . . . . . . . . . . . . . . . . . . . . . . . . . . . 4–6
    Total 20:1 . . . . . . . . . . . . . . . . . . . . . 15–22

**References** *Lipids 10 (1):* 33 (1975)
    *Fette Seifen Austrichm. 73 (10):* 639 (1971)

# Soapberry Seed Oil
*Cupania anacardioides*

Specific Gravity (SG)
    15.5/15.5°C
    25/25°C
    Other SG
Refractive Index (RI)
    25°C
    40°C
    Other RI
Iodine Value
Saponification Value
Titer °C
% Unsaponifiable
Melting Point °C

**Fatty Acid Composition (%)**
    16:0 . . . . . . . . . . . . . . . . . . . . . . . . . . . 12
    9c-16:1 . . . . . . . . . . . . . . . . . . . . . . . . . . 8
    18:0 . . . . . . . . . . . . . . . . . . . . . . . . . . . . 6
    Total 18:1 . . . . . . . . . . . . . . . . . . . . . . . 10
    9c,12c-18:2 . . . . . . . . . . . . . . . . . . . . . . 16
    20:0 . . . . . . . . . . . . . . . . . . . . . . . . . . . . 2
    Total 20:1 . . . . . . . . . . . . . . . . . . . . . . . 46

**References** *J. Am. Oil Chem. Soc. 63:* 671 (1986)

# Soapberry Seed Oil
*Paullinia elegans*

Specific Gravity (SG)
    15.5/15.5°C
    25/25°C
    Other SG
Refractive Index (RI)
    25°C
    40°C
    Other RI
Iodine Value
Saponification Value
Titer °C
% Unsaponifiable
Melting Point °C

**Fatty Acid Composition (%)**
    16:0 . . . . . . . . . . . . . . . . . . . . . . . . . . . 2.2
    9c-16:1 . . . . . . . . . . . . . . . . . . . . . . . . . 3.6
    18:0 . . . . . . . . . . . . . . . . . . . . . . . . . . . 1.7
    9c-18:1 . . . . . . . . . . . . . . . . . . . . . . . . 12.2
    11c-18:1 . . . . . . . . . . . . . . . . . . . . . . . 19.8
    9c,12c-18:2 . . . . . . . . . . . . . . . . . . . . . . 3.1
    Undefined 18:3 . . . . . . . . . . . . . . . . . . . 1.8
    20:0 . . . . . . . . . . . . . . . . . . . . . . . . . . . . .5
    Total 20:1 . . . . . . . . . . . . . . . . . . . . . . 48.7
    22:0 . . . . . . . . . . . . . . . . . . . . . . . . . . . 0.3
    Unidentified 22:1 . . . . . . . . . . . . . . . . . 1.2

**References**

# Solanum Melongena Seed Oil
*Solanum melongena*

Specific Gravity (SG)
    15.5/15.5°C
    25/25°C
    Other SG
Refractive Index (RI)
    25°C
    40°C
    Other RI
Iodine Value
Saponification Value

Titer °C
% Unsaponifiable
Melting Point °C

**Fatty Acid Composition (%)**
- 14:0 .......................... 0.12
- 16:0 .......................... 9.49
- 9c-16:1 ....................... 0.22
- 18:0 .......................... 3.22
- 9c-18:1 ...................... 14.53
- 7c-18:1 ......................... 1
- 9c,12c-18:2 .................. 68.95
- 9c,12c,15c-18:3 ............... 1.49
- 20:0 .......................... 0.23
- 11c-20:1 ...................... 0.08
- 22:0 .......................... 0.12
- 22:2 .......................... 0.03
- 24:0 .......................... 0.15

Tocopherol Composition, mg/kg
- α-Tocopherol .................. 56
- β-Tocopherol .................. 35
- γ-Tocopherol ................. 372
- δ-Tocopherol .................. 39
- Total, mg/kg

**References** *JAOCS 80:* 1013–1020 (2003)

## Sorghum Seed Oil
*Sorghum bicolor*

Specific Gravity (SG)
  15.5/15.5°C
  25/25°C
  Other SG
Refractive Index (RI)
  25°C
  40°C
  Other RI
Iodine Value ................... 108–126
Saponification Value
Titer °C
% Unsaponifiable
Melting Point °C

**Fatty Acid Composition (%)**
- 16:0 ......................... 15–25
- 9c-16:1 ....................... tr-1.1
- 18:0 ......................... 1.0–1.4
- Total 18:1 .................... 30–42

9c,12c-18:2 .................... 36–51
Undefined 18:3 .................. 1.6–2.3

Sterol Composition, %
- Cholesterol ................... 0.8–2.1
- Brassicasterol
- Campesterol .................. 19–29
- Stigmasterol ................. 14–21
- Stigmasta-8,22-dien-3β-ol
- 5α-Stigmasta-7,22-dien-3β-ol
- Δ7,25-Stigmastadienol
- β-Sitosterol .................. 44–58
- Δ5-Avenasterol ................ 4.1–7.4
- Δ7-Stigmasterol ................ tr-2.5
- Δ7-Avenasterol
- Δ7-Campesterol
- Δ7-Ergosterol
- Δ7,25-Stigmasterol
- Sitostanol
- Spinasterol
- Squalene
- 24-Methylene Cholesterol
- Other
- % sterols in oil
- Total Sterols, mg/kg

**References** *J. Sci. Food Agric. 70:* 334 (1996)

## Sorghum Seed Oil
*Sorghum vulgare*

Specific Gravity (SG)
  15.5/15.5°C
  25/25°C
  Other SG
Refractive Index (RI)
  25°C ................... 1.4686–1.4720
  40°C
  Other RI
Iodine Value ................... 108–122
Saponification Value ........... 181–191
Titer °C
% Unsaponifiable ............... 1.7–3.2
Melting Point °C

**Fatty Acid Composition (%)**
- 14:0 .......................... 0–1
- 16:0 ......................... 6–10

9c-16:1 . . . . . . . . . . . . . . . . . . . . . . . 0–1
18:0. . . . . . . . . . . . . . . . . . . . . . . . . . 3–6
Total 18:1 . . . . . . . . . . . . . . . . . . . 30–47
9c,12c-18:2 . . . . . . . . . . . . . . . . . . 40–55
Undefined 18:3 . . . . . . . . . . . . . . . . . 0–1

**References**

## Soybean Oil
*Glycine max*
Specific Gravity (SG)
  15.5/15.5°C
  25/25°C
  Other SG . . . . . . . . . . (20/20) 0.919–0.925
Refractive Index (RI)
  25°C
  40°C . . . . . . . . . . . . . . . . . . . 1.466–1.470
  Other RI
Iodine Value . . . . . . . . . . . . . . . . . . 118–139
Saponification Value . . . . . . . . . . . 189–195
Titer °C
% Unsaponifiable . . . . . . . . . . . . . . . . . 0–1.5
Melting Point -16°C

**Fatty Acid Composition (%)**
  12:0. . . . . . . . . . . . . . . . . . . . . . . . 0–0.1
  14:0. . . . . . . . . . . . . . . . . . . . . . . . 0–0.2
  16:0. . . . . . . . . . . . . . . . . . . . . . . 9.7–13.3
  16:1. . . . . . . . . . . . . . . . . . . . . . . . . . . .tr
  9c-16:1 . . . . . . . . . . . . . . . . . . . . . . 0–0.2
  18:0. . . . . . . . . . . . . . . . . . . . . . . . 3.0–5.4
  Total 18:1 . . . . . . . . . . . . . . . . . . . 17.7–28.5
  Undefined 18:2 . . . . . . . . . . . . . . . . . 53.7
  9c,12c-18:2 . . . . . . . . . . . . . . . . . . 49.8–57.1
  Undefined 18:3 . . . . . . . . . . . . . . . . 5.5–9.5
  20:0. . . . . . . . . . . . . . . . . . . . . . . . 0.1–0.6
  Total 20:1 . . . . . . . . . . . . . . . . . . . . 0–0.3
  20:2. . . . . . . . . . . . . . . . . . . . . . . . 0–0.1
  22:0. . . . . . . . . . . . . . . . . . . . . . . . 0.3–0.7
  Unidentified 22:1 . . . . . . . . . . . . . . 0–0.3
  24:0. . . . . . . . . . . . . . . . . . . . . . . . 0–0.4

Sterol Composition, %
  Cholesterol . . . . . . . . . . . . . . . . . . 0.6–1.4
  Brassicasterol . . . . . . . . . . . . . . . . . 0–0.3
  Campesterol . . . . . . . . . . . . . . 15.8–24.2
  Stigmasterol . . . . . . . . . . . . . . 15.9–19.1
  Stigmasta-8,22-dien-3β-ol

  5α-Stigmasta-7,22-dien-3β-ol
  Δ7,25-Stigmastadienol
  β-Sitosterol . . . . . . . . . . . . . . . . 51.7–57.6
  Δ5-Avenasterol . . . . . . . . . . . . . . . 1.9–3.7
  Δ7-Stigmasterol . . . . . . . . . . . . . 1.4–5.2
  Δ7-Avenasterol . . . . . . . . . . . . . . . 1.0–4.6
  Δ7-Campesterol
  Δ7-Ergosterol
  Δ7,25-Stigmasterol
  Sitostanol
  Spinasterol
  Squalene
  24-Methylene Cholesterol
  Other. . . . . . . . . . . . . . . . . . . . . . . 0–1.8
% sterols in oil
Total Sterols, mg/kg . . . . . . . . 1840–4090

Tocopherol Composition, mg/kg
  α-Tocopherol . . . . . . . . . . . . . . . . . 9–352
  β-Tocopherol. . . . . . . . . . . . . . . . . . 0–40
  γ-Tocopherol. . . . . . . . . . . . . . . 89–2400
  δ-Tocopherol. . . . . . . . . . . . . . . 150–932
  Total, mg/kg . . . . . . . . . . . . . . 573–3363

Tocotrienols Composition, mg/kg
  α-Tocotrienol . . . . . . . . . . . . . . . . . 0–69
  β-Tocotrienol
  γ-Tocotrienol. . . . . . . . . . . . . . . . . 0–103
  δ-Tocotrienol
  Total Tocotrienols, mg/kg

**References** *Codex* 1997/17
*J. Sci. Food Aric. 72:* 403 (1996)
*JAOCS 74:* 375–380 (1997)

## Soybean Oil (high palmitic; HP)
*GMO*
Specific Gravity (SG)
  15.5/15.5°C
  25/25°C
  Other SG
Refractive Index (RI)
  25°C
  40°C
  Other RI
Iodine Value
Saponification Value
Titer °C

% Unsaponifiable
Melting Point °C

**Fatty Acid Composition (%)**
- 14:0 .......................... 0.1
- 16:0 .......................... 23.8
- 9c-16:1 ....................... 0.7
- 18:0 .......................... 3.8
- Total 18:1 .................... 15.4
- 9c,12c-18:2 ................... 44.1
- Undefined 18:3 ................ 11.0
- 20:0 .......................... 0.4
- Total 20:1 .................... 0.1
- 22:0 .......................... 0.6
- 24:0 .......................... 0.1

**References** *J. Am. Oil Chem. Soc. 70:* 983 (1993)
*J. Am. Oil Chem. Soc. 74:* 989 (1997)

## Soybean Oil (high saturate; Hsat)
*GMO*

Specific Gravity (SG)
  15.5/15.5°C
  25/25°C
  Other SG
Refractive Index (RI)
  25°C
  40°C
  Other RI
Iodine Value
Saponification Value
Titer °C
% Unsaponifiable
Melting Point °C

**Fatty Acid Composition (%)**
- 14:0 .......................... 0.1
- 16:0 .......................... 21.9
- 9c-16:1 ....................... 0.3
- 18:0 .......................... 17.5
- Total 18:1 .................... 9.4
- 9c,12c-18:2 ................... 37.5
- Undefined 18:3 ................ 11.0
- 20:0 .......................... 1.3
- Total 20:1 .................... 0.1
- 22:0 .......................... 1.0
- 24:0 .......................... 0.2

**References** *J. Am. Oil Chem. Soc. 70:* 983 (1993)
*J. Am. Oil Chem. Soc. 74:* 989 (1997)

## Soybean Oil (high stearic; HS)
*GMO*

Specific Gravity (SG)
  15.5/15.5°C
  25/25°C
  Other SG
Refractive Index (RI)
  25°C
  40°C
  Other RI
Iodine Value
Saponification Value
Titer °C
% Unsaponifiable
Melting Point °C

**Fatty Acid Composition (%)**
- 16:0 .......................... 8.0
- 9c-16:1 ....................... 0.1
- 18:0 .......................... 24.7
- Total 18:1 .................... 17.2
- 9c,12c-18:2 ................... 39.2
- Undefined 18:3 ................ 8.3
- 20:0 .......................... 1.5
- Total 20:1 .................... 0.1
- 22:0 .......................... 0.7
- 24:0 .......................... 0.1

**References** *J. Am. Oil Chem. Soc. 70:* 983 (1993)
*J. Am. Oil Chem. Soc. 74:* 989 (1997)

## Soybean Oil (HP/LLn)
*GMO*

Specific Gravity (SG)
  15.5/15.5°C
  25/25°C
  Other SG

Refractive Index (RI)
 25°C
 40°C
 Other RI
Iodine Value
Saponification Value
Titer °C
% Unsaponifiable
Melting Point °C

**Fatty Acid Composition (%)**
 14:0 . . . . . . . . . . . . . . . . . . . . . . . . . . . 0.1
 16:0 . . . . . . . . . . . . . . . . . . . . . . . . . . . 19.2
 9c-16:1 . . . . . . . . . . . . . . . . . . . . . . . . 0.8
 18:0 . . . . . . . . . . . . . . . . . . . . . . . . . . . 4.1
 Total 18:1 . . . . . . . . . . . . . . . . . . . . . 23.2
 9c,12c-18:2 . . . . . . . . . . . . . . . . . . . . 48.2
 Undefined 18:3 . . . . . . . . . . . . . . . . 3.3
 20:0 . . . . . . . . . . . . . . . . . . . . . . . . . . . 0.4
 Total 20:1 . . . . . . . . . . . . . . . . . . . . . 0.1
 22:0 . . . . . . . . . . . . . . . . . . . . . . . . . . . 0.5
 24:0 . . . . . . . . . . . . . . . . . . . . . . . . . . . 0.1

**References** *J. Am. Oil Chem. Soc. 70:* 983 (1993)
*J. Am. Oil Chem. Soc. 74:* 989 (1997)

# Soybean Oil (low linolenic; LLn)
*GMO*

Specific Gravity (SG)
 15.5/15.5°C
 25/25°C
 Other SG
Refractive Index (RI)
 25°C
 40°C
 Other RI
Iodine Value . . . . . . . . . . . . . . . . . . . . . . 126
Saponification Value
Titer °C
% Unsaponifiable
Melting Point °C

**Fatty Acid Composition (%)**
 14:0 . . . . . . . . . . . . . . . . . . . . . . . . . . . 0.1
 16:0 . . . . . . . . . . . . . . . . . . . . . . . . . . . 10.9–11

 9c-16:1 . . . . . . . . . . . . . . . . . . . . . . . . 0.1
 18:0 . . . . . . . . . . . . . . . . . . . . . . . . . . . 4–5.7
 Total 18:1 . . . . . . . . . . . . . . . . . . . . . 25–27.5
 9c,12c-18:2 . . . . . . . . . . . . . . . . . . . . 51.5–55
 Undefined 18:3 . . . . . . . . . . . . . . . . 3.0–4.0
 20:0 . . . . . . . . . . . . . . . . . . . . . . . . . . . 0.5
 Total 20:1 . . . . . . . . . . . . . . . . . . . . . 0.2
 22:0 . . . . . . . . . . . . . . . . . . . . . . . . . . . 0.4
 24:0 . . . . . . . . . . . . . . . . . . . . . . . . . . . 0.1

**References** *J. Am. Oil Chem. Soc. 70:* 983 (1993)
*J. Am. Oil Chem. Soc. 74:* 989 (1997)

# Soybean Oil (low saturate; Lsat)
*GMO*

Specific Gravity (SG)
 15.5/15.5°C
 25/25°C
 Other SG
Refractive Index (RI)
 25°C
 40°C
 Other RI
Iodine Value
Saponification Value
Titer °C
% Unsaponifiable
Melting Point °C

**Fatty Acid Composition (%)**
 16:0 . . . . . . . . . . . . . . . . . . . . . . . . . . . 3.5
 9c-16:1 . . . . . . . . . . . . . . . . . . . . . . . . 0.1
 18:0 . . . . . . . . . . . . . . . . . . . . . . . . . . . 2.8
 Total 18:1 . . . . . . . . . . . . . . . . . . . . . 22.7
 9c,12c-18:2 . . . . . . . . . . . . . . . . . . . . 60.3
 Undefined 18:3 . . . . . . . . . . . . . . . . 9.8
 20:0 . . . . . . . . . . . . . . . . . . . . . . . . . . . 0.2
 Total 20:1 . . . . . . . . . . . . . . . . . . . . . 0.3
 22:0 . . . . . . . . . . . . . . . . . . . . . . . . . . . 0.2
 24:0 . . . . . . . . . . . . . . . . . . . . . . . . . . . 0.1

**References** *J. Am. Oil Chem. Soc. 70:* 983 (1993)
*J. Am. Oil Chem. Soc. 74:* 989 (1997)

## Soybean Oil (Lsat/LLn)
*GMO*

Specific Gravity (SG)
  15.5/15.5°C
  25/25°C
  Other SG
Refractive Index (RI)
  25°C
  40°C
  Other RI
Iodine Value
Saponification Value
Titer °C
% Unsaponifiable
Melting Point °C

**Fatty Acid Composition (%)**
  16:0 .......................... 4.1
  9c-16:1 ....................... 0.1
  18:0 .......................... 3.4
  Total 18:1 .................... 28.3
  9c,12c-18:2 ................... 60.6
  Undefined 18:3 ................ 2.7
  20:0 .......................... 0.2
  Total 20:1 .................... 0.3
  22:0 .......................... 0.3
  24:0 .......................... 0.1

**References** *J. Am. Oil Chem. Soc. 70:* 983 (1993)
*J. Am. Oil Chem. Soc. 74:* 989 (1997)

## Soybean Oil (Tropical area)

Specific Gravity (SG)
  15.5/15.5°C
  25/25°C
  Other SG
Refractive Index (RI)
  25°C
  40°C
  Other RI
Iodine Value
Saponification Value
Titer °C
% Unsaponifiable
Melting Point °C

**Fatty Acid Composition (%)**
  16:0 .......................... 10
  16:2 .......................... 1.9
  18:0 .......................... 3.4
  9c-18:1 ....................... 21.2
  11c-18:1 ...................... 1.3
  9c,12c-18:2 ................... 40.5
  6c,9c,12c-18:3 ................ 0.4
  9c,12c,15c-18:3 ............... 5.5
  20:0 .......................... 0.4
  Total 20:1 .................... 0.2
  22:0 .......................... 0.4

**References** *J. Am. Oil Chem. Soc. 75:* 1031 (1998)

## Spicebush Kernel Fat
*Lindera benzoin*

Specific Gravity (SG)
  15.5/15.5°C
  25/25°C
  Other SG
Refractive Index (RI)
  25°C .......................... 1.4553
  40°C
  Other RI
Iodine Value ..................... 14
Saponification Value ............. 284
Titer °C
% Unsaponifiable
Melting Point °C

**Fatty Acid Composition (%)**
  8:0 ........................... 0.2
  10:0 .......................... 42–47
  12:0 .......................... 45–47
  14:0 .......................... 2–3
  16:0 .......................... 0.4–1.0
  18:0 .......................... 0–0.1
  Total 18:1 .................... 2–4
  9c,12c-18:2 ................... 2–3
  Undefined 18:3 ................ 0.1
  Total 20:1 .................... 0.1

**References** *Lipids 1:* 118 (1966)
*Lipids 2:* 345 (1967)

## Sterculia Tomentosa
*Sterculia tomentosa*

Specific Gravity (SG)
  15.5/15.5°C
  25/25°C
  Other SG
Refractive Index (RI)
  25°C
  40°C
  Other RI
Iodine Value
Saponification Value
Titer °C
% Unsaponifiable
Melting Point °C

**Fatty Acid Composition (%)**
  14:0 .......................... 0.5
  16:0 .......................... 20.5
  9c-16:1 ....................... 0.5
  17:0 .......................... 0.7
  18:0 .......................... 5.7
  9c-18:1 ....................... 20.5
  9c,12c-18:2 ................... 29.8
  9c,12c,15c-18:3 ............... 2.1
  20:0 .......................... 0.5
  22:0 .......................... 0.3

**References** *JAOCS 70:* 205 (1993)

## Sterculia tragacantha
*Sterculia tragacantha*

Specific Gravity (SG)
  15.5/15.5°C
  25/25°C
  Other SG
Refractive Index (RI)
  25°C
  40°C
  Other RI
Iodine Value
Saponification Value
Titer °C
% Unsaponifiable
Melting Point °C

**Fatty Acid Composition (%)**
  14:0 .......................... 0.2
  16:0 .......................... 23.6
  9c-16:1 ....................... 0.6
  17:0 .......................... 0.2
  18:0 .......................... 5.64
  9c-18:1 ....................... 14.8
  7c-18:1 ....................... 0.9
  9c,12c-18:2 ................... 15.9
  9c,12c,15c-18:3 ............... 1.8
  20:0 .......................... 0.9
  22:0 .......................... 0.6

**References** *JAOCS 70:* 205 (1993)

## Stillingia Seed Kernel Oil (Chinese Tallow Tree)
*Sapium sebiferum*

Specific Gravity (SG)
  15.5/15.5°C
  25/25°C .................. 0.936–0.944
  Other SG
Refractive Index (RI)
  25°C .................... 1.4817–1.484
  40°C
  Other RI
Iodine Value ................... 169–191
Saponification Value ........... 202–212
Titer °C
% Unsaponifiable ................. 0.5–3
Melting Point °C

**Fatty Acid Composition (%)**
  16:0 .......................... 6–9
  18:0 .......................... 3–5
  Total 18:1 .................... 7–10
  7c-18:1 ....................... tr
  9c,12c-18:2 ................... 24–30
  Undefined 18:3 ................ 41–54
  Other: 10:2, 4–5

**References**

## Sunflower Seed Oil

*Helianthus annuus*

Specific Gravity (SG)
   15.5/15.5°C
   25/25°C
   Other SG . . . . . . . . . (20/20) 0.918–0.923
Refractive Index (RI)
   25°C . . . . . . . . . . . . . . . . . . 1.472–1.476
   40°C . . . . . . . . . . . . . . . . . . 1.467–1.469
   Other RI
Iodine Value . . . . . . . . . . . . . . . . . . 118–145
Saponification Value . . . . . . . . . . . . 188–194
Titer °C
% Unsaponifiable . . . . . . . . . . . . . . . . . 0–1.5
Melting Point -17°C

**Fatty Acid Composition (%)**
   12:0. . . . . . . . . . . . . . . . . . . . . . . . 0–0.1
   14:0. . . . . . . . . . . . . . . . . . . . . . . . 0–0.2
   16:0. . . . . . . . . . . . . . . . . . . . . . . . . . 5–8
   9c-16:1 . . . . . . . . . . . . . . . . . . . . . 0–0.3
   18:0. . . . . . . . . . . . . . . . . . . . . . . 2.5–7.0
   Total 18:1 . . . . . . . . . . . . . . . . . . . 13–40
   9c,12c-18:2 . . . . . . . . . . . . . . . . . . 48–74
   Undefined 18:3 . . . . . . . . . . . . . . . 0–0.3
   20:0. . . . . . . . . . . . . . . . . . . . . . . 0.2–0.5
   Total 20:1 . . . . . . . . . . . . . . . . . . . . 0–0.5
   22:0. . . . . . . . . . . . . . . . . . . . . . . 0.5–1.3
   Unidentified 22:1 . . . . . . . . . . . . . . 0–0.5
   24:0. . . . . . . . . . . . . . . . . . . . . . . . 0–0.4

Sterol Composition, %
   Cholesterol . . . . . . . . . . . . . . . . . . 0–0.7
   Brassicasterol . . . . . . . . . . . . . . . . 0–0.2
   Campesterol . . . . . . . . . . . . . . . . . . 7–13
   Stigmasterol . . . . . . . . . . . . . . . . . . 7–12
   Stigmasta-8,22-dien-3β-ol
   5α-Stigmasta-7,22-dien-3β-ol
   Δ7,25-Stigmastadienol
   β-Sitosterol . . . . . . . . . . . . . . . . . . 56–65
   Δ5-Avenasterol . . . . . . . . . . . . . . . . 1.5–7
   Δ7-Stigmasterol . . . . . . . . . . . . . . . . 7–24
   Δ7-Avenasterol . . . . . . . . . . . . . . . . 3–6.5
   Δ7-Campesterol
   Δ7-Ergosterol
   Δ7,25-Stigmasterol
   Sitostanol
   Spinasterol
   Squalene
   24-Methylene Cholesterol
   Other: Δ7-Campesterol, 2–3, clerosterol,
     0.7–1, other, 0–5.3
% sterols in oil
Total Sterols, mg/kg . . . . . . . . 2440–4550

Tocopherol Composition, mg/kg
   α-Tocopherol . . . . . . . . . . . . . . . 400–950
   β-Tocopherol. . . . . . . . . . . . . . . . . . . 0–50
   γ-Tocopherol. . . . . . . . . . . . . . . . . . . 0–50
   δ-Tocopherol. . . . . . . . . . . . . . . . . . . 0–10
   Total, mg/kg

**References** *Codex* 1997/17
*J. Am. Oil Chem. Soc. 60:* 387 (1983)
*J. Am. Oil Chem. Soc. 74:* 989 (1997)
*JAOCS 74:* 375–380 (1997)

## Sunflower Seed Oil (high linoleic; HL)

*GMO*

Specific Gravity (SG)
   15.5/15.5°C
   25/25°C
   Other SG
Refractive Index (RI)
   25°C
   40°C
   Other RI
Iodine Value
Saponification Value
Titer °C
% Unsaponifiable
Melting Point °C

**Fatty Acid Composition (%)**
   14:0. . . . . . . . . . . . . . . . . . . . . . . . . . 0.1
   16:0. . . . . . . . . . . . . . . . . . . . . . . . . . 7.5
   9c-16:1 . . . . . . . . . . . . . . . . . . . . . . . 0.1
   18:0. . . . . . . . . . . . . . . . . . . . . . . . . . 1.9
   Total 18:1 . . . . . . . . . . . . . . . . . . . . 13.3
   9c,12c-18:2 . . . . . . . . . . . . . . . . . . . 76.0
   Undefined 18:3 . . . . . . . . . . . . . . . . . 0.1
   20:0. . . . . . . . . . . . . . . . . . . . . . . . . . 0.1
   Total 20:1 . . . . . . . . . . . . . . . . . . . . . 0.2

22:0 . . . . . . . . . . . . . . . . . . . . . . . . . . 0.4
24:0 . . . . . . . . . . . . . . . . . . . . . . . . . . 0.2

**References** *J. Am. Oil Chem. Soc. 74:* 989 (1997)

## Sunflower Seed Oil (high oleic; HO)

Specific Gravity (SG)
  15.5/15.5°C
  25/25°C . . . . . . . . . . . . . . . . . 0.912–0.915
  Other SG . . . . . . . . . (20/20) 0.915–0.920
Refractive Index (RI)
  25°C . . . . . . . . . . . . . . . . . . . 1.467–1.469
  40°C
  Other RI
Iodine Value . . . . . . . . . . . . . . . . . . . . 75–90
Saponification Value
Titer °C
% Unsaponifiable . . . . . . . . . . . . . . . 0.8–2.0
Melting Point °C

**Fatty Acid Composition (%)**
  16:0 . . . . . . . . . . . . . . . . . . . . . . . . 3–5.2
  16:1 . . . . . . . . . . . . . . . . . . . . . . . . . . 0.1
  9c-16:1 . . . . . . . . . . . . . . . . . . . . 0.1–0.2
  18:0 . . . . . . . . . . . . . . . . . . . . . . . . . . 3–5
  Total 18:1 . . . . . . . . . . . . . . . . . . . 70–92
  Undefined 18:2 . . . . . . . . . . . . . . . . 56.5
  9c,12c-18:2 . . . . . . . . . . . . . . . . . . . 2–20
  Undefined 18:3 . . . . . . . . . . . . . . . . . . tr
  20:0 . . . . . . . . . . . . . . . . . . . . . . . . . . 0.3
  Total 20:1 . . . . . . . . . . . . . . . . . . . . . 0.2
  22:0 . . . . . . . . . . . . . . . . . . . . . . . . . . . . 1
  Unidentified 22:1 . . . . . . . . . . . . . . . 0.1
  24:0 . . . . . . . . . . . . . . . . . . . . . . . . . . 0.4

Sterol Composition, %
  Cholesterol . . . . . . . . . . . . . . . . . . 0–0.5
  Brassicasterol . . . . . . . . . . . . . . . . 0–0.1
  Campesterol . . . . . . . . . . . . . . . . . . 7–12
  Stigmasterol . . . . . . . . . . . . . . . . . . 8–13
  Stigmasta-8,22-dien-3β-ol
  5α-Stigmasta-7,22-dien-3β-ol
  Δ7,25-Stigmastadienol
  β-Sitosterol . . . . . . . . . . . . . . . . . . 53–61
  Δ5-Avenasterol . . . . . . . . . . . . . . . . 1.5–5
  Δ7-Stigmasterol . . . . . . . . . . . . . . . 7–21
  Δ7-Avenasterol . . . . . . . . . . . . . . . . . 3–6
  Δ7-Campesterol
  Δ7-Ergosterol
  Δ7,25-Stigmasterol
  Sitostanol . . . . . . . . . . . . . . . . . . 0.3 –1.5
  Spinasterol
  Squalene
  24-Methylene Cholesterol
  Other Δ7-Campesterol, 1–3
  % sterols in oil
  Total Sterols, mg/kg

Tocopherol Composition, mg/kg
  α-Tocopherol . . . . . . . . . . . . . . . 94–430
  β-Tocopherol . . . . . . . . . . . . . . . . . . . . . 2
  γ-Tocopherol . . . . . . . . . . . . . . . . . . . . . 1
  δ-Tocopherol
  Total, mg/kg . . . . . . . . . . . . . . . . . . . 450

**References** *J. Am. Oil Chem. Soc. 74:* 989 (1997)
*J. Am. Oil Chem. Soc. 63:* 1062 (1986)
*Riv. Ital. Sost. Grasse 71:* 171 (1994)
*J. Am. Oil Chem. Soc. 72:* 1513 (1995)
*J. Chromatogr. 630:* 213 (1993)

## Sunflower Seed Oil (high palmitic/high linoleic; HP/HL)
*GMO*

Specific Gravity (SG)
  15.5/15.5°C
  25/25°C
  Other SG
Refractive Index (RI)
  25°C
  40°C
  Other RI
Iodine Value
Saponification Value
Titer °C
% Unsaponifiable
Melting Point °C

**Fatty Acid Composition (%)**
  14:0 . . . . . . . . . . . . . . . . . . . . . . . . . . 0.1
  16:0 . . . . . . . . . . . . . . . . . . . . . . . . . 27.3
  9c-16:1 . . . . . . . . . . . . . . . . . . . . . . . . 4.4

18:0 . . . . . . . . . . . . . . . . . . . . . . . . . 2.7
Total 18:1 . . . . . . . . . . . . . . . . . . . 17.1
9c,12c-18:2 . . . . . . . . . . . . . . . . . . 46.8
Undefined 18:3 . . . . . . . . . . . . . . . . 0.1
20:0 . . . . . . . . . . . . . . . . . . . . . . . . . 0.3
Total 20:1 . . . . . . . . . . . . . . . . . . . . 0.1
22:0 . . . . . . . . . . . . . . . . . . . . . . . . . 0.8
Unidentified 22:1 . . . . . . . . . . . . . . . 0.1
24:0 . . . . . . . . . . . . . . . . . . . . . . . . . 0.3

**References** *J. Am. Oil Chem. Soc. 74:* 989 (1997)

## Sunflower Seed Oil (high palmitic/high oleic; HP/HO)
*GMO*

Specific Gravity (SG)
  15.5/15.5°C
  25/25°C
  Other SG
Refractive Index (RI)
  25°C
  40°C
  Other RI
Iodine Value
Saponification Value
Titer °C
% Unsaponifiable
Melting Point °C

**Fatty Acid Composition (%)**
16:0 . . . . . . . . . . . . . . . . . . . . . . . . 24.6
9c-16:1 . . . . . . . . . . . . . . . . . . . . . . 6.1
18:0 . . . . . . . . . . . . . . . . . . . . . . . . . 2.9
Total 18:1 . . . . . . . . . . . . . . . . . . . 59.8
9c,12c-18:2 . . . . . . . . . . . . . . . . . . . 3.5
Undefined 18:3 . . . . . . . . . . . . . . . . 0.1
20:0 . . . . . . . . . . . . . . . . . . . . . . . . . 0.4
Total 20:1 . . . . . . . . . . . . . . . . . . . . 0.2
22:0 . . . . . . . . . . . . . . . . . . . . . . . . . 1.8
Unidentified 22:1 . . . . . . . . . . . . . . . 0.1
24:0 . . . . . . . . . . . . . . . . . . . . . . . . . 0.6

**References** *J. Am. Oil Chem. Soc. 74:* 989 (1997)

## Sunflower Seed Oil (high stearic/high oleic; HS/HO)
*GMO*

Specific Gravity (SG)
  15.5/15.5°C
  25/25°C
  Other SG
Refractive Index (RI)
  25°C
  40°C
  Other RI
Iodine Value
Saponification Value
Titer °C
% Unsaponifiable
Melting Point °C

**Fatty Acid Composition (%)**
16:0 . . . . . . . . . . . . . . . . . . . . . . . . . 4.6
9c-16:1 . . . . . . . . . . . . . . . . . . . . . . 0.1
18:0 . . . . . . . . . . . . . . . . . . . . . . . . 11.0
Total 18:1 . . . . . . . . . . . . . . . . . . . 79.1
9c,12c-18:2 . . . . . . . . . . . . . . . . . . . 2.0
Undefined 18:3 . . . . . . . . . . . . . . . . 0.1
20:0 . . . . . . . . . . . . . . . . . . . . . . . . . 0.9
Total 20:1 . . . . . . . . . . . . . . . . . . . . 0.2
22:0 . . . . . . . . . . . . . . . . . . . . . . . . . 1.8
24:0 . . . . . . . . . . . . . . . . . . . . . . . . . 0.3

**References** *J. Am. Oil Chem. Soc. 74:* 989 (1997)

## Sweet Rocket Oil (Dame's Violet)
*Hesperis matronalis*

Specific Gravity (SG)
  15.5/15.5°C
  25/25°C
  Other SG
Refractive Index (RI)
  25°C
  40°C
  Other RI

Iodine Value . . . . . . . . . . . . . . . . . . . . . . . 190
Saponification Value . . . . . . . . . . . . . . . . 193
Titer °C
% Unsaponifiable . . . . . . . . . . . . . . . . . . . . 2
Melting Point °C

**Fatty Acid Composition (%)**
    Total 18:1 . . . . . . . . . . . . . . . . . . . . . . . 11
    9c,12c-18:2 . . . . . . . . . . . . . . . . . . . . . . 35
    Undefined 18:3 . . . . . . . . . . . . . . . . . . . 46
    Other: saturate, 8–9

**References**

## Tabebuia argentia
*Tabebuia argentia*

Specific Gravity (SG)
    15.5/15.5°C
    25/25°C
    Other SG
Refractive Index (RI)
    25°C
    40°C
    Other RI
Iodine Value . . . . . . . . . . . . . . . . . . . . . . 116.3
Saponification Value . . . . . . . . . . . . . . . 203.1
Titer °C
% Unsaponifiable . . . . . . . . . . . . . . . . . . . 2.9
Melting Point °C

**Fatty Acid Composition (%)**
    16:0 . . . . . . . . . . . . . . . . . . . . . . . . . . . 21.7
    18:0 . . . . . . . . . . . . . . . . . . . . . . . . . . . . 3.8
    9c-18:1 . . . . . . . . . . . . . . . . . . . . . . . . . 9.8
    9c,12c-18:2 . . . . . . . . . . . . . . . . . . . . . . 52.7
    Undefined 18:3 . . . . . . . . . . . . . . . . . . . . 3

**References** *JAOCS 688:* 520–521 (1991)

## Tall Oil (Crude from pine wood pulping)

Specific Gravity (SG)
    15.5/15.5°C
    25/25°C . . . . . . . . . . . . . . . . 0.968–0.976
    Other SG

Refractive Index (RI)
    25°C . . . . . . . . . . . . . . . . . . . . . . . . . 1.494
    40°C
    Other RI
Iodine Value . . . . . . . . . . . . . . . . . . . 140–180
Saponification Value . . . . . . . . . . . . 154–180
Titer °C
% Unsaponifiable . . . . . . . . . . . . . . . . . 9–23
Melting Point °C

**Fatty Acid Composition (%)**
    16:0 . . . . . . . . . . . . . . . . . . . . . . . . . . . 5–6
    18:0 . . . . . . . . . . . . . . . . . . . . . . . . . . . 2–3
    Total 18:1 . . . . . . . . . . . . . . . . . . . . . 41–48
    9c,12c-18:2 . . . . . . . . . . . . . . . . . . . . 41–52
    Other: 18:2 conjugated, 7–13

**References**

## Tamarind Kernel Oil
*Tamarindus indica*

Specific Gravity (SG)
    15.5/15.5°C
    25/25°C . . . . . . . . . . . . . . . . . . . . . . 0.920
    Other SG
Refractive Index (RI)
    25°C . . . . . . . . . . . . . . . . . . . . . . . . 1.4750
    40°C
    Other RI
Iodine Value . . . . . . . . . . . . . . . . . . . . . . 118
Saponification Value . . . . . . . . . . . . . . . 190
Titer °C
% Unsaponifiable . . . . . . . . . . . . . . . . . . 2.5
Melting Point °C

**Fatty Acid Composition (%)**
    16:0 . . . . . . . . . . . . . . . . . . . . . . . . . . . . 15
    18:0 . . . . . . . . . . . . . . . . . . . . . . . . . . . . . 6
    Total 18:1 . . . . . . . . . . . . . . . . . . . . . . . 27
    9c,12c-18:2 . . . . . . . . . . . . . . . . . . . . . . . 8
    Undefined 18:3 . . . . . . . . . . . . . . . . . . . . 6
    20:0 . . . . . . . . . . . . . . . . . . . . . . . . . . . . . 5
    22:0 . . . . . . . . . . . . . . . . . . . . . . . . . . . . 12
    24:0 . . . . . . . . . . . . . . . . . . . . . . . . . . . . 22

**References** *J. Am. Oil Chem. Soc. 54:* 592 (1977)

## Tanacetum Seed Oil

*Chrysanthemum corymbosum*

Specific Gravity (SG)
  15.5/15.5°C
  25/25°C
  Other SG
Refractive Index (RI)
  25°C
  40°C
  Other RI
Iodine Value
Saponification Value
Titer °C
% Unsaponifiable
Melting Point °C

**Fatty Acid Composition (%)**
  16:0 . . . . . . . . . . . . . . . . . . . . . . . . . . . . 4
  17:0 . . . . . . . . . . . . . . . . . . . . . . . . . . 0.3
  18:0 . . . . . . . . . . . . . . . . . . . . . . . . . . 1–2
  Total 18:1 . . . . . . . . . . . . . . . . . . . . . 0.6
  Undefined 18:2 . . . . . . . . . . . . . . . . . . 58
  9c,12a-18:2 . . . . . . . . . . . . . . . . . . . . . 10
  Undefined 18:3 . . . . . . . . . . . . . . . . . . 0.2
  8t,10t, 12a-18:3 . . . . . . . . . . . . . . . . . . 18
  20:0 . . . . . . . . . . . . . . . . . . . . . . . . . . 0.3

**References** *Lipids 33:* 723 (1998)


## Taramira Seed Oil (Rocket Salad)

*Eruca sativa*

Specific Gravity (SG)
  15.5/15.5°C . . . . . . . . . . . . . 0.914–0.920
  25/25°C . . . . . . . . . . . . . . . . . . . . 0.910
  Other SG
Refractive Index (RI)
  25°C
  40°C . . . . . . . . . . . . . . . . . . . . . . 1.4680
  Other RI . . . . . . . . . . . . .(20) 1.472–1.475
Iodine Value . . . . . . . . . . . . . . . . . 130–137
Saponification Value . . . . . . . . . . . 168–176
Titer °C
% Unsaponifiable . . . . . . . . . . . . . . . 0.7–1.5
Melting Point °C

**Fatty Acid Composition (%)**
  16:0 . . . . . . . . . . . . . . . . . . . . . . . . . 2–10
  9c-16:1 . . . . . . . . . . . . . . . . . . . . . . . 0.2
  18:0 . . . . . . . . . . . . . . . . . . . . . . . . . 1–2
  Total 18:1 . . . . . . . . . . . . . . . . . . . . 8–24
  9c,12c-18:2 . . . . . . . . . . . . . . . . . . . 8–16
  Undefined 18:3 . . . . . . . . . . . . . . . 10–36
  20:0 . . . . . . . . . . . . . . . . . . . . . . . . . 1–2
  20:2 . . . . . . . . . . . . . . . . . . . . . . . . . 0–1
  22:0 . . . . . . . . . . . . . . . . . . . . . . . . . 1–2
  Unidentified 22:1 . . . . . . . . . . . . . 10–58
  24:0 . . . . . . . . . . . . . . . . . . . . . . . . . 0–1
  15c-24:1 . . . . . . . . . . . . . . . . . . . . . 0.4
  Other: 13c-20:1, 8–13; 5c,16c-22:2, 0.5

**References** *Indust. Crops Prod. 1:* 52 (1992)
*J. Am. Oil Chem. Soc. 62:* 1134 (1985)
*J. Am. Oil Chem. Soc. 66:* 139 (1989)
*J. Sci. Food Agric. 27:* 373 (1976)


## Taxus Baccata Seed Oil

*Taxus baccata*

Specific Gravity (SG)
  15.5/15.5°C
  25/25°C
  Other SG
Refractive Index (RI)
  25°C
  40°C
  Other RI
Iodine Value
Saponification Value
Titer °C
% Unsaponifiable
Melting Point °C

**Fatty Acid Composition (%)**
  14:0 . . . . . . . . . . . . . . . . . . . . . . . . . . . tr
  16:0 . . . . . . . . . . . . . . . . . . . . . . . 3–3.03
  16:1 . . . . . . . . . . . . . . . . . . . . . . . . . 0.06
  9c-16:1 . . . . . . . . . . . . . . . . . . . . . . . 0.1
  17:0 . . . . . . . . . . . . . . . . . . . . . . . . . 0.05
  18:0 . . . . . . . . . . . . . . . . . . . . . . 2.47–2.5
  Total 18:1 . . . . . . . . . . . . . . . . . . . . . . 56
  9c-18:1 . . . . . . . . . . . . . . . . . . . . . . 54.78
  11c-18:1 . . . . . . . . . . . . . . . . . . . . . . 0.33
  5,9–18:2 . . . . . . . . . . . . . . . . . . . . . . . 9.5

9c,12a-18:2 .................... 23.08
9c,12c-18:2 ....................... 23
Undefined 18:3 ..................... 2
9c,12c,15c-18:3 ................. 1.27
5c,9c,12c-18:3 .................. 0.33
5,9c,12c,15c-18:4 .................. tr
20:0 ........................ tr-0.06
Total 20:1 ......................... 1
11c-20:1 ....................... 1.33
20:2 ........................ 0.27–0.6
5,11c,14c-20:3 .................. 1.64
5c,11c,14c, 17–20:4 ............. 0.28
Other: 5c,9c-18:2, 10; 5c,11c-20:2, 0.2; 5c,11c,14c-20:3, 1.5; 5c,11c,14c,17c-20:4, 0.2

**References** *INFORM 8:* 116 (1997)
*JAOCS 75:* 1637–1641 (1998)

## Taxus Cuspidata Seed Oil
*Taxus cuspidata*
Specific Gravity (SG)
  15.5/15.5°C
  25/25°C
  Other SG
Refractive Index (RI)
  25°C
  40°C
  Other RI
Iodine Value
Saponification Value
Titer °C
% Unsaponifiable
Melting Point °C

**Fatty Acid Composition (%)**
14:0 .............................. tr
16:0 ........................... 3.18
16:1 .............................. tr
17:0 ........................... 0.06
18:0 ........................... 0.87
9c-18:1 ........................ 39.21
11c-18:1 ....................... 0.62
5,9–18:2 ....................... 16.16
9c,12a-18:2 .................... 29.35
9c,12c,15c-18:3 .................... 2

5c,9c,12c-18:3 .................. 2.66
5,9c,12c,15c-18:4 ............... 0.25
20:0 ........................... 0.06
11c-20:1 ....................... 1.49
20:2 ........................ 0.21–0.65
5,11c,14c-20:3 .................. 2.16
5c,11c,14c, 17–20:4 ............. 0.08

**References** *JAOCS 75:* 1637–1641 (1998)

## Taxus Chinensis Seed Oil
*Taxus chinensis*
Specific Gravity (SG)
  15.5/15.5°C
  25/25°C
  Other SG
Refractive Index (RI)
  25°C
  40°C
  Other RI
Iodine Value
Saponification Value
Titer °C
% Unsaponifiable
Melting Point °C

**Fatty Acid Composition (%)**
14:0 ........................... 0.07
16:0 ........................... 3.23
16:1 ........................... 0.09
17:0 ........................... 0.05
18:0 .............................. 1
9c-18:1 ........................ 34.31
11c-18:1 ....................... 0.54
5,9–18:2 ....................... 16.08
9c,12a-18:2 .................... 34.22
9c,12c,15c-18:3 ................. 2.09
5c,9c,12c-18:3 .................. 3.31
5,9c,12c,15c-18:4 ............... 0.28
20:0 ........................... 0.06
11c-20:1 ....................... 1.46
20:2 ............................ 0.7
5,11c,14c-20:3 .................. 2.13
5c,11c,14c, 17–20:4 ............... tr

**References** *JAOCS 75:* 1637–1641 (1998)

## Taxus Grandis Seed Oil
*Taxus grandis*

Specific Gravity (SG)
 15.5/15.5°C
 25/25°C
 Other SG
Refractive Index (RI)
 25°C
 40°C
 Other RI
Iodine Value
Saponification Value
Titer °C
% Unsaponifiable
Melting Point °C

**Fatty Acid Composition (%)**
 14:0 . . . . . . . . . . . . . . . . . . . . . . . . . 0.02
 16:0 . . . . . . . . . . . . . . . . . . . . . . . . . 5.49
 16:1 . . . . . . . . . . . . . . . . . . . . . . . . . 0.05
 17:0 . . . . . . . . . . . . . . . . . . . . . . . . . 0.05
 18:0 . . . . . . . . . . . . . . . . . . . . . . . . . 2.22
 9c-18:1 . . . . . . . . . . . . . . . . . . . . . . 40.39
 11c-18:1 . . . . . . . . . . . . . . . . . . . . . . 0.54
 5,9-18:2 . . . . . . . . . . . . . . . . . . . . . . . . tr
 9c,12a-18:2 . . . . . . . . . . . . . . . . . . 32.05
 9c,12c,15c-18:3 . . . . . . . . . . . . . . . . 0.50
 5c,9c,12c-18:3 . . . . . . . . . . . . . . . . . 0.04
 20:0 . . . . . . . . . . . . . . . . . . . . . . . . . 0.07
 11c-20:1 . . . . . . . . . . . . . . . . . . . . . 2.12
 20:2 . . . . . . . . . . . . . . . . . . . . . 0.82–3.21
 5,11c,14c-20:3 . . . . . . . . . . . . . . . 11.20
 5c,11c,14c,17–20:4 . . . . . . . . . . . . . 0.19

**References** *JAOCS 75:* 1637–1641 (1998)

## Teaseed Oil
*Camilla sasanqua*

Specific Gravity (SG)
 15.5/15.5°C
 25/25°C . . . . . . . . . . . . . . . . 0.909–0.920
 Other SG
Refractive Index (RI)
 25°C . . . . . . . . . . . . . . . . . . . 1.466–1.470
 40°C . . . . . . . . . . . . . . . . . . . 1.460–1.464
 Other RI

Iodine Value . . . . . . . . . . . . . . . . . . . . 80–92
Saponification Value . . . . . . . . . . . 188–196
Titer °C
% Unsaponifiable . . . . . . . . . . . . . . . . 0.1–1
Melting Point °C

**Fatty Acid Composition (%)**
 16:0 . . . . . . . . . . . . . . . . . . . . . . . . . . . 12
 18:0 . . . . . . . . . . . . . . . . . . . . . . . . . . . . 1
 Total 18:1 . . . . . . . . . . . . . . . . . . . . 72–74
 9c,12c-18:2 . . . . . . . . . . . . . . . . . . . 14–15

**References**

## Teaseed Oil
*Thea sinensis*

Specific Gravity (SG)
 15.5/15.5°C
 25/25°C
 Other SG
Refractive Index (RI)
 25°C . . . . . . . . . . . . . . . . . . . 1.466–1.469
 40°C . . . . . . . . . . . . . . . . . . . 1.462–1.464
 Other RI
Iodine Value . . . . . . . . . . . . . . . . . . . . 84–91
Saponification Value . . . . . . . . . . . 188–195
Titer °C
% Unsaponifiable . . . . . . . . . . . . . . . . 0.1–1
Melting Point °C

**Fatty Acid Composition (%)**
 14:0 . . . . . . . . . . . . . . . . . . . . . . . . . . 0–2
 16:0 . . . . . . . . . . . . . . . . . . . . . . . . . 5–17
 18:0 . . . . . . . . . . . . . . . . . . . . . . . . . 0.3–4
 Total 18:1 . . . . . . . . . . . . . . . . . . . . 58–87
 9c,12c-18:2 . . . . . . . . . . . . . . . . . . . . 7–24
 20:0 . . . . . . . . . . . . . . . . . . . . . . . . . 0–0.6
 Total 20:1 . . . . . . . . . . . . . . . . . . . . . 0–2
 20:2 . . . . . . . . . . . . . . . . . . . . . . . . . . 0–2

Sterol Composition, %
 Cholesterol
 Brassicasterol
 Campesterol
 Stigmasterol
 Stigmasta-8,22-dien-3β-ol
 5α-Stigmasta-7,22-dien-3β-ol

Δ7,25-Stigmastadienol
β-Sitosterol
Δ5-Avenasterol
Δ7-Stigmasterol .................. 34
Δ7-Avenasterol .................... 2
Δ7-Campesterol
Δ7-Ergosterol
Δ7,25-Stigmasterol
Sitostanol
Spinasterol ...................... 60
Squalene
24-Methylene Cholesterol
Other: 24-Methyl-cholest-7-enol, 4 % sterols in oil
Total Sterols, mg/kg

Tocopherol Composition, mg/kg
    α-Tocopherol
    β-Tocopherol
    γ-Tocopherol
    δ-Tocopherol
    Total, mg/kg ................... 1020

**References**  *J. Am. Dietetic Assn. 73:* 39 (1978)
*J. Sci. Food Agric. 27:* 1115 (1976)

## Teaseed Oil (Tsubaki Oil)
*Camellia japonica*

Specific Gravity (SG)
    15.5/15.5°C
    25/25°C
    Other SG
Refractive Index (RI)
    25°C .................. 1.465–1.468
    40°C
    Other RI
Iodine Value .................... 78–81
Saponification Value ............ 189–197
Titer °C
% Unsaponifiable
Melting Point °C

**Fatty Acid Composition (%)**
    16:0 ............................ 9
    18:0 ............................ 2
    Total 18:1 ...................... 87
    9c,12c-18:2 ..................... 2

**References**

## Teaseed Oil (Turkish)

Specific Gravity (SG)
    15.5/15.5°C
    25/25°C .................... 0.9180
    Other SG
Refractive Index (RI)
    25°C ....................... 1.4692
    40°C
    Other RI
Iodine Value .................... 91
Saponification Value ............ 193
Titer °C
% Unsaponifiable ................ 1.1
Melting Point °C

**Fatty Acid Composition (%)**
    16:0 ........................... 16
    18:0 ............................ 2
    Total 18:1 ..................... 59
    9c,12c-18:2 .................... 22
    20:0 ............................ 1

**References**  *Fette Seifen Anstrichm. 79:* 115 (1977)

## Thumba Oil
*Citrulus colocynthis*

Specific Gravity (SG)
    15.5/15.5°C
    25/25°C .................... 0.9200
    Other SG
Refractive Index (RI)
    25°C ....................... 1.4741
    40°C
    Other RI
Iodine Value .................... 129
Saponification Value ............ 192
Titer °C
% Unsaponifiable
Melting Point °C

**Fatty Acid Composition (%)**
- 14:0 .............................. 1
- 16:0 ............................ 9–14
- 16:1 .............................. 1
- 18:0 ............................ 6–9
- Total 18:1 .................... 17–26
- Undefined 18:2 ............... 50–65

**References** *INFORM 13:* 151 (2002)

## Tobacco Seed Oil

Specific Gravity (SG)
- 15.5/15.5°C .............. 0.920–0.925
- 25/25°C ................ 0.9186–0.9196
- Other SG

Refractive Index (RI)
- 25°C ................. 1.4715–1.4728
- 40°C ................. 1.4678–1.4717
- Other RI

Iodine Value .................. 187–200
Saponification Value ............ 112–145
Titer °C
% Unsaponifiable .................. 0–3
Melting Point °C

**Fatty Acid Composition (%)**
- 16:0 ............................ 3–12
- 18:0 ............................ 3–12
- Total 18:1 .................... 8–40
- 9c,12c-18:2 .................. 52–80
- Undefined 18:3 ................ 0.6–3
- 20:0 ........................... 0.1–1

Sterol Composition, %
- Cholesterol ...................... 16
- Brassicasterol
- Campesterol ...................... 7
- Stigmasterol ..................... 13
- Stigmasta-8,22-dien-3β-ol
- 5α-Stigmasta-7,22-dien-3β-ol
- Δ7,25-Stigmastadienol
- β-Sitosterol ...................... 60
- Δ5-Avenasterol
- Δ7-Stigmasterol
- Δ7-Avenasterol
- Δ7-Campesterol
- Δ7-Ergosterol
- Δ7,25-Stigmasterol

- Sitostanol
- Spinasterol
- Squalene
- 24-Methylene Cholesterol
- Other
- % sterols in oil
- Total Sterols, mg/kg ............. 1500

**References** *J. Am. Oil Chem. Soc. 53:* 680 (1976)
*J. Am. Dietetic Assn. 73:* 39 (1978)

## Tomato Seed Oil
*Lycopersicon lycopersicum*

Specific Gravity (SG)
- 15.5/15.5°C .............. 0.920–0.925
- 25/25°C .................. 0.918–0.920
- Other SG

Refractive Index (RI)
- 25°C ................... 1.472–1.473
- 40°C ................... 1.466–1.468
- Other RI

Iodine Value .................. 112–125
Saponification Value ............ 186–198
Titer °C
% Unsaponifiable ............... 0.4–1.4
Melting Point °C

**Fatty Acid Composition (%)**
- 14:0 ........................... 0–0.2
- 16:0 ............................ 12–16
- 9c-16:1 ........................ 0–0.6
- 18:0 ............................ 4–7
- Total 18:1 .................... 16–25
- 9c,12c-18:2 .................. 50–60
- Undefined 18:3 ............... 1.5–2.5
- 20:0 ........................... 0–0.6
- Total 20:1 ..................... 0–0.2
- 22:0 ........................... 0–0.2

Sterol Composition, %
- Cholesterol .................... 7–27
- Brassicasterol
- Campesterol ..................... 4–7
- Stigmasterol .................... 7–17
- Stigmasta-8,22-dien-3β-ol
- 5α-Stigmasta-7,22-dien-3β-ol

Δ7,25-Stigmastadienol
β-Sitosterol . . . . . . . . . . . . . . . . . . . 54–73
Δ5-Avenasterol . . . . . . . . . . . . . . . . . 0–8
Δ7-Stigmasterol . . . . . . . . . . . . . . . . . . 1
Δ7-Avenasterol
Δ7-Campesterol
Δ7-Ergosterol
Δ7,25-Stigmasterol
Sitostanol
Spinasterol
Squalene
24-Methylene Cholesterol
Other: Δ7-cholesterol, 1.5–4
% sterols in oil
Total Sterols, mg/kg

**References** *Riv. Ital. Sost. Grasse 65:* 43 (1988)
*Riv. Ital. Sost. Grasse 52:* 79 (1975)

## Tonka Bean Oil
*Dipteryx odorata*
*(Erythrina spp)*
Specific Gravity (SG)
  15.5/15.5°C . . . . . . . . . . . . . . . . . . . 0.923
  25/25°C . . . . . . . . . . . . . . . . . . . . . . 0.916
  Other SG
Refractive Index (RI)
  25°C
  40°C . . . . . . . . . . . . . . . . . 1.457–1.468
  Other RI
Iodine Value . . . . . . . . . . . . . . . . . . 72–79
Saponification Value . . . . . . . . . . . 183–198
Titer °C
% Unsaponifiable . . . . . . . . . . . . . . . . . 0–1
Melting Point °C

**Fatty Acid Composition (%)**
  16:0 . . . . . . . . . . . . . . . . . . . . . . . . . 5–13
  18:0 . . . . . . . . . . . . . . . . . . . . . . . . . 4–11
  Total 18:1 . . . . . . . . . . . . . . . . . . . 47–61
  9c,12c-18:2 . . . . . . . . . . . . . . . . . . . 7–15
  20:0 . . . . . . . . . . . . . . . . . . . . . . . . . . 3–4
  22:0 . . . . . . . . . . . . . . . . . . . . . . . . 11–15
  24:0 . . . . . . . . . . . . . . . . . . . . . . . . . . 0–1

**References** *J. Am. Oil Chem. Soc. 37:* 440 (1960)

## Trichosanthes Kirilowii Seed Oil
*Trichosanthes kirilowii*
Specific Gravity (SG)
  15.5/15.5°C
  25/25°C
  Other SG
Refractive Index (RI)
  25°C
  40°C
  Other RI
Iodine Value
Saponification Value
Titer °C
% Unsaponifiable
Melting Point °C

**Fatty Acid Composition (%)**
  16:0 . . . . . . . . . . . . . . . . . . . . . . . . . . . . 5
  18:0 . . . . . . . . . . . . . . . . . . . . . . . . . . . . 3
  9c-18:1 . . . . . . . . . . . . . . . . . . . . . . . . 12
  11c-18:1 . . . . . . . . . . . . . . . . . . . . . . 0.7
  9c,12c-18:2 . . . . . . . . . . . . . . . . . . . . 38
  α-Eleostearic 9c,11t,13t-18:3 . . . . . . . . 3
  11c-20:1 . . . . . . . . . . . . . . . . . . . . . . 0.2
  Other: 9c,11t,13c-18:3, 38 (punicic);
    9t,11t,13c-18:3, 2;

**References** *J. Am. Oil Chem. Soc. 72:* 1037 (1995)

## Tucum (Aoiara) Kernel Oil
*Astrocarpum spp*
Specific Gravity (SG)
  15.5/15.5°C
  25/25°C
  Other SG . . . . . . . . . (100/15) 0.865–0.867
Refractive Index (RI)
  25°C
  40°C . . . . . . . . . . . . . . . . . 1.449–1.451
  Other RI
Iodine Value . . . . . . . . . . . . . . . . . . 10–14
Saponification Value . . . . . . . . . . . 240–250
Titer °C
% Unsaponifiable . . . . . . . . . . . . . . . . . 0.3
Melting Point °C . . . . . . . . . . . . . . . 30–33

**Fatty Acid Composition (%)**
- 8:0 .................................. 1
- 10:0 ................................. 4
- 12:0 ................................ 49
- 14:0 ................................ 22
- 16:0 ................................. 6
- 18:0 ................................. 2
- Total 18:1 .......................... 13
- 9c,12c-18:2 .......................... 2

**References**

## Tucum Pulp Oil
*Astrocarpum vulgare*
Specific Gravity (SG)
 15.5/15.5°C
 25/25°C
 Other SG
Refractive Index (RI)
 25°C
 40°C
 Other RI
Iodine Value ........................ 64
Saponification Value ............... 189
Titer °C
% Unsaponifiable
Melting Point °C

**Fatty Acid Composition (%)**
- 12:0 ............................... 0.2
- 14:0 ............................... 0.5
- 16:0 ................................ 30
- 18:0 ................................. 2
- Total 18:1 .......................... 60
- 9c,12c-18:2 .......................... 3
- 20:0 ................................. 4

**References** *Riv. Ital. Sost. Grasse 71:* 425 (1994)
*Food Chem. 30:* 277 (1988)

## Tung Oil
*Aleurites fordii (aleurites montana; Vernicia montana)*
Specific Gravity (SG)
 15.5/15.5°C
 25/25°C .................... 0.913–0.917
 Other SG
Refractive Index (RI)
 25°C ...................... 1.514–1.520
 40°C
 Other RI
Iodine Value .................. 160–175
Saponification Value ........... 189–195
Titer °C
% Unsaponifiable ................. 0–1
Melting Point °C

**Fatty Acid Composition (%)**
- 16:0 ................................. 2
- 18:0 ................................. 3
- Total 18:1 ........................ 4–10
- 9c,12c-18:2 ...................... 8–15
- Undefined 18:3 ...................... 2
- α-Eleostearic 9c,11t,13t-18:3 .... 71–82

**References** FDA Tech. Report SCI-025–67

## Ucuhuba Butter Oil
*Virola surinamensis*
Specific Gravity (SG)
 15.5/15.5°C
 25/25°C
 Other SG .............. (100/15) 0.871
Refractive Index (RI)
 25°C
 40°C
 Other RI .......... (50) 1.4502–1.4525;
                    (70) 1.4431–1.4446
Iodine Value .................... 11–17
Saponification Value ........... 221–229
Titer °C
% Unsaponifiable ................. 1–4
Melting Point °C ................... 47

**Fatty Acid Composition (%)**
- 10:0 ................................. 1
- 12:0 .............................. 13–15
- 14:0 .............................. 64–73
- 16:0 ............................... 3–9
- 18:0 ................................. 1
- Total 18:1 ........................ 6–8
- 9c,12c-18:2 ....................... 3–5

References *J. Am. Dietetic Assn. 68:* 224 (1976)

## Vernonia Seed Oil
*Vernonia anthelmintica*
Specific Gravity (SG)
  15.5/15.5°C
  25/25°C
  Other SG . . . . . . . . . . . . . .(30/30) 0.9050
Refractive Index (RI)
  25°C
  40°C
  Other RI . . . . . . . . . . . . . . . . . (32) 1.4860
Iodine Value . . . . . . . . . . . . . . . . . . . . . . 55
Saponification Value . . . . . . . . . . . . . . . 176
Titer °C
% Unsaponifiable . . . . . . . . . . . . . . . . . . 1–2
Melting Point °C

**Fatty Acid Composition (%)**
  16:0. . . . . . . . . . . . . . . . . . . . . . . . . . . 3–7
  18:0. . . . . . . . . . . . . . . . . . . . . . . . . . . 2–6
  Total 18:1 . . . . . . . . . . . . . . . . . . . . . . 1–6
  9c,12c-18:2. . . . . . . . . . . . . . . . . . . . 9–17
  Other: 12,13-Epoxy-octadeca-9-enoic, 62–72 (vernolic)

**References**

## Vernonia Seed Oil
*Vernonia galamensis*
Specific Gravity (SG)
  15.5/15.5°C
  25/25°C
  Other SG
Refractive Index (RI)
  25°C
  40°C
  Other RI
Iodine Value
Saponification Value
Titer °C
% Unsaponifiable
Melting Point °C

**Fatty Acid Composition (%)**
  16:0. . . . . . . . . . . . . . . . . . . . . . . . 2.7–3.3
  18:0. . . . . . . . . . . . . . . . . . . . . . . . 2.7–3.9
  Total 18:1 . . . . . . . . . . . . . . . . . . . 3.6–5.6
  9c,12c-18:2. . . . . . . . . . . . . . . . . . . . 9–14
  Undefined 18:3 . . . . . . . . . . . . . . . . 0–0.3
  20:0. . . . . . . . . . . . . . . . . . . . . . . . 0.2–0.5
  Total 20:1 . . . . . . . . . . . . . . . . . . . 0.2–0.4
  Other: 12,13-Epoxy-octadeca-9-enoic, 72–78 (vernolic)

**References** E.H. Pryde, *et al.*, eds., *New Sources Of Fats and Oils,* AOCS Press, Champaign, 1981, pp. 55; *JAOCS 65:* 942 (1988); *J. Liq. Chromatogr. 18:* 4165 (1995)

## Walnut Oil
*Juglans regia*
Specific Gravity (SG)
  15.5/15.5°C. . . . . . . . . . . . . . 0.927–0.930
  25/25°C. . . . . . . . . . . . . . . . . 0.923–0.925
  Other SG
Refractive Index (RI)
  25°C . . . . . . . . . . . . . . . . . . . 1.472–1.475
  40°C . . . . . . . . . . . . . . . . . . . 1.469–1.471
  Other RI
Iodine Value . . . . . . . . . . . . . . . . . . 138–162
Saponification Value . . . . . . . . . . . . 189–197
Titer °C
% Unsaponifiable . . . . . . . . . . . . . . . . . . 0.5
Melting Point °C

**Fatty Acid Composition (%)**
  16:0. . . . . . . . . . . . . . . . . . . . . . . . 6.22–8
  9c-16:1 . . . . . . . . . . . . . . . . . . . . . 0.1–0.2
  18:0. . . . . . . . . . . . . . . . . . . . . . . . 1.8–2.2
  Total 18:1 . . . . . . . . . . . . . . . . . . . . 17–19
  9c,12c-18:2. . . . . . . . . . . . . . . . . . . 56–60
  Undefined 18:3 . . . . . . . . . . . . . . . . 13–14
  20:0. . . . . . . . . . . . . . . . . . . . . . . . . . . 0.1
  Total 20:1 . . . . . . . . . . . . . . . . . . . . . . 0.2
  22:0. . . . . . . . . . . . . . . . . . . . . . . . . . . 0.1

Sterol Composition, %
  Cholesterol

Brassicasterol
Campesterol . . . . . . . . . . . . . . . . . . . . . . 5
Stigmasterol
Stigmasta-8,22-dien-3β-ol
5α-Stigmasta-7,22-dien-3β-ol
Δ7,25-Stigmastadienol
β-Sitosterol . . . . . . . . . . . . . . . . . . . . . . 89
Δ5-Avenasterol . . . . . . . . . . . . . . . . . . . . 5
Δ7-Stigmasterol
Δ7-Avenasterol
Δ7-Campesterol
Δ7-Ergosterol
Δ7,25-Stigmasterol
Sitostanol
Spinasterol
Squalene
24-Methylene Cholesterol
Other
% sterols in oil
Total Sterols, mg/kg . . . . . . . . . . . . . . 1760

Tocopherol Composition, mg/kg
   α-Tocopherol . . . . . . . . . . . . . . . 10–28.7
   β-Tocopherol . . . . . . . . . . . . . . . . . 1–8.2
   γ-Tocopherol . . . . . . . . . . . . . . 206.9–355
   δ-Tocopherol . . . . . . . . . . . . . . . 29.6–62.1
   Total, mg/kg . . . . . . . . . . . . . . . . 309–455

**References** *JAOAC 48:* 902 (1965)
   *J. Am. Dietetic Assn. 73:* 39 (1978)
   *J. Am. Oil Chem. Soc. 53:* 732 (1976)
   *Fat Sci. Technol. 93:* 519 (1991)
   *JAOCS 76:* 1059–1063 (1999)
   *J. Korean Soc. Food Nutr. 13:* 263 (1984)

# Watermelon Seed Oil
*Citrullus lanatus/vulgaris*

Specific Gravity (SG)
   15.5/15.5°C
   25/25°C . . . . . . . . . . . . . . . . . 0.919–0.922
   Other SG
Refractive Index (RI)
   25°C
   40°C
   Other RI . . . . . . . . . . .(20) 1.4669–1.4748
Iodine Value . . . . . . . . . . . . . . . . . . 124–134
Saponification Value . . . . . . . . . . . 190  198

Titer °C
% Unsaponifiable . . . . . . . . . . . . . . 0.5–1.2
Melting Point °C

**Fatty Acid Composition (%)**
   14:0 . . . . . . . . . . . . . . . . . . . . . . . . . . . 0.1
   15:0 . . . . . . . . . . . . . . . . . . . . . . . . . . . . tr
   16:0 . . . . . . . . . . . . . . . . . . . . . . . . 9–11.2
   9c-16:1 . . . . . . . . . . . . . . . . . . . . . . 0.1–1
   18:0 . . . . . . . . . . . . . . . . . . . . . . . . . . 5–10
   Total 18:1 . . . . . . . . . . . . . . . . . . . . 13–19
   9c-18:1 . . . . . . . . . . . . . . . . . . . . . . . 10.2
   9c,12c-18:2 . . . . . . . . . . . . . . . . . 62–71.3
   9c,12c,15c-18:3 . . . . . . . . . . . . . . . . . 0.2
   20:0 . . . . . . . . . . . . . . . . . . . . . . . . . . . 0–1
   11c-20:1 . . . . . . . . . . . . . . . . . . . . . . . 0.1

**References** *JAOCS 69:* 314–315 (1992)


# Welwitschia Mirabilis Seed Oil
*Welwitschia mirabilis*

Specific Gravity (SG)
   15.5/15.5°C
   25/25°C
   Other SG
Refractive Index (RI)
   25°C
   40°C
   Other RI
Iodine Value
Saponification Value
Titer °C
% Unsaponifiable
Melting Point °C

**Fatty Acid Composition (%)**
   14:0 . . . . . . . . . . . . . . . . . . . . . . . . . . . 0.2
   15:0 . . . . . . . . . . . . . . . . . . . . . . . . . . . 0.1
   16:0 . . . . . . . . . . . . . . . . . . . . . . . . . . . 1.6
   9c-16:1 . . . . . . . . . . . . . . . . . . . . . . . . .01
   11c-16:1 . . . . . . . . . . . . . . . . . . . . . . . 0.7
   17:0 . . . . . . . . . . . . . . . . . . . . . . . . . . . 0.6
   18:0 . . . . . . . . . . . . . . . . . . . . . . . . . . . 2.7
   9,10 epoxy-18:0 . . . . . . . . . . . . . . . . 0.2
   9c-18:1 . . . . . . . . . . . . . . . . . . . . . . . 11.1
   11c-18:1 . . . . . . . . . . . . . . . . . . . . . . . 1.7
   5,9–18:2 . . . . . . . . . . . . . . . . . . . . . . . 0.3

9c,12c-18:2 .................... 15.1
9c,12c,15c-18:3 ............... 33.5
5,9c,12c,15c-18:4 ............... 0.1
19:0 .............................. tr
20:0 .............................. 0.2
11c-20:1 .......................... 0.1
22:0 .............................. 0.2
24:0 .............................. 0.3

**References** *JAOCS 75:* 1761–1765 (1998)

## Western Soapberry Seed Fat (Wild Chinaberry)
*Sapindus drummondii*

Specific Gravity (SG)
  15.5/15.5°C
  25/25°C ................. 0.903–0.917
  Other SG
Refractive Index (RI)
  25°C ................. 1.4686–1.4722
  40°C
  Other RI
Iodine Value .................... 82–89
Saponification Value ........... 192–219
Titer °C
% Unsaponifiable .................... 1
Melting Point °C

**Fatty Acid Composition (%)**
  16:0 ............................... 5
  Total 18:1 ....................... 55
  9c,12c-18:2 ...................... 16
  Undefined 18:3 .................... 4
  20:0 .............................. 3
  Total 20:1 ....................... 17

**References** *Fat Sci. Technol. 96:* 69 (1994)
*Lipids 2:* 258 (1967)

## Wheat Germ Oil
*Triticum aestinum/durum*

Specific Gravity (SG)
  15.5/15.5°C ............. 0.928–0.938
  25/25°C ................. 0.925–0.933
  Other SG

Refractive Index (RI)
  25°C .................. 1.474–1.483
  40°C .................. 1.469–1.478
  Other RI
Iodine Value .................. 100–128
Saponification Value ........... 179–217
Titer °C
% Unsaponifiable ............. 1.59–7.28
Melting Point °C

**Fatty Acid Composition (%)**
  14:0 ........................... 0–0.2
  16:0 ........................... 12–20
  9c-16:1 ....................... 0.2–0.5
  18:0 ............................ 0.3–3
  Total 18:1 ..................... 13–23
  9c,12c-18:2 .................... 50–59
  Undefined 18:3 .................. 2–9
  20:0 ............................. 0.3
  Total 20:1 ....................... 0.3
  22:0 ........................... 0–0.1
  Unidentified 22:1 ................ 0.3
  24:0 ............................. 0–1

Sterol Composition, %
  Cholesterol
  Brassicasterol .................. 0–0.4
  Campesterol .................... 19–29
  Stigmasterol ................... 0.3–4
  Stigmasta-8,22-dien-3β-ol
  5α-Stigmasta-7,22-dien-3β-ol
  Δ7,25-Stigmastadienol
  β-Sitosterol .................... 56–67
  Δ5-Avenasterol .................. 2–6
  Δ7-Stigmasterol ................. 1–4
  Δ7-Avenasterol .................... 2
  Δ7-Campesterol
  Δ7-Ergosterol
  Δ7,25-Stigmasterol
  Sitostanol
  Spinasterol
  Squalene
  24-Methylene Cholesterol
  Other: Cholestanol, 0.1–0.4;
    campestanol, 1–2
% sterols in oil
Total Sterols, mg/kg ............. 5500

Tocopherol Composition, mg/kg
  α-Tocopherol ............. 166–3100

β-Tocopherol. . . . . . . . . . . . . . 66.6–1150
γ-Tocopherol. . . . . . . . . . . . . . . . 18–950
δ-Tocopherol. . . . . . . . . . . . . . . . 20–100
Total, mg/kg . . . . . . . . . . . . . 1350–2500

Tocotrienols Composition, mg/kg
  α-Tocotrienol . . . . . . . . . . . . . . . 10–200
  β-Tocotrienol . . . . . . . . . . . . . . . 10–200
  γ-Tocotrienol
  δ-Tocotrienol
  Total Tocotrienols, mg/kg . . . . . . . 20–400

**References** *Riv. Ital. Sost. Grasse 60:* 195 (1983)
*J. Am. Oil Chem. Soc. 63:* 328 (1986)
*Riv. Ital. Sost. Grasse 54:* 177 (1977)
*J. Sci. Food Agric. 31:* 997 (1980)
*Wheat: Chemistry and Technology,*
Vol. 1, Y. Pomeraz, ed., AACC, 1998, St. Paul, MN
*JAOCS 77:* 969–974 (2000)

## Winged Bean Oil
*Psophocarpus tetragonolobus*
Specific Gravity (SG)
  15.5/15.5°C
  25/25°C
  Other SG . . . . . . . . (20/20) 0.911–0.914;
                       (40/20) 0.897–0.900
Refractive Index (RI)
  25°C
  40°C . . . . . . . . . . . . . . . . 1.4628–1.4633
  Other RI . . . . . . . . . . .(20) 1.4699–1.4703
Iodine Value . . . . . . . . . . . . . . . . . . . 81–86
Saponification Value . . . . . . . . . . . 186–187
Titer °C
% Unsaponifiable . . . . . . . . . . . . . . . 0.3–1
Melting Point °C

**Fatty Acid Composition (%)**
  14:0. . . . . . . . . . . . . . . . . . . . . . . . . . 0.1
  16:0. . . . . . . . . . . . . . . . . . . . . . . . . 10–12
  9c-16:1 . . . . . . . . . . . . . . . . . . . . . . . 0.2
  17:0. . . . . . . . . . . . . . . . . . . . . . . . . . 0.1
  18:0. . . . . . . . . . . . . . . . . . . . . . . . .4–8
  Total 18:1 . . . . . . . . . . . . . . . . . . . 35–41
  9c,12c-18:2. . . . . . . . . . . . . . . . . . 15–32

Undefined 18:3 . . . . . . . . . . . . . . . . . . . . 1
20:0. . . . . . . . . . . . . . . . . . . . . . . . . . . 2–4
22:0. . . . . . . . . . . . . . . . . . . . . . . . . 13–20
Unidentified 22:1 . . . . . . . . . . . . 0.1–0.7
24:0. . . . . . . . . . . . . . . . . . . . . . . . . . . 3–5

Sterol Composition, %
  Cholesterol
  Brassicasterol
  Campesterol . . . . . . . . . . . . . . . . . . . . . 6
  Stigmasterol . . . . . . . . . . . . . . . . . . 34–38
  Stigmasta-8,22-dien-3β-ol
  5α-Stigmasta-7,22-dien-3β-ol
  Δ7,25-Stigmastadienol
  β-Sitosterol . . . . . . . . . . . . . . . . . . . 40–42
  Δ5-Avenasterol
  Δ7-Stigmasterol
  Δ7-Avenasterol
  Δ7-Campesterol
  Δ7-Ergosterol
  Δ7,25-Stigmasterol
  Sitostanol
  Spinasterol
  Squalene . . . . . . . . . . . . . . . . . . . . . . 0.3
  24-Methylene Cholesterol
  Other: C31,C32 hydrocarbons, 2–4;
    Triterpenes, 4–6
% sterols in oil
Total Sterols, mg/kg

**References** *J. Am. Oil Chem. Soc. 56:* 931 (1979)
*J. Am. Oil Chem. Soc. 59:* 523 (1982)
*J. Nutr. Sci. Vitaminol. 33:* 49 (1987)

## Yam Bean
*Pachyrhizus spp.*
Specific Gravity (SG)
  15.5/15.5°C
  25/25°C. . . . . . . . . . . . . . . . . . . . . 1.4700
  Other SG
Refractive Index (RI)
  25°C
  40°C
  Other RI
Iodine Value . . . . . . . . . . . . . . . . . . . . . 92

Saponification Value . . . . . . . . . . . . . . . . 183
Titer °C
% Unsaponifiable
Melting Point °C

**Fatty Acid Composition (%)**
  16:0. . . . . . . . . . . . . . . . . . . . . . . 24–31.4
  18:0. . . . . . . . . . . . . . . . . . . . . . . . 3.9–7.5
  Total 18:1 . . . . . . . . . . . . . . . . . 12.66–27
  9c-18:1 . . . . . . . . . . . . . . . . . . . 21–28.6
  11c-18:1 . . . . . . . . . . . . . . . . . . 0.58–0.85
  Undefined 18:2 . . . . . . . . . . . . . 36–62.48
  9c,12c-18:2 . . . . . . . . . . . . . . . . 34.9–40.7
  Undefined 18:3 . . . . . . . . . . . . . 0.9–16.17
  Total 20:1 . . . . . . . . . . . . . . . . . 0.06–0.14

Tocopherol Composition, mg/kg
  α-Tocopherol . . . . . . . . . . . . . . . . 0.4–5
  β-Tocopherol
  γ-Tocopherol . . . . . . . . . . . . . . 94.5–98.5
  δ-Tocopherol. . . . . . . . . . . . . . . . 0.2–1.4
  Total, mg/kg . . . . . . . . . . . . . . . 285–684

Tocotrienols Composition, mg/kg
  α-Tocotrienol . . . . . . . . . . . . . . . . 0–9.7
  β-Tocotrienol
  γ-Tocotrienol. . . . . . . . . . . . . . . 90.1–100
  δ-Tocotrienol. . . . . . . . . . . . . . . . . 0–2.5
  Total Tocotrienols, mg/kg . . . 249.3–896.2

**References** *JAOCS 76:* 1309 (1999)

# Characteristics of Oils and Fats of Animal Origin

## Anchovy Oil

Specific Gravity (SG)
   15.5/15.5°C
   25/25°C
   Other SG
Refractive Index (RI)
   25°C
   40°C
   Other RI
Iodine Value .................. 163–199
Saponification Value ............ 191–194
Titer °C
% Unsaponifiable ................ 0.3–3
Melting Point °C
% hydrocarbons
% sterols
% Squalene
% Pristane

**Fatty Acid Composition (%)**
   14:0 ........................ 0.4–1.3
   16:0 ........................ 16–20
   16:1 ........................ 8.12
   17:0 ........................ 0.5–2
   18:0 ........................ 3–7
   Unassigned 18:1 ............... 9–14
   18:2 ........................ 1–3
   18:3 ........................ 0.3–1.3
   20:1 ........................ 2–8
   5c,8c,11c,14c-20:4 (n-6) ........ 0.3–1
   6c,9c,12c,15c,17c-20:5 .......... 10–20
   22:1 ........................ 2–4
   7c,10c,13c,16c,19c-22:5 (n-3) ...... 1–2
   4c,7c,10c,13c,16c,19c-22:6 ....... 4–11
   Other ........................ 3–10

**References**

## Alewife Oil
*Alosa pseudoharengus*

Specific Gravity (SG)
   15.5/15.5°C
   25/25°C
   Other SG
Refractive Index (RI)
   25°C
   40°C
   Other RI
Iodine Value
Saponification Value
Titer °C
% Unsaponifiable
Melting Point °C
% hydrocarbons
% sterols
% Squalene
% Pristane

**Fatty Acid Composition (%)**
   16:1 ........................ 16
   Unassigned 18:1 ............... 24
   18:2 ........................ 2
   18:3 ........................ 2
   20:4 ........................ 4
   20:5 ........................ 9
   22:6 ........................ 9

**References** *J. Food Comp. Anal. 2:* 13 (1989)

## Bass, Rock Oil
*Ambloplites rupestris*

Specific Gravity (SG)
   15.5/15.5°C
   25/25°C
   Other SG
Refractive Index (RI)
   25°C
   40°C
   Other RI
Iodine Value
Saponification Value
Titer °C
% Unsaponifiable
Melting Point °C
% hydrocarbons
% sterols
% Squalene
% Pristane

**Fatty Acid Composition (%)**
   14:0 ........................ 2
   16:0 ........................ 19
   16:1 ........................ 9
   18:0 ........................ 4–5

Unassigned 18:1 .................. 18
18:2 .......................... 2
18:3 .......................... 2
20:4 (n-6) ..................... 8
20:5 (n-3) ..................... 4
22:5 .............. (n-3), 21; (n-6), 2

**References** *JAOCS 54:* 424 (1977)

## Bass, Sea

Specific Gravity (SG)
   15.5/15.5°C
   25/25°C
   Other SG
Refractive Index (RI)
   25°C
   40°C
   Other RI
Iodine Value
Saponification Value
Titer °C
% Unsaponifiable
Melting Point °C
% hydrocarbons
% sterols
% Squalene
% Pristane

**Fatty Acid Composition (%)**
   14:0 ........................... 2
   15:0 ........................... 1
   16:0 .......................... 17
   16:1 ........................... 7
   18:0 ........................... 4
   Unassigned 18:1 .................. 23
   18:2 ........................... 1
   20:1 ........................... 3
   20:4 (n-6) ..................... 2
   20:5 (n-3) ..................... 9
   22:5 (n-3) ..................... 2
   22:6 (n-3) ..................... 24
   Other  24:0, 2

**References**

## Beef Flank Fat

Specific Gravity (SG)
   15.5/15.5°C
   25/25°C
   Other SG
Refractive Index (RI)
   25°C
   40°C
   Other RI
Iodine Value
Saponification Value
Titer °C
% Unsaponifiable
Melting Point °C
% hydrocarbons
% sterols
% Squalene
% Pristane

**Fatty Acid Composition (%)**
   14:0 ......................... 1–5
   14:1 ........................ 0–1.6
   16:0 ........................ 14–45
   16:1 ......................... 2–8
   18:0 ......................... 9–20
   Unassigned 18:1 .............. 28–64
   18:2 ......................... 1–3
   18:3 ........................ 0–1.4
   5c,8c,11c,14c-20:4 (n-6)......... 0–0.8
   Other: 16:1t, 0–0.8; 18:1t, 0.8–5;
     other 1–4.3

Cholesterol ................. (600 mg/kg)

**References** USDA, *Agriculture Handbook 8* (1989)
*J. Am Oil Chem. Soc. 75:* 1001 (1998)

## Butterfat

Specific Gravity (SG)
   15.5/15.5°C ............. 0.930–0.940
   25/25°C
   Other SG
Refractive Index (RI)
   25°C
   40°C ................... 1.453–1.457
   Other RI
Iodine Value .................... 26–40
Saponification Value .......... 210–232
Titer °C ........................ 33–38
% Unsaponifiable ................ 0–0.5

Melting Point °C
% hydrocarbons
% sterols
% Squalene
% Pristane

**Fatty Acid Composition (%)**
10:0 . . . . . . . . . . . . . . . . . . . . . . . 1.7–3.2
12:0 . . . . . . . . . . . . . . . . . . . . . . . 2.2–4.5
14:0 . . . . . . . . . . . . . . . . . . . . . . . 5.4–14.6
14:1 . . . . . . . . . . . . . . . . . . . . . . . 0.6–1.6
16:0 . . . . . . . . . . . . . . . . . . . . . . . 25–41
16:1 . . . . . . . . . . . . . . . . . . . . . . . 2–6
18:0 . . . . . . . . . . . . . . . . . . . . . . . 6–12
Unassigned 18:1 . . . . . . . . . . . 18.7–33.4
18:2 . . . . . . . . . . . . . . . . . . . . . . . 0.9–3.7
20:0 . . . . . . . . . . . . . . . . . . . . . . . 1.2–2.4
Other: 4:0, 2.8–4; 6:0, 1.4–3.0; 8:0, 0.5–1.7

**References** *J. Am. Oil Chem. Soc. 52:* 154 (1975)
*J. Am. Oil Chem. Soc. 70:* 1161 (1993)

## Capelin Oil
*Mallotus villosus*

Specific Gravity (SG)
  15.5/15.5°C
  25/25°C
  Other SG
Refractive Index (RI)
  25°C
  40°C
Other RI . . . . . . . . . . . . . (50) 1.4620–1.4645
Iodine Value . . . . . . . . . . . . . . . . . . . . 94–164
Saponification Value . . . . . . . . . . . . 185–202
Titer °C
% Unsaponifiable . . . . . . . . . . . . . . . . . . 1–5
Melting Point °C
% hydrocarbons
% sterols
% Squalene
% Pristane

**Fatty Acid Composition (%)**
14:0 . . . . . . . . . . . . . . . . . . . . . . . . . . . . 5–9
16:0 . . . . . . . . . . . . . . . . . . . . . . . . . . . 8–12
16:1 . . . . . . . . . . . . . . . . . . . . . . . . . . . 8–12

18:0 . . . . . . . . . . . . . . . . . . . . . . . . . . . . 1–2
Unassigned 18:1 . . . . . . . . . . . . . . . . 12–18
18:2 . . . . . . . . . . . . . . . . . . . . . . . . . . . . 1–2
18:3 . . . . . . . . . . . . . . . . . . . . . . . . . . . . 0–1
18:4 . . . . . . . . . . . . . . . . . . . . . . . . . . . . 0–7
20:1 . . . . . . . . . . . . . . . . . . . . . . . . . . . 9–27
20:2 . . . . . . . . . . . . . . . . . . . . . . . . . . . . 0–1
5c,8c,11c,14c-20:4 (n-6) . . . . . . . . . . 0–2
6c,9c,12c,15c,17c-20:5 . . . . . . . . . . 3–12
22:1 . . . . . . . . . . . . . . . . . . . . . . . . . . . 9–25
22:2 . . . . . . . . . . . . . . . . . . . . . . . . . . . . 0–1
7c,10c,13c,16c,19c-22:5 (n-3) . . . . . . 0–1
4c,7c,10c,13c,16c,19c-22:6 . . . . . . . 1–11
Other . . . . . . . . . . . . . . . . . . . . . . . . . . . 3–6

**References**

## Carp Lipids
*Cyprinus carpio*

Specific Gravity (SG)
  15.5/15.5°C
  25/25°C
  Other SG
Refractive Index (RI)
  25°C
  40°C
  Other RI
Iodine Value
Saponification Value
Titer °C
% Unsaponifiable
Melting Point °C
% hydrocarbons
% sterols
% Squalene
% Pristane

**Fatty Acid Composition (%)**
14:0 . . . . . . . . . . . . . . . . . . . . . . . . . . . . 2–3
16:0 . . . . . . . . . . . . . . . . . . . . . . . . . . 17–20
16:1 . . . . . . . . . . . . . . . . . . . . . . . . . . . 9–17
17:0 . . . . . . . . . . . . . . . . . . . . . . . . . . . . 1–2
18:0 . . . . . . . . . . . . . . . . . . . . . . . . . . . . 4–5
Unassigned 18:1 . . . . . . . . . . . . . . . . 23–28
18:2 . . . . . . . . . . . . . . . . . . . . . . . . . . . 4–13
18:3 . . . . . . . . . . . . . . . . . . . . . . . . . . . . 2–6
6c,9c,12c,15c-18:4 (n-3) . . . . . . . . . . . 0.2

20:1 . . . . . . . . . . . . . . . . . . . . . . . . . . 1–4
20:4 (n-6) . . . . . . . . . . . . . . . . . . . . . 3–4
5c,8c,11c,14c-20:4 (n-6). . . . . . . . . . . . 3
6c,9c,12c,15c,17c-20:5. . . . . . . . . . . 3–6
7c, 10c, 13c, 16c-22:4. . . . . . . . . . . . . . 2
7c,10c,13c,16c,19c-22:5 (n-3) . . . . . . . . 1
4c,7c,10c,13c,16c,19c-22:6 . . . . . . . . . . 5

**References** Ackman, R.G. in *Objective Methods for Food Analysis,* National Academy of Sciences, Washington, DC, 1976
*Bull. Korean Fish Soc. 19:* 195 (1986)
*Prog. Lipid Res. 26:* 281 (1987)

## Catfish Lipids

Specific Gravity (SG)
 15.5/15.5°C
 25/25°C
 Other SG
Refractive Index (RI)
 25°C
 40°C
 Other RI
Iodine Value
Saponification Value
Titer °C
% Unsaponifiable
Melting Point °C
% hydrocarbons
% sterols
% Squalene
% Pristane

**Fatty Acid Composition (%)**
 14:0. . . . . . . . . . . . . . . . . . . . . . . . . . 1–2
 16:0. . . . . . . . . . . . . . . . . . . . . . . . 15–22
 16:1. . . . . . . . . . . . . . . . . . . . . . . . . 3–6
 18:0. . . . . . . . . . . . . . . . . . . . . . . . . 4–9
 Unassigned 18:1 . . . . . . . . . . . . . . 30–50
 18:2. . . . . . . . . . . . . . . . . . . . . . . 10–16
 18:3. . . . . . . . . . . . . . . . . . . . . . . 0.5–3
 18:4. . . . . . . . . . . . . . . . . . . . . . . 0.4–1
 20:1 . . . . . . . . . . . . . . . . . . . . . . . . . 1–2
 5c,8c,11c,14c-20:4 (n-6). . . . . . . . . . . 1–6
 6c,9c,12c,15c,17c-20:5. . . . . . . . . 0.2–2.5
 7c,10c,13c,16c,19c-22:5 (n-3) . . . 0.2–1.3
 4c,7c,10c,13c,16c,19c-22:6 . . . . . . 0.6–6

Cholesterol . . . . 1–5 (30–56 mg% in muscle)

**References** Ackman, R.G. in *Objective Methods for Food Analysis,* National Academy of Sciences, Washington, DC, 1976

## Chicken Egg Lipids, whole egg

Specific Gravity (SG)
 15.5/15.5°C
 25/25°C
 Other SG
Refractive Index (RI)
 25°C
 40°C
 Other RI
Iodine Value
Saponification Value
Titer °C
% Unsaponifiable
Melting Point °C
% hydrocarbons
% sterols
% Squalene
% Pristane

**Fatty Acid Composition (%)**
 16:0. . . . . . . . . . . . . . . . . . . . . . . . . . 2.5
 18:0. . . . . . . . . . . . . . . . . . . . . . . . . . 0.9
 Unassigned 18:1 . . . . . . . . . . . . . . . . . . 4
 18:2. . . . . . . . . . . . . . . . . . . . . . . . . . 1.2
 18:3. . . . . . . . . . . . . . . . . . . . . . . . . 0.03
 5c,8c,11c,14c-20:4 (n-6). . . . . . . . . . . 0.1

Cholesterol . . . . . . . . . . . . . 99 (5480 mg/kg)

**References** USDA, *Agriculture Handbook 8-1* (1976)

## Chicken Egg Lipids, yolk

Specific Gravity (SG)
 15.5/15.5°C
 25/25°C
 Other SG
Refractive Index (RI)
 25°C
 40°C

Other RI
Iodine Value
Saponification Value
Titer °C
% Unsaponifiable
Melting Point °C
% hydrocarbons
% sterols
% Squalene
% Pristane

**Fatty Acid Composition (%)**
    14:0 . . . . . . . . . . . . . . . . . . . . . . . . . . 0.1
    16:0 . . . . . . . . . . . . . . . . . . . . . . . . . . 7.3
    16:1 . . . . . . . . . . . . . . . . . . . . . . . . . . 1.1
    18:0 . . . . . . . . . . . . . . . . . . . . . . . . . . 2.5
    Unassigned 18:1 . . . . . . . . . . . . . . . . 12.1
    18:2 . . . . . . . . . . . . . . . . . . . . . . . . . . 3.7
    18:3 . . . . . . . . . . . . . . . . . . . . . . . . . . 0.1
    5c,8c,11c,14c-20:4 (n-6) . . . . . . . . . . 0.3

Cholesterol . . . . . . . . . . . 99 (16000 mg/kg)

**References** USDA, *Agriculture Handbook 8-1* (1976)


## Chicken Fat

Specific Gravity (SG)
    15.5/15.5°C . . . . . . . . . . . . . 0.914–0.924
    25/25°C
    Other SG
Refractive Index (RI)
    25°C
    40°C . . . . . . . . . . . . . . . . . . . 1.452–1.460
    Other RI
Iodine Value . . . . . . . . . . . . . . . . . . . . 76–80
Saponification Value
Titer °C . . . . . . . . . . . . . . . . . . . . . . . 32–36
% Unsaponifiable
Melting Point °C . . . . . . . . . . . . . . . . 30–34
% hydrocarbons
% sterols
% Squalene
% Pristane

**Fatty Acid Composition (%)**
    12:0 . . . . . . . . . . . . . . . . . . . . . . . . . . 0.1
    14:0 . . . . . . . . . . . . . . . . . . . . . . . . . . 0.9
    16:0 . . . . . . . . . . . . . . . . . . . . . . . . . . . 22
    16:1 . . . . . . . . . . . . . . . . . . . . . . . . . . . . 6
    18:0 . . . . . . . . . . . . . . . . . . . . . . . . . . . . 6
    Unassigned 18:1 . . . . . . . . . . . . . . . . . 37
    18:2 . . . . . . . . . . . . . . . . . . . . . . . . . . . 20
    18:3 . . . . . . . . . . . . . . . . . . . . . . . . . . . . 1
    20:1 . . . . . . . . . . . . . . . . . . . . . . . . . . . . 1

Cholesterol . . . . . . . . . . . . . 99 (850 mg/kg)

**References** USDA, *Agriculture Handbook 8-4* (1979)


## Chinook Salmon Lipids

Specific Gravity (SG)
    15.5/15.5°C
    25/25°C
    Other SG
Refractive Index (RI)
    25°C
    40°C
    Other RI
Iodine Value
Saponification Value
Titer °C
% Unsaponifiable
Melting Point °C
% hydrocarbons
% sterols
% Squalene
% Pristane

**Fatty Acid Composition (%)**
    14:0 . . . . . . . . . . . . . . . . . . . . . . . . 5.0–5.4
    16:0 . . . . . . . . . . . . . . . . . . . . . . 20.7–21.3
    16:1 . . . . . . . . . . . . . . . . . . . . . . . . 8.2–8.8
    18:0 . . . . . . . . . . . . . . . . . . . . . . . . 4.6–4.8
    9c-18:1 . . . . . . . . . . . . . . . . . . . . 23.2–24.3
    11c-18:1 . . . . . . . . . . . . . . . . . . . . . 5.9–6.5
    18:2 . . . . . . . . . . . . . . . . . . . . . . . . 0.7–1.3
    18:3 . . . . . . . . . . . . . . . . . . . . . . . . 0.3–1.1
    8c,11c,14c-20:3 . . . . . . . . . . . . . . . 0.1–0.2
    5c,8c,11c,14c-20:4 (n-6) . . . . . . . . 5.6–5.9
    6c,9c,12c,15c,17c-20:5 . . . . . . . . . . 7.6–8.0
    7c,10c,13c,16c,19c-22:5 (n-3) . . . 2.3–3.7
    4c,7c,10c,13c,16c,19c-22:6 . . . . . . . . 9.5
    24:0 . . . . . . . . . . . . . . . . . . . . . . . . 0–0.1
    24:1 . . . . . . . . . . . . . . . . . . . . . . . . 0.4–1.0

## Cod, Atlantic

Specific Gravity (SG)
  15.5/15.5°C
  25/25°C
  Other SG
Refractive Index (RI)
  25°C
  40°C
  Other RI
Iodine Value
Saponification Value
Titer °C
% Unsaponifiable
Melting Point °C
% hydrocarbons
% sterols
% Squalene
% Pristane

**Fatty Acid Composition (%)**
  14:0 . . . . . . . . . . . . . . . . . . . . . . . . . . . . 0.5
  16:0 . . . . . . . . . . . . . . . . . . . . . . . . . . . . . 22
  16:1 . . . . . . . . . . . . . . . . . . . . . . . . . . . . . . 2
  18:0 . . . . . . . . . . . . . . . . . . . . . . . . . . . . . . 5
  Unassigned 18:1 . . . . . . . . . . . . . . . . . 10
  18:2 . . . . . . . . . . . . . . . . . . . . . . . . . . . . . . 1
  20:1 . . . . . . . . . . . . . . . . . . . . . . . . . . . . . . 2
  20:2 . . . . . . . . . . . . . . . . . . . . . . . . . . . . . 0.8
  5c,8c,11c,14c-20:4 (n-6) . . . . . . . . . . . 1.5
  6c,9c,12c,15c,17c-20:5 . . . . . . . . . . . . 16
  22:1 . . . . . . . . . . . . . . . . . . . . . . . . . . . . . . 2
  7c,10c,13c,16c,19c-22:5 (n-3) . . . . . . 0.7
  4c,7c,10c,13c,16c,19c-22:6 . . . . . . . . . 36

**References** *J. Food Sci.* 52: 1209 (1987)

## Cod Liver Oil

Specific Gravity (SG)
  15.5/15.5°C . . . . . . . . . . . . . . 0.922–0.928
  25/25°C
  Other SG
Refractive Index (RI)
  25°C . . . . . . . . . . . . . . . . . . . 1.478–1.485
  40°C
  Other RI
Iodine Value . . . . . . . . . . . . . . . . . . . 142–176
Saponification Value . . . . . . . . . . . . 180–192
Titer °C . . . . . . . . . . . . . . . . . . . . . . . . 18–24
% Unsaponifiable . . . . . . . . . . . . . . . . . . 0–2
Melting Point °C
% hydrocarbons
% sterols
% Squalene
% Pristane

**Fatty Acid Composition (%)**
  14:0 . . . . . . . . . . . . . . . . . . . . . . . . . . . . 3–5
  15:0 . . . . . . . . . . . . . . . . . . . . . . . . . 0.3–0.5
  16:0 . . . . . . . . . . . . . . . . . . . . . . . . . . 10–14
  16:1 . . . . . . . . . . . . . . . . . . . . . . . . . . . 6–12
  16:2 . . . . . . . . . . . . . . . . . . . . . . . . . . 0.3–1
  17:0 . . . . . . . . . . . . . . . . . . . . . . . . . . 0.1–1
  18:0 . . . . . . . . . . . . . . . . . . . . . . . . . . . . 1–4
  Unassigned 18:1 . . . . . . . . . . . . . . . . 19–27
  9c-18:1 . . . . . . . . . . . . . . . . . . . . . . . 14–20
  11c-18:1 . . . . . . . . . . . . . . . . . . . . . . . . 5–7
  18:2 . . . . . . . . . . . . . . . . . . . . . . . . . . . . 1–2
  18:3 . . . . . . . . . . . . . . . . . . . . . . . . . . 0.2–1
  6c,9c,12c,15c-18:4 (n-3) . . . . . . . . 0.4–2.4
  18:4 . . . . . . . . . . . . . . . . . . . . . . . . . 0.4–2.4
  20:1 . . . . . . . . . . . . . . . . . . . . . . . . . . . 7–15
  20:2 . . . . . . . . . . . . . . . . . . . . . . . . . 0.1–0.4
  5c,8c,11c,14c-20:4 (n-6) . . . . . . . . . 0.4–2
  6c,9c,12c,15c,17c-20:5 . . . . . . . . . . . 8–14
  22:1 . . . . . . . . . . . . . . . . . . . . . . . . . . . 4–13
  22:4 . . . . . . . . . . . . . . . . . . . . . . . . . . . . 0.5
  7c, 10c, 13c, 16c-22:4 . . . . . . . . . . . . . . 0.5
  7c,10c,13c,16c,19c-22:5 (n-3) . . . . . . 1–3
  4c,7c,10c,13c,16c,19c-22:6 . . . . . . . 6–17
  24:1 . . . . . . . . . . . . . . . . . . . . . . . . . 0.2–0.7
  Other: 16:3, 0.2–0.6; 17:1, 0.1–0.6;
    8c,11,14c,17c-20:4, 0.3–0.6;
    4c,7c,10c,13c,16c-22:5, 0.4–1

**References** *J. Fisheries Res. Bd. Can.* 24: 613 (1967)
*JAOCS* 72: 575 (1995)

## Cod, Pacific

Specific Gravity (SG)
  15.5/15.5°C
  25/25°C
  Other SG

Refractive Index (RI)
  25°C
  40°C
  Other RI
Iodine Value
Saponification Value
Titer °C
% Unsaponifiable
Melting Point °C
% hydrocarbons
% sterols
% Squalene
% Pristane

**Fatty Acid Composition (%)**
  14:0 . . . . . . . . . . . . . . . . . . . . . . . . . . . 1
  16:0 . . . . . . . . . . . . . . . . . . . . . . . . . . 18
  16:1 . . . . . . . . . . . . . . . . . . . . . . . . . . . 4
  18:0 . . . . . . . . . . . . . . . . . . . . . . . . . . . 5
  Unassigned 18:1 . . . . . . . . . . . . . . . . . 13
  18:2 . . . . . . . . . . . . . . . . . . . . . . . . . . . 2
  20:1 . . . . . . . . . . . . . . . . . . . . . . . . . . . 2
  20:2 . . . . . . . . . . . . . . . . . . . . . . . . . . . 1
  20:4 . . . . . . . . . . . . . . . . . . . . . . . . . . . 1
  6c,9c,12c,15c,17c-20:5 . . . . . . . . . . . . . 16
  7c,10c,13c,16c,19c-22:5 (n-3) . . . . . . . . 2
  4c,7c,10c,13c,16c,19c-22:6 . . . . . . . . . 28

**References** *J. Food Sci. 52:* 1209 (1987)

# Crab Lipids, King

Specific Gravity (SG)
  15.5/15.5°C
  25/25°C
  Other SG
Refractive Index (RI)
  25°C
  40°C
  Other RI
Iodine Value
Saponification Value
Titer °C
% Unsaponifiable
Melting Point °C
% hydrocarbons
% sterols
% Squalene
% Pristane

**Fatty Acid Composition (%)**
  14:0 . . . . . . . . . . . . . . . . . . . . . . . . . . . 1
  16:0 . . . . . . . . . . . . . . . . . . . . . . . . . . . 9
  16:1 . . . . . . . . . . . . . . . . . . . . . . . . . . . 5
  18:0 . . . . . . . . . . . . . . . . . . . . . . . . . . . 4
  Unassigned 18:1 . . . . . . . . . . . . . . . . . 15
  18:2 . . . . . . . . . . . . . . . . . . . . . . . . . . . 3
  18:3 . . . . . . . . . . . . . . . . . . . . . . . . . . . 3
  18:4 . . . . . . . . . . . . . . . . . . . . . . . . . . . 2
  20:1 . . . . . . . . . . . . . . . . . . . . . . . . . . . 4
  5c,8c,11c,14c-20:4 (n-6) . . . . . . . . . . . 0.6
  6c,9c,12c,15c,17c-20:5 . . . . . . . . . . . . . 22
  7c,10c,13c,16c,19c-22:5 (n-3) . . . . . . . . 1
  4c,7c,10c,13c,16c,19c-22:6 . . . . . . . . . 10

**References** Ackman, R.G. in *Objective Methods for Food Analysis,* National Academy of Sciences, Washington, DC, 1976

# Crab Lipids, Queen

Specific Gravity (SG)
  15.5/15.5°C
  25/25°C
  Other SG
Refractive Index (RI)
  25°C
  40°C
  Other RI
Iodine Value
Saponification Value
Titer °C
% Unsaponifiable
Melting Point °C
% hydrocarbons
% sterols
% Squalene
% Pristane

**Fatty Acid Composition (%)**
  14:0 . . . . . . . . . . . . . . . . . . . . . . . . . . 0.5
  16:0 . . . . . . . . . . . . . . . . . . . . . . . . . . 14
  16:1 . . . . . . . . . . . . . . . . . . . . . . . . . . . 6
  18:0 . . . . . . . . . . . . . . . . . . . . . . . . . . . 2
  Unassigned 18:1 . . . . . . . . . . . . . . . . . 22
  18:2 . . . . . . . . . . . . . . . . . . . . . . . . . . . 1
  18:3 . . . . . . . . . . . . . . . . . . . . . . . . . . 0.2
  18:4 . . . . . . . . . . . . . . . . . . . . . . . . . . 0.1

20:1 .......................... 3
5c,8c,11c,14c-20:4 (n-6). ........... 4
6c,9c,12c,15c,17c-20:5. ............ 31
7c,10c,13c,16c,19c-22:5 (n-3) ....... 1
4c,7c,10c,13c,16c,19c-22:6 ......... 13

**References** Ackman, R.G. in *Objective Methods for Food Analysis,* National Academy of Sciences, Washington, DC, 1976

## Dolphin Dorsal Blubber
*Inia geoffrensis*
Specific Gravity (SG)
  15.5/15.5°C
  25/25°C
  Other SG
Refractive Index (RI)
  25°C
  40°C
  Other RI
Iodine Value
Saponification Value
Titer °C
% Unsaponifiable
Melting Point °C
% hydrocarbons
% sterols
% Squalene
% Pristane

**Fatty Acid Composition (%)**
  10:0 .......................... 0.2
  12:0 .......................... 1.4
  12:1 .......................... 0.5
  14:0 .......................... 6
  14:1 .......................... 3
  16:0 .......................... 22
  16:1 .......................... 20
  16:2 .......................... 0.4
  17:0 .......................... 0.8
  18:0 .......................... 3
  Unassigned 18:1 ................ 24
  18:2 .......................... 6
  20:1 .......................... 0.4
  20:2 .......................... 0.2
  8c,11c,14c-20:3 ................ 1

5c,8c,11c,14c-20:4 (n-6). ........... 1
6c,9c,12c,15c,17c-20:5. ............ 0.1
7c,10c,13c,16c,19c-22:5 (n-3) ...... 0.1
4c,7c,10c,13c,16c,19c-22:6 ......... 0.1
Other: 19:0, 0.1: See ref. For ISO + ANTISO content and Others, 4.5

**References** *Lipids 6:* 69 (1971)

## Dogfish, Birdbeak Muscle Oil
*Deania calceus*
Specific Gravity (SG)
  15.5/15.5°C
  25/25°C
  Other SG
Refractive Index (RI)
  25°C
  40°C
  Other RI
Iodine Value
Saponification Value
Titer °C
% Unsaponifiable
Melting Point °C
% hydrocarbons
% sterols
% Squalene
% Pristane

**Fatty Acid Composition (%)**
  14:0 .......................... 0.6
  16:0 .......................... 13
  16:1 .......................... 2
  16:2 (n-6) ..................... 0.6
  16:2 (n-4) ..................... 0.3
  16:3 (n-4) ..................... 0.4
  18:0 .......................... 5
  Unassigned 18:1 ................ 17
  18:2 .......................... 2
  18:3 .......................... 0.3
  6c,9c,12c,15c-18:4 (n-3). ....... 0.6
  20:1 (n-9) ..................... 6
  20:2 .......................... 0.4
  5c,8c,11c,14c-20:4 (n-6). ....... 3
  20:5 (n-3) ..................... 1
  22:1 .......................... 6

4c,7c,10c,13c,16c,19c-22:6 ......... 30
24:1 ........................... 0.2

**References** *JAOCS 70:* 1081 (1993)

## Dogfish, Sping Liver Oil
*Centroscyllium ritteri*
Specific Gravity (SG)
  15.5/15.5°C
  25/25°C
  Other SG ............... (15/4) 0.8875
Refractive Index (RI)
  25°C
  40°C
  Other RI .................. (20) 1.4765
Iodine Value
Saponification Value
Titer °C
% Unsaponifiable ................... 56
Melting Point °C
% hydrocarbons .................... 63
% sterols .......................... 3
% Squalene ...................... 96.8
% Pristane

**Fatty Acid Composition (%)**
  14:0 ............................. 4
  14:1 ........................... 0.7
  15:0 ........................... 0.7
  16:0 ............................ 19
  16:1 ............................. 7
  17:0 ........................... 1.5
  18:0 ............................. 5
  Unassigned 18:1 ................. 48
  18:2 ........................... 1.2
  19:0 ............................. 3
  20:0 ........................... 1.6
  20:1 ........................... 0.2

**References** *INFORM 9:* 794 (1998)

## Dogfish, Spur Liver Oil
*Squalus acanthias*
Specific Gravity (SG)
  15.5/15.5°C
  25/25°C
  Other SG ............... (15/4) 0.9191
Refractive Index (RI)
  25°C
  40°C
  Other RI .................. (20) 1.4763
Iodine Value
Saponification Value
Titer °C
% Unsaponifiable ................ 8–13
Melting Point °C
% hydrocarbons .................... 3
% sterols .......................... 8
% Squalene .................... 0.3–65
% Pristane .................... 0.2–35

**Fatty Acid Composition (%)**
  8.0 ............................ 0.1
  10:1 ........................... 0.2
  11:1 ........................... 0.1
  12:1 ........................... 0.1
  14:0 ............................. 4
  14:1 ........................... 0.8
  15:0 ......................... 0.6–1
  16:0 ......................... 7–18
  16:1 ......................... 6–17
  17:0 ........................... 0.5
  18:0 ......................... 0.4–1
  Unassigned 18:1 .............. 22–54
  18:2 ........................... 1–3
  18:3 ........................... 0.5
  18:4 ........................... 0.5
  19:0 ............................. 1
  20:1 ......................... 7–14
  20:5 ............................. 4
  21:0 ........................... 0.3
  22:1 ............................ 17
  22:4 ........................... 0.5
  22:5 ........................... 0.5
  22:6 ............................. 9

**References** *INFORM 9:* 794 (1998)
*JAOCS 46:* 554 (1969)

## Dover Sole Lipids
Specific Gravity (SG)
  15.5/15.5°C
  25/25°C

Other SG
Refractive Index (RI)
  25°C
  40°C
  Other RI
Iodine Value
Saponification Value
Titer °C
% Unsaponifiable
Melting Point °C
% hydrocarbons
% sterols
% Squalene
% Pristane

**Fatty Acid Composition (%)**
14:0........................ 4.6–5.2
16:0........................ 20–21
16:1........................ 4.3–4.7
18:0........................ 6.3–6.7
9c-18:1 .................... 8.0–8.4
11c-18:1 ................... 4.2–4.6
18:2........................ 0.2–1
18:3........................ 0–1
20:0........................ 0–1.4
20:1........................ 1.4–1.5
8c,11c,14c-20:3 ............ 0.2
5c,8c,11c,14c-20:4 (n-6).... 5.3–5.7
6c,9c,12c,15c,17c-20:5...... 16.5–17.4
7c,10c,13c,16c,19c-22:5 (n-3) ... 5.1–5.9
4c,7c,10c,13c,16c,19c-22:6 ... 17.3–18.3
24:1........................ 1–4

**References** *J. Food Comp. Anal. 4:* 128 (1991)


# Eel Lipids

Specific Gravity (SG)
  15.5/15.5°C
  25/25°C
  Other SG
Refractive Index (RI)
  25°C
  40°C
  Other RI
Iodine Value
Saponification Value
Titer °C
% Unsaponifiable
Melting Point °C
% hydrocarbons
% sterols
% Squalene
% Pristane

**Fatty Acid Composition (%)**
14:0........................ 6
16:0........................ 14
16:1........................ 12
18:0........................ 1
Unassigned 18:1 ............ 28
18:2........................ 1
20:1........................ 28
5c,8c,11c,14c-20:4 (n-6).... 0.5
6c,9c,12c,15c,17c-20:5...... 1
7c,10c,13c,16c,19c-22:5 (n-3) ... 1
4c,7c,10c,13c,16c,19c-22:6 ... 0.5

**References** Ackman, R.G. in *Objective Methods for Food Analysis,* National Academy of Sciences, Washington, DC, 1976


# Emu Oil

Specific Gravity (SG)
  15.5/15.5°C ............... 0.89–0.95
  25/25°C
  Other SG
Refractive Index (RI)
  25°C .................... 1.410–1.470
  40°C
  Other RI
Iodine Value ................. 40–80
Saponification Value ......... 175–210
Titer °C
% Unsaponifiable
Melting Point °C
% hydrocarbons
% sterols
% Squalene
% Pristane

**Fatty Acid Composition (%)**
14:0........................ 0.3–0.6
16:0........................ 19–25
16:1........................ 2–5

17:0 . . . . . . . . . . . . . . . . . . . . . . . . . . 0.1
18:0 . . . . . . . . . . . . . . . . . . . . . . . . . 8–11
Unassigned 18:1 . . . . . . . . . . . . . . 41–54
18:2 . . . . . . . . . . . . . . . . . . . . . . . . . 9–22
18:3 . . . . . . . . . . . . . . . . . . . . . . . . 0.2–1.5
20:2 . . . . . . . . . . . . . . . . . . . . . . . . . . 0.1
5c,8c,11c,14c-20:4 (n-6) . . . . . . . . . . 0.2

**References** *Ostrich News* (1997), p. 43

## Ghee (Buffalo Milk) Butter

Specific Gravity (SG)
   15.5/15.5°C
   25/25°C
   Other SG
Refractive Index (RI)
   25°C
   40°C
   Other RI
Iodine Value . . . . . . . . . . . . . . . . . . . . 28–32
Saponification Value . . . . . . . . . . . . 225–235
Titer °C
% Unsaponifiable . . . . . . . . . . . . . . . . . . . 5
Melting Point °C
% hydrocarbons
% sterols
% Squalene
% Pristane

**Fatty Acid Composition (%)**
   Other Saturates, 62–69%

**References**

## Goose Fat

Specific Gravity (SG)
   15.5/15.5°C . . . . . . . . . . . . . . 0.923–0.930
   25/25°C
   Other SG
Refractive Index (RI)
   25°C . . . . . . . . . . . . . . . . . . . 1.458–1.463
   40°C . . . . . . . . . . . . . . . . . . . 1.459–1.466
   Other RI
Iodine Value . . . . . . . . . . . . . . . . . . . . 66–73
Saponification Value . . . . . . . . . . . . 193–198
Titer °C . . . . . . . . . . . . . . . . . . . . . . . 34–41

% Unsaponifiable . . . . . . . . . . . . . . . . . . . 1
Melting Point °C . . . . . . . . . . . . . . . . 28–34
% hydrocarbons
% sterols
% Squalene
% Pristane

**Fatty Acid Composition (%)**
14:0 . . . . . . . . . . . . . . . . . . . . . . . . . . 0.5
16:0 . . . . . . . . . . . . . . . . . . . . . . . . . . . 21
16:1 . . . . . . . . . . . . . . . . . . . . . . . . . . . . 3
18:0 . . . . . . . . . . . . . . . . . . . . . . . . . . . . 6
Unassigned 18:1 . . . . . . . . . . . . . . . . . 54
18:2 . . . . . . . . . . . . . . . . . . . . . . . . . . . 10
18:3 . . . . . . . . . . . . . . . . . . . . . . . . . . 0.5
20:1 . . . . . . . . . . . . . . . . . . . . . . . . . . 0.1

Cholesterol . . . . . . . . . . . . 99 (1000 mg/kg)

**References** USDA, *Agriculture Handbook 8-4* (1979)

## Guinea Fowl Egg Fat
*Numida meleagris*

Specific Gravity (SG)
   15.5/15.5°C
   25/25°C
   Other SG
Refractive Index (RI)
   25°C
   40°C
   Other RI
Iodine Value
Saponification Value
Titer °C
% Unsaponifiable
Melting Point °C
% hydrocarbons
% sterols
% Squalene
% Pristane

**Fatty Acid Composition (%)**
14:0 . . . . . . . . . . . . . . . . . . . . . . . . 0.1–0.5
16:0 . . . . . . . . . . . . . . . . . . . . . . . . . 32–34
16:1 . . . . . . . . . . . . . . . . . . . . . . . . . . 1–2
18:0 . . . . . . . . . . . . . . . . . . . . . . . . . 15–16
Unassigned 18:1 . . . . . . . . . . . . . . . 28–29

18:2 . . . . . . . . . . . . . . . . . . . . . . . . 16–18
5c,8c,11c,14c-20:4 (n-6) . . . . . . . . . . 3–5

Cholesterol . . . (whole egg, 550–560 mg/kg; yolk, 1530–1830 mg/kg)

**References** *Food Chem. 30:* 211 (1988)

## Haddock

Specific Gravity (SG)
  15.5/15.5°C
  25/25°C
  Other SG
Refractive Index (RI)
  25°C
  40°C
  Other RI
Iodine Value
Saponification Value
Titer °C
% Unsaponifiable
Melting Point °C
% hydrocarbons
% sterols
% Squalene
% Pristane

**Fatty Acid Composition (%)**
  14:0 . . . . . . . . . . . . . . . . . . . . . . . . . . . 1
  16:0 . . . . . . . . . . . . . . . . . . . . . . . . . . . 22
  16:1 . . . . . . . . . . . . . . . . . . . . . . . . . . . 5
  18:0 . . . . . . . . . . . . . . . . . . . . . . . . . . . 5
  Unassigned 18:1 . . . . . . . . . . . . . . . . . . 14
  18:2 . . . . . . . . . . . . . . . . . . . . . . . . . . . 2
  6c,9c,12c,15c-18:4 (n-3) . . . . . . . . . . . . . 3
  20:1 . . . . . . . . . . . . . . . . . . . . . . . . . . . 3
  5c,8c,11c,14c-20:4 (n-6) . . . . . . . . . . . . . 2
  6c,9c,12c,15c,17c-20:5 . . . . . . . . . . . . . 15
  22:5 (n-3) . . . . . . . . . . . . . . . . . . . . . . . 1
  4c,7c,10c,13c,16c,19c-22:6 . . . . . . . . . 25

**References** *J. Food Sci 52:* 1209 (1987)

## Halibut, Greenland Oil
*Reinhardtius hippoglossoides*

Specific Gravity (SG)
  15.5/15.5°C
  25/25°C
  Other SG
Refractive Index (RI)
  25°C
  40°C
  Other RI
Iodine Value
Saponification Value
Titer °C
% Unsaponifiable
Melting Point °C
% hydrocarbons
% sterols
% Squalene
% Pristane

**Fatty Acid Composition (%)**
  14:0 . . . . . . . . . . . . . . . . . . . . . . . . . . . 5
  16:0 . . . . . . . . . . . . . . . . . . . . . . . . . . . 21
  16:1 . . . . . . . . . . . . . . . . . . . . . . . . . . . 10
  18:0 . . . . . . . . . . . . . . . . . . . . . . . . . . . 2
  Unassigned 18:1 . . . . . . . . . . . . . . . . . . 15
  18:2 . . . . . . . . . . . . . . . . . . . . . . . . . . . 1
  18:3 . . . . . . . . . . . . . . . . . . . . . . . . . . . 1
  6c,9c,12c,15c-18:4 (n-3) . . . . . . . . . . . . . 2
  20:1 (n-9) . . . . . . . . . . . . . . . . . . . . . . 11
  20:2 . . . . . . . . . . . . . . . . . . . . . . . . . . . 0.3
  6c,9c,12c,15c,17c-20:5 . . . . . . . . . . . . . 6
  22:1 . . . . . . . . . . . . . . . . . . . . . . . . . . . 9
  7c,10c,13c,16c,19c-22:5 (n-3) . . . . . . . . 1
  4c,7c,10c,13c,16c,19c-22:6 . . . . . . . . . . 6

**References** *JAOCS 70:* 1081 (1993)

## Herring Oil

Specific Gravity (SG)
  15.5/15.5°C
  25/25°C
  Other SG . . . . . . . . . . . . . . (20/20) 0.9162
Refractive Index (RI)
  25°C . . . . . . . . . . . . . . . . . 1.4730–1.4750
  40°C
  Other RI
Iodine Value . . . . . . . . . . . . . . . . . 115–160
Saponification Value . . . . . . . . . . . 161–192
Titer °C
% Unsaponifiable . . . . . . . . . . . . . . . 0.5–2.5
Melting Point °C

% hydrocarbons
% sterols
% Squalene
% Pristane

**Fatty Acid Composition (%)**
  14:0 . . . . . . . . . . . . . . . . . . . . . . . . 3–10
  16:0 . . . . . . . . . . . . . . . . . . . . . . . . 8–25
  16:1 . . . . . . . . . . . . . . . . . . . . . . . . 3–12
  18:0 . . . . . . . . . . . . . . . . . . . . . . . . . 1–4
  Unassigned 18:1 . . . . . . . . . . . . . . . 5–22
  18:2 . . . . . . . . . . . . . . . . . . . . . . . . 0.1–2
  18:3 . . . . . . . . . . . . . . . . . . . . . . . . . 0–2
  18:4 . . . . . . . . . . . . . . . . . . . . . . . . . 1–5
  20:1 . . . . . . . . . . . . . . . . . . . . . . . . 6–20
  20:2 . . . . . . . . . . . . . . . . . . . . . . . 0.5–0.7
  20:4 . . . . . . . . . . . . . . . . . . . . . . . 0.3–0.5
  5c,8c,11c,14c-20:4 (n-6) . . . . . . . 0.3–0.5
  20:5 . . . . . . . . . . . . . . . . . . . . . . . . 4–15
  6c,9c,12c,15c,17c-20:5 . . . . . . . . 4–15
  22:1 . . . . . . . . . . . . . . . . . . . . . . . . 4–31
  22:2 . . . . . . . . . . . . . . . . . . . . . . . . 0.4–1
  22:5 . . . . . . . . . . . . . . . . . . . . . . . 0.5–1.3
  7c,10c,13c,16c,19c-22:5 (n-3) . . . 0.5–1.3
  22:6 . . . . . . . . . . . . . . . . . . . . . . . . 2–10
  24:1 . . . . . . . . . . . . . . . . . . . . . . . 0.2–1.3
  Other: 24:5, 0–0.5

**References** *J. Am. Oil Chem. Soc. 75:* 581 (1998)
Stansby, M.E., *et al.* in *Fish Oils in Nutrition,* (Stansby, M.E., ed.) van Nostrand Reinhold, N.Y., 1990, p. 30
Enser, M. in *Analysis of Oilseeds, Fats and Fatty Foods,* (Pritchard, J.L.R., ed.) Elsevier Applied Science, New York, N.Y. 1991, p. 376.

## Horse Fat

Specific Gravity (SG)
  15.5/15.5°C . . . . . . . . . . . . . . 0.918–0.922
  25/25°C
  Other SG
Refractive Index (RI)
  25°C . . . . . . . . . . . . . . . . . . . 1.465–1.470
  40°C . . . . . . . . . . . . . . . . . . . . . . . . 1.462
  Other RI
Iodine Value . . . . . . . . . . . . . . . . . . 72–84
Saponification Value . . . . . . . . . . . 195–199
Titer °C . . . . . . . . . . . . . . . . . . . . . . 34–38
% Unsaponifiable
Melting Point °C . . . . . . . . . . . . . . . 36–48
% hydrocarbons
% sterols
% Squalene
% Pristane

**Fatty Acid Composition (%)**
  14:0 . . . . . . . . . . . . . . . . . . . . . . . . . . . 2
  14:1 . . . . . . . . . . . . . . . . . . . . . . . . . . . 1
  16:0 . . . . . . . . . . . . . . . . . . . . . . . . . . 30
  18:0 . . . . . . . . . . . . . . . . . . . . . . . . . . . 4
  Unassigned 18:1 . . . . . . . . . . . . . . . . . 33
  18:2 . . . . . . . . . . . . . . . . . . . . . . . . . . . 4
  18:3 . . . . . . . . . . . . . . . . . . . . . . . . . . 16
  20:0 . . . . . . . . . . . . . . . . . . . . . . . . . . 0.2
  Other . . . . . . . . . . . . . . . . . . . . . . . . . . 3

**References** Pearson, A.M. in *Advances in Food Research,* Vol. 23 (Chichester, C.O., ed.) Academic Press, N.Y., 1977, p. 28

## Lamb Shoulder Fat

Specific Gravity (SG)
  15.5/15.5°C
  25/25°C
  Other SG
Refractive Index (RI)
  25°C
  40°C
  Other RI
Iodine Value
Saponification Value
Titer °C
% Unsaponifiable
Melting Point °C
% hydrocarbons
% sterols
% Squalene
% Pristane

**Fatty Acid Composition (%)**
  12:0 . . . . . . . . . . . . . . . . . . . . . . . . 0–1.1
  14:0 . . . . . . . . . . . . . . . . . . . . . . . 2.1–10.4
  16:0 . . . . . . . . . . . . . . . . . . . . . . . 12.2–63.1
  16:1 . . . . . . . . . . . . . . . . . . . . . . . . 0.8–6.5

18:0. . . . . . . . . . . . . . . . . . . . . . . . . 8–42
Unassigned 18:1 . . . . . . . . . . . . . . 16–90
18:2. . . . . . . . . . . . . . . . . . . . . . . . . . 1–4
18:3. . . . . . . . . . . . . . . . . . . . . . . . . . 0–2
Other: 16:1t, 0–1.4; 18:1t, 2–8; 18:2c,t,
0–0.5

Cholesterol . . . . . . . . . . . (700–800 mg/kg)

**References** USDA, *Agriculture Handbook 8* (1989)
*J. Am. Oil Chem. Soc 75:* 1001 (1998)

## Lard (Rendered Pork Fat)

Specific Gravity (SG)
 15.5/15.5°C
 25/25°C
 Other SG . . . . . . . . . . (20/20) 0.894–0.906
Refractive Index (RI)
 25°C
 40°C
 Other RI . . . . . . . . . . . . . . . (20) 1.448–1.461
Iodine Value . . . . . . . . . . . . . . . . . . . . 45–70
Saponification Value . . . . . . . . . . . . 192–203
Titer °C . . . . . . . . . . . . . . . . . . . . . . . . 32–45
% Unsaponifiable . . . . . . . . . . . . . . . . . . 12
Melting Point °C
% hydrocarbons
% sterols
% Squalene
% Pristane

**Fatty Acid Composition (%)**
 12:0 . . . . . less than 0.5 for C12 and lower
 14:0. . . . . . . . . . . . . . . . . . . . . . . . . 0.5–2.5
 14:1. . . . . . . . . . . . . . . . . . . . . . . . . . 0–0.2
 15:0. . . . . . . . . . . . . . . . . . . . . . . . . . 0–0.1
 16:0. . . . . . . . . . . . . . . . . . . . . . . . . 20–32
 16:1. . . . . . . . . . . . . . . . . . . . . . . . . 1.7–5.0
 17:0. . . . . . . . . . . . . . . . . . . . . . . . . . 0–0.5
 18:0. . . . . . . . . . . . . . . . . . . . . . . . . . 5–24
 Unassigned 18:1 . . . . . . . . . . . . . . . 35–62
 18:2. . . . . . . . . . . . . . . . . . . . . . . . . . 3–16
 18:3. . . . . . . . . . . . . . . . . . . . . . . . . . 0–0.5
 20:0. . . . . . . . . . . . . . . . . . . . . . . . . . 0–1.0
 20:1. . . . . . . . . . . . . . . . . . . . . . . . . . 0–1.0
 20:2. . . . . . . . . . . . . . . . . . . . . . . . . . 0–1.0
 5c,8c,11c,14c-20:4 (n-6). . . . . . . . . 0–1.0

Other 15:ISO, 0–0.1; 16:ISO, 0–0.1; 22:0, 0–1.0

Cholesterol . . . . . . . . . . . . . . . . . (950 mg/kg)

**References** *Codex* Alinorm 97/17
USDA, *Agriculture Handbook 8-4* (1979)
*Rev. Franc. Corps Gras 33:* 437 (1996)

## Maasbanker Oil

*Trachurus trachurus*

Specific Gravity (SG)
 15.5/15.5°C
 25/25°C
 Other SG . . . . . . . . . . . . . . (20/20) 0.9227
Refractive Index (RI)
 25°C
 40°C
 Other RI
Iodine Value
Saponification Value . . . . . . . . . . . . . . . . 194
Titer °C
% Unsaponifiable . . . . . . . . . . . . . . . . . . 1–4
Melting Point °C
% hydrocarbons
% sterols
% Squalene
% Pristane

**Fatty Acid Composition (%)**
 14:0. . . . . . . . . . . . . . . . . . . . . . . . . . . 4–9
 15:0. . . . . . . . . . . . . . . . . . . . . . . . . . 0–0.3
 16:0. . . . . . . . . . . . . . . . . . . . . . . . . 14–22
 16:1. . . . . . . . . . . . . . . . . . . . . . . . . . . 5–9
 17:0. . . . . . . . . . . . . . . . . . . . . . . . . . . 0–1
 18:0. . . . . . . . . . . . . . . . . . . . . . . . . . . 3–6
 Unassigned 18:1 . . . . . . . . . . . . . . . . 6–13
 18:2. . . . . . . . . . . . . . . . . . . . . . . . . . . 1–2
 18:4. . . . . . . . . . . . . . . . . . . . . . . . . . . 1–2
 20:1. . . . . . . . . . . . . . . . . . . . . . . . . . . 5–9
 20:4. . . . . . . . . . . . . . . . . . . . . . . . . . . 1–2
 5c,8c,11c,14c-20:4 (n-6). . . . . . . . . . 1–2
 20:5. . . . . . . . . . . . . . . . . . . . . . . . . . 8–13
 6c,9c,12c,15c,17c-20:5. . . . . . . . . . . 8–13
 22:1. . . . . . . . . . . . . . . . . . . . . . . . . . 6–18
 22:5. . . . . . . . . . . . . . . . . . . . . . . . . . . . 2
 7c,10c,13c,16c,19c-22:5 (n-3) . . . . . . . 2

| | |
|---|---|
| 22:6 | 6–23 |
| 4c,7c,10c,13c,16c,19c-22:6 | 6–23 |
| Other | 1–11 |

**References** Enser, M. in *Analysis of Oilseeds, Fats and Fatty Foods,* (Pritchard, J.L.R., ed.) Elsevier Applied Science, New York, N.Y., 1991, p. 377.

## Mackerel Oil

Specific Gravity (SG)
 15.5/15.5°C
 25/25°C
 Other SG ..............(15/15) 0.9301
Refractive Index (RI)
 25°C
 40°C
 Other RI ................... (20) 1.4811
Iodine Value .................. 136–167
Saponification Value ........... 136–167
Titer °C
% Unsaponifiable ............... 0.4–1.4
Melting Point °C
% hydrocarbons
% sterols
% Squalene
% Pristane

**Fatty Acid Composition (%)**

| | |
|---|---|
| 14:0 | 7–8 |
| 16:0 | 13–16 |
| 16:1 | 4–9 |
| 18:0 | 2–3 |
| Unassigned 18:1 | 13–14 |
| 18:2 | 1–2 |
| 18:3 | 1–2 |
| 18:4 | 2–5 |
| 20:1 | 12 |
| 20:2 | 0.2 |
| 20:5 | 6–8 |
| 6c,9c,12c,15c,17c-20:5 | 6–8 |
| 22:1 | 14–16 |
| 22:5 | 1 |
| 7c,10c,13c,16c,19c-22:5 (n-3) | 1 |
| 22:6 | 8–9 |
| 4c,7c,10c,13c,16c,19c-22:6 | 8–9 |
| Other | 5–8 |
| Cholesterol | 0.2–0.3 |

**References** Enser, M. in *Analysis of Oilseeds, Fats and Fatty Foods,* (Pritchard, J.L.R., ed.) Elsevier Applied Science, New York, N.Y., 1991, p. 378.
Ackman, R.G. in *Objective*

## Mackerel, Atlantic Oil
*Scomber scombrus*

Specific Gravity (SG)
 15.5/15.5°C
 25/25°C
 Other SG
Refractive Index (RI)
 25°C
 40°C
 Other RI
Iodine Value
Saponification Value
Titer °C
% Unsaponifiable
Melting Point °C
% hydrocarbons
% sterols
% Squalene
% Pristane

**Fatty Acid Composition (%)**

| | |
|---|---|
| 10:0 | 0.1 |
| 12:0 | 0.3 |
| 13:0 | 0.06 |
| 14:0 | 6 |
| 14:1 | 0.06 |
| 15:0 | 0.6 |
| 16:0 | 15 |
| 16:1 | 0.4 |
| 16:1 (n-9) | 0.4 |
| 16:1 (n-5) | 0.8 |
| 16:2 (n-4) | 0.6 |
| 16:3 (n-3) | 0.2 |
| 16:3 (n-4) | 0.4 |
| 17:0 | 0.2 |
| 18:0 | 2 |
| 9c-18:1 | 10 |
| 11c-18:1 | 4 |
| 13c-18:1 | 0.8 |
| 18:2 | 1 |
| 18:2 (n-4) | 0.2 |
| 18:3 | 0.04 |

18:3 (n-3) . . . . . . . . . . . . . . . . . . . . . . . 1.4
6c,9c,12c,15c-18:4 (n-3) . . . . . . . . . . . . . 3
20:0 . . . . . . . . . . . . . . . . . . . . . . . . . . . . 0.2
20:1 (n-5) . . . . . . . . . . . . . . . . . . . . . . . 0.4
20:1 (n-7) . . . . . . . . . . . . . . . . . . . . . . . . 2
20:1 (n-9) . . . . . . . . . . . . . . . . . . . . . . . . 8
20:2 . . . . . . . . . . . . . . . . . . . . . . . . . . . . 0.2
11c,14c,17c-20:3 (n-3) . . . . . . . . . . . . 0.2
20:4 . . . . . . . . . . . . . . . . . . . . . . . . . . . . 0.8
5c,8c,11c,14c-20:4 (n-6) . . . . . . . . . . . 0.5
21:5 . . . . . . . . . . . . . . . . . . . . . . . . . . . . 0.3
22:0 . . . . . . . . . . . . . . . . . . . . . . . . . . . . 0.2
22:1 . . . . . . . . . . . . . . . . . . . . . . . . . . . . . 11
22:1 (n-7) . . . . . . . . . . . . . . . . . . . . . . . 0.7
22:1 (n-9) . . . . . . . . . . . . . . . . . . . . . . . 2.5
7c,10c,13c,16c,19c-22:5 (n-3) . . . . . . 0.6
4c,7c,10c,13c,16c,19c-22:6 . . . . . . . 10.5
24:0 . . . . . . . . . . . . . . . . . . . . . . . . . . . . 0.2
24:1 . . . . . . . . . . . . . . . . . . . . . . . . . . . . . 1

**References** *JAOCS 63:* 324 (1986)

# Menhaden Oil

Specific Gravity (SG)
  15.5/15.5°C . . . . . . . . . . . . . . 0.912–0.930
  25/25°C
  Other SG
Refractive Index (RI)
  25°C
  40°C
  Other RI . . . . . . . . . . . . . . . (65) 1.490–1.423
Iodine Value . . . . . . . . . . . . . . . . . . 150–200
Saponification Value . . . . . . . . . . . . 192–199
Titer °C
% Unsaponifiable . . . . . . . . . . . . . . . . 0.6–1.6
Melting Point °C
% hydrocarbons
% sterols
% Squalene
% Pristane

**Fatty Acid Composition (%)**
  14:0 . . . . . . . . . . . . . . . . . . . . . . . . . . 6–12
  14:1 . . . . . . . . . . . . . . . . . . . . . . . . . 0.2–0.4
  15:0 . . . . . . . . . . . . . . . . . . . . . . . . . 0.4–1.1
  16:0 . . . . . . . . . . . . . . . . . . . . . . . . . . 14–23
  16:1 . . . . . . . . . . . . . . . . . . . . . . . . . . 7–15
  16:2 . . . . . . . . . . . . . . . . . . . . . . . . . . . 1–2

17:0 . . . . . . . . . . . . . . . . . . . . . . . . . . . 0.3–2.5
18:0 . . . . . . . . . . . . . . . . . . . . . . . . . . . . . 2–4
Unassigned 18:1 . . . . . . . . . . . . . . . . . 6–16
18:2 . . . . . . . . . . . . . . . . . . . . . . . . . . . . . 1–2
18:3 . . . . . . . . . . . . . . . . . . . . . . . . . . . . . 1–2
18:4 . . . . . . . . . . . . . . . . . . . . . . . . . . . . . 1–5
20:0 . . . . . . . . . . . . . . . . . . . . . . . . . . . . . 0.2
20:1 . . . . . . . . . . . . . . . . . . . . . . . . . . . 0.5–2
20:2 . . . . . . . . . . . . . . . . . . . . . . . . . . . . . 0.2
5c,8c,11c,14c-20:4 (n-6) . . . . . . . . . . 0.9–4
6c,9c,12c,15c,17c-20:5 . . . . . . . . . . . 12–18
22:1 . . . . . . . . . . . . . . . . . . . . . . . . . . 0.1–0.4
7c,10c,13c,16c,19c-22:5 (n-3) . . . . . . 2–4
4c,7c,10c,13c,16c,19c-22:6 . . . . . . . 4–15
Other: 16:3, 1–3; 16:4, 0.5–2; 17:1,
  1.8–1.9; 21:5, 0.5–1; 22:0, 0.1

**References**

# Mullet Oil

Specific Gravity (SG)
  15.5/15.5°C
  25/25°C
  Other SG
Refractive Index (RI)
  25°C
  40°C
  Other RI
Iodine Value
Saponification Value
Titer °C
% Unsaponifiable
Melting Point °C
% hydrocarbons
% sterols
% Squalene
% Pristane

**Fatty Acid Composition (%)**
  14:0 . . . . . . . . . . . . . . . . . . . . . . . . . . 5–12
  15:0 . . . . . . . . . . . . . . . . . . . . . . . . . . 3–12
  15:1 . . . . . . . . . . . . . . . . . . . . . . . . . . . 0–1
  16:0 . . . . . . . . . . . . . . . . . . . . . . . . . . 20–34
  16:1 . . . . . . . . . . . . . . . . . . . . . . . . . . 13–29
  17:0 . . . . . . . . . . . . . . . . . . . . . . . . . . . 0–3
  17:1 . . . . . . . . . . . . . . . . . . . . . . . . . . . 2–8
  17:2 . . . . . . . . . . . . . . . . . . . . . . . . . . . 0–4
  18:0 . . . . . . . . . . . . . . . . . . . . . . . . . . . 2–5

Unassigned 18:1 . . . . . . . . . . . . . . . . . 7–14
18:2. . . . . . . . . . . . . . . . . . . . . . . . . . . 0.7–3
18:3. . . . . . . . . . . . . . . . . . . . . . . . . . . 0.3–1
6c,9c,12c,15c-18:4 (n-3). . . . . . . . . 0.7–2
18:4 (n-6) . . . . . . . . . . . . . . . . . . . . . 0.1–2
19:1. . . . . . . . . . . . . . . . . . . . . . . . . . . 0–2.5
20:0. . . . . . . . . . . . . . . . . . . . . . . . . . . . . 0–4
20:3. . . . . . . . . . . (n-3) 0.1–0.8; (n-6) 0–2
8c,11c,14c-20:3 . . . . . . . . . . . . . . . . . . 0–2
8c,11c,14c,17c-20:4 (n-3). . . . . . . 0.3–0.6
5c,8c,11c,14c-20:4 (n-6). . . . . . . . . . . 2–4
20:5. . . . . . . . . . . . . . . . . . . . . . . . . . . . 5–8
6c,9c,12c,15c,17c-20:5. . . . . . . . . . . . 5–8
22:3. . . . . . . . . . . . . . . . . . . . . . . . . . . 0–0.2
22:4. . . . . . . . . . . . . . . . . . . . . . . . . . 0.2–0.6
22:5. . . . . . . . . . . . . . . . . . . . . . . . . . . . 1–4
7c,10c,13c,16c,19c-22:5 (n-3) . . . . . . 1–4
22:6. . . . . . . . . . (n-6) 0.4–1; (n-3) 0.7–4
4c,7c,10c,13c,16c,19c-22:6 . . . . . . 0.7–4
Other. . . . . . . . . . . . . . . . . . . . . . . . . . 3–21

**References** Stansby, M.E. *et al.* in *fish oils in Nutrition*, (Stansby, M.E., ed.) Von Nostrand Reinhold, N.Y., 1990, p. 31

# Norway Pout Oil

Specific Gravity (SG)
  15.5/15.5°C
  25/25°C
  Other SG
Refractive Index (RI)
  25°C
  40°C
  Other RI
Iodine Value . . . . . . . . . . . . . . . . . . . . . 141
Saponification Value
Titer °C
% Unsaponifiable . . . . . . . . . . . . . . . . . 5–6
Melting Point °C
% hydrocarbons
% sterols
% Squalene
% Pristane

**Fatty Acid Composition (%)**
  14:0. . . . . . . . . . . . . . . . . . . . . . . . . . 4–6
  16:0. . . . . . . . . . . . . . . . . . . . . . . . . 9–17
  16:1. . . . . . . . . . . . . . . . . . . . . . . . . . 4–8

18:0. . . . . . . . . . . . . . . . . . . . . . . . . . . . . 2–3
Unassigned 18:1 . . . . . . . . . . . . . . . . 10–20
18:2. . . . . . . . . . . . . . . . . . . . . . . . . . . . . 1–2
18:3. . . . . . . . . . . . . . . . . . . . . . . . . . . . . 1–2
18:4. . . . . . . . . . . . . . . . . . . . . . . . . . . . . 2–7
20:1. . . . . . . . . . . . . . . . . . . . . . . . . . . . 8–13
20:2. . . . . . . . . . . . . . . . . . . . . . . . . . . 0.3–1
5c,8c,11c,14c-20:4 (n-6). . . . . . . . . . . 1–4
6c,9c,12c,15c,17c-20:5. . . . . . . . . . . 5–10
22:1. . . . . . . . . . . . . . . . . . . . . . . . . . . 9–15
22:2. . . . . . . . . . . . . . . . . . . . . . . . . . 0.5–1
7c,10c,13c,16c,19c-22:5 (n-3) . . . . . . 1–2
4c,7c,10c,13c,16c,19c-22:6 . . . . . 11–20
Other. . . . . . . . . . . . . . . . . . . . . . . . . . 4–18

**References**

# Orange Roughy
*Hoplostethus atlanticus*

Specific Gravity (SG)
  15.5/15.5°C
  25/25°C
  Other SG
Refractive Index (RI)
  25°C
  40°C
  Other RI
Iodine Value
Saponification Value
Titer °C
% Unsaponifiable
Melting Point °C
% hydrocarbons
% sterols
% Squalene
% Pristane

**Fatty Acid Composition (%)**
  14:0. . . . . . . . . . . . . . . . . . . . . . . . . . . . 1
  16:0. . . . . . . . . . . . . . . . . . . . . . . . . . . . 3
  16:1. . . . . . . . . . . . . . . . . . . . . . . . . . . . 7
  16:3 (n-4) . . . . . . . . . . . . . . . . . . . . . 0.5
  18:0. . . . . . . . . . . . . . . . . . . . . . . . . . . . 1
  Unassigned 18:1 . . . . . . . . . . . . . . . . . 34
  18:2. . . . . . . . . . . . . . . . . . . . . . . . . . . . 1
  18:2 (n-4) . . . . . . . . . . . . . . . . . . . . . 0.3
  18:3. . . . . . . . . . . . . . . . . . . . . . . . . . 0.7

6c,9c,12c,15c-18:4 (n-3). . . . . . . . . . . 0.6
20:1 (n-9) . . . . . . . . . . . . . . . . . . . . . . 27
20:2. . . . . . . . . . . . . . . . . . . . . . . . . . . . 1
20:3. . . . . . . . . . . . . . . . . . . . . . . . . . . 0.1
8c,11c,14c,17c-20:4 (n-3). . . . . . . . . . 0.1
5c,8c,11c,14c-20:4 (n-6). . . . . . . . . . . 0.2
6c,9c,12c,15c,17c-20:5. . . . . . . . . . . . . . 1
22:1 (n-9) . . . . . . . . . . . . . . . . . . . . . . 14
22:2 (n-6) . . . . . . . . . . . . . . . . . . . . . . . 3
22:6 (n-3) . . . . . . . . . . . . . . . . . . . . . . . 2
24:1. . . . . . . . . . . . . . . . . . . . . . . . . . . 0.5

**References**

## Oyster Lipids (American)
Specific Gravity (SG)
  15.5/15.5°C
  25/25°C
  Other SG
Refractive Index (RI)
  25°C
  40°C
  Other RI
Iodine Value
Saponification Value
Titer °C
% Unsaponifiable
Melting Point °C
% hydrocarbons
% sterols
% Squalene
% Pristane

**Fatty Acid Composition (%)**
  14:0. . . . . . . . . . . . . . . . . . . . . . . . . . . . 4
  16:0. . . . . . . . . . . . . . . . . . . . . . . . . . . 29
  16:1. . . . . . . . . . . . . . . . . . . . . . . . . . . . 4
  18:0. . . . . . . . . . . . . . . . . . . . . . . . . . . . 4
  Unassigned 18:1 . . . . . . . . . . . . . . . . . 8
  18:2. . . . . . . . . . . . . . . . . . . . . . . . . . . . 2
  18:3. . . . . . . . . . . . . . . . . . . . . . . . . . . . 3
  20:1. . . . . . . . . . . . . . . . . . . . . . . . . . . . 5
  6c,9c,12c,15c,17c-20:5. . . . . . . . . . . . . . 2
  22:1. . . . . . . . . . . . . . . . . . . . . . . . . . . 0.3
  7c,10c,13c,16c,19c-22:5 (n-3) . . . . . . 0.3
  4c,7c,10c,13c,16c,19c-22:6 . . . . . . . . . 10

**References** Ackman, R.G. in *Objective Methods for Food Analysis,* National Academy of Sciences, Washington, DC, 1976

## Oyster Lipids (European)
Specific Gravity (SG)
  15.5/15.5°C
  25/25°C
  Other SG
Refractive Index (RI)
  25°C
  40°C
  Other RI
Iodine Value
Saponification Value
Titer °C
% Unsaponifiable
Melting Point °C
% hydrocarbons
% sterols
% Squalene
% Pristane

**Fatty Acid Composition (%)**
  14:0. . . . . . . . . . . . . . . . . . . . . . . . . . . . 9
  16:0. . . . . . . . . . . . . . . . . . . . . . . . . . . 34
  16:1. . . . . . . . . . . . . . . . . . . . . . . . . . . . 6
  18:0. . . . . . . . . . . . . . . . . . . . . . . . . . . 10
  Unassigned 18:1 . . . . . . . . . . . . . . . . . 7
  18:2. . . . . . . . . . . . . . . . . . . . . . . . . . . . 1
  18:3. . . . . . . . . . . . . . . . . . . . . . . . . . . . 4
  18:4. . . . . . . . . . . . . . . . . . . . . . . . . . . . 1
  20:1. . . . . . . . . . . . . . . . . . . . . . . . . . . . 3
  5c,8c,11c,14c-20:4 (n-6). . . . . . . . . . . . . 1
  6c,9c,12c,15c,17c-20:5. . . . . . . . . . . . . . 3
  22:1. . . . . . . . . . . . . . . . . . . . . . . . . . . . 1
  7c,10c,13c,16c,19c-22:5 (n-3) . . . . . . 0.1
  4c,7c,10c,13c,16c,19c-22:6 . . . . . . . . . . 1

**References** Ackman, R.G. in *Objective Methods for Food Analysis,* National Academy of Sciences, Washington, DC, 1976

## Perch, White Oil
*Morone americanus*
Specific Gravity (SG)
  15.5/15.5°C

25/25°C
Other SG
Refractive Index (RI)
  25°C
  40°C
  Other RI
Iodine Value
Saponification Value
Titer °C
% Unsaponifiable
Melting Point °C
% hydrocarbons
% sterols
% Squalene
% Pristane

**Fatty Acid Composition (%)**
  14:0 . . . . . . . . . . . . . . . . . . . . . . . . . . . 3
  16:0 . . . . . . . . . . . . . . . . . . . . . . . . . . 19
  16:1 . . . . . . . . . . . . . . . . . . . . . . . . . . 14
  18:0 . . . . . . . . . . . . . . . . . . . . . . . . . . . 3
  Unassigned 18:1 . . . . . . . . . . . . . . . . . . 25
  18:2 . . . . . . . . . . . . . . . . . . . . . . . . . . . 4
  18:3 . . . . . . . . . . . . . . . . . . . . . . . . . . . 3
  6c,9c,12c,15c-18:4 (n-3) . . . . . . . . . . . . 2
  18:4 . . . . . . . . . . . . . . . . . . . . . . . . . . . 2
  20:1 . . . . . . . . . . . . . . . . . . . . . . . . . . . 1
  20:4 . . . . . . . . . . . . . . . . . . . . . . . . . . . 5
  8c,11c,14c,17c-20:4 (n-3) . . . . . . . . . . . 5
  20:5 . . . . . . . . . . . . . . . . . . . . . . . . . . 11
  6c,9c,12c,15c,17c-20:5 . . . . . . . . . . . . 11
  22:5 . . . . . . . . . . . . . . . . . . . . . . . . . . . 2
  7c,10c,13c,16c,19c-22:5 (n-3) . . . . . . . 2
  22:6 . . . . . . . . . . . . . . . . . . . . . . . . . . . 4
  4c,7c,10c,13c,16c,19c-22:6 . . . . . . . . . 4

**References** *JAOCS 54:* 424 (1977)

## Pike, Northern Oil
*Esox lucius*
Specific Gravity (SG)
  15.5/15.5°C
  25/25°C
  Other SG
Refractive Index (RI)
  25°C
  40°C

Other RI
Iodine Value
Saponification Value
Titer °C
% Unsaponifiable
Melting Point °C
% hydrocarbons
% sterols
% Squalene
% Pristane

**Fatty Acid Composition (%)**
  14:0 . . . . . . . . . . . . . . . . . . . . . . . . . . . 2
  16:0 . . . . . . . . . . . . . . . . . . . . . . . . . . 16
  16:1 . . . . . . . . . . . . . . . . . . . . . . . . . . . 6
  18:0 . . . . . . . . . . . . . . . . . . . . . . . . . . . 4
  Unassigned 18:1 . . . . . . . . . . . . . . . . . . 13
  18:2 . . . . . . . . . . . . . . . . . . . . . . . . . . . 4
  18:3 . . . . . . . . . . . . . . . . . . . . . . . . . . . 3
  20:4 . . . . . . . . . . . . . . . . . . . . . . . . . . . 8
  5c,8c,11c,14c-20:4 (n-6) . . . . . . . . . . . . 8
  20:5 . . . . . . . . . . . . . . . . . . . . . . . . . . . 6
  6c,9c,12c,15c,17c-20:5 . . . . . . . . . . . . . 6
  7c, 10c, 13c, 16c-22:4 . . . . . . . . . . . . . . 1
  22:5 . . . . . . . . . . . . . . . . . . . . . . . . . . . 4
  4c, 7c,10c,13c,16c-22:5 (n-6) . . . . . . . . 1
  7c,10c,13c,16c,19c-22:5 (n-3) . . . . . . . 3
  22:6 . . . . . . . . . . . . . . . . . . . . . . . . . . 31
  4c,7c,10c,13c,16c,19c-22:6 . . . . . . . . 31

**References** *JAOCS 54:* 424 (1977)

## Pompano
Specific Gravity (SG)
  15.5/15.5°C
  25/25°C
  Other SG
Refractive Index (RI)
  25°C
  40°C
  Other RI
Iodine Value
Saponification Value
Titer °C
% Unsaponifiable
Melting Point °C
% hydrocarbons

% sterols
% Squalene
% Pristane

**Fatty Acid Composition (%)**
14:0 . . . . . . . . . . . . . . . . . . . . . . . . . . 3
15:0 . . . . . . . . . . . . . . . . . . . . . . . . . 2.5
16:0 . . . . . . . . . . . . . . . . . . . . . . . . . . 14
16:1 . . . . . . . . . . . . . . . . . . . . . . . . . . . 9
18:0 . . . . . . . . . . . . . . . . . . . . . . . . . . 16
Unassigned 18:1 . . . . . . . . . . . . . . . . . 10
18:2 . . . . . . . . . . . . . . . . . . . . . . . . . . . 1
6c,9c,12c,15c-18:4 (n-3) . . . . . . . . . . 0.6
18:4 . . . . . . . . . . . . . . . . . . . . . . . . . . 0.6
20:1 . . . . . . . . . . . . . . . . . . . . . . . . . . . 3
20:4 . . . . . . . . . . . . . . . . . . . . . . . . . . . 7
5c,8c,11c,14c-20:4 (n-6) . . . . . . . . . . . . 7
20:5 . . . . . . . . . . . . . . . . . . . . . . . . . . . 4
6c,9c,12c,15c,17c-20:5 . . . . . . . . . . . . . 4
7c, 10c, 13c, 16c-22:4 . . . . . . . . . . . . . . 1
22:5 . . . . . . . . . . . . . . . . . . . . . . . . . . . 7
4c, 7c,10c,13c,16c-22:5 (n-6) . . . . . . . . 2
7c,10c,13c,16c,19c-22:5 (n-3) . . . . . . . 5
4c,7c,10c,13c,16c,19c-22:6 . . . . . . . . 20

**References** *J. Food Sci. 52:* 1209 (1987)

## Pout, Norway Oil
*Trisopterus esmarki*

Specific Gravity (SG)
 15.5/15.5°C
 25/25°C
 Other SG
Refractive Index (RI)
 25°C
 40°C
 Other RI
Iodine Value
Saponification Value
Titer °C
% Unsaponifiable
Melting Point °C
% hydrocarbons
% sterols
% Squalene
% Pristane

**Fatty Acid Composition (%)**
14:0 . . . . . . . . . . . . . . . . . . . . . . . . . . 2
16:0 . . . . . . . . . . . . . . . . . . . . . . . . . . 17
16:1 . . . . . . . . . . . . . . . . . . . . . . . . . . . 3
18:0 . . . . . . . . . . . . . . . . . . . . . . . . . . . 4
Unassigned 18:1 . . . . . . . . . . . . . . . . . 15
18:2 . . . . . . . . . . . . . . . . . . . . . . . . . . . 1
18:3 . . . . . . . . . . . . . . . . . . . . . . . . . . 0.5
6c,9c,12c,15c-18:4 (n-3) . . . . . . . . . . . . 1
20:1 . . . . . . . . . . . . . . . . . . . . . . . . . . . 3
20:2 . . . . . . . . . . . . . . . . . . . . . . . . . . 0.2
20:4 . . . . . . . . . . . . . . . . . . . . . . . . . . . 1
6c,9c,12c,15c,17c-20:5 . . . . . . . . . . . . 14
22:1 (n-9) . . . . . . . . . . . . . . . . . . . . . . . 2
7c,10c,13c,16c,19c-22:5 (n-3) . . . . . . . 1
4c,7c,10c,13c,16c,19c-22:6 . . . . . . . . 33

**References** *JAOCS 70:* 1081 (1993)

## Premier Jus (Beef/sheep tallow)

Specific Gravity (SG)
 15.5/15.5°C
 25/25°C
 Other SG . . . . . . . . . . (20/20) 0.893–0.904
Refractive Index (RI)
 25°C
 40°C
 Other RI . . . . . . . . . . . . . (20) 1.448–1.460
Iodine Value . . . . . . . . . . . . . . . . . . 32–50
Saponification Value . . . . . . . . . . . 190–202
Titer °C . . . . . . . . . . . . . . . . . . . . . . 40–49
% Unsaponifiable . . . . . . . . . . . . . . . . . . 12
Melting Point °C
% hydrocarbons
% sterols
% Squalene
% Pristane

**Fatty Acid Composition (%)**
12:0 . . . . . less than 2.5 for C12 and lower
14:0 . . . . . . . . . . . . . . . . . . . . . . . . . 2–6
14:1 . . . . . . . . . . . . . . . . . . . . . . . 0.5–1.5
15:0 . . . . . . . . . . . . . . . . . . . . . . . 0.5–1.0
16:0 . . . . . . . . . . . . . . . . . . . . . . . . 20–30
16:1 . . . . . . . . . . . . . . . . . . . . . . . . . 1–5
16:2 . . . . . . . . . . . . . . . . . . . . . . . . 0–1.0

17:0 . . . . . . . . . . . . . . . . . . . . . . . . . 0.5–2
18:0 . . . . . . . . . . . . . . . . . . . . . . . . . 6–30
Unassigned 18:1 . . . . . . . . . . . . . . . 30–45
18:2 . . . . . . . . . . . . . . . . . . . . . . . . . 1–6
18:3 . . . . . . . . . . . . . . . . . . . . . . . . . <1.5
20:0 . . . . . . . . . . . . . . . . . . . . . . . . . 0–0.5
20:1 . . . . . . . . . . . . . . . . . . . . . . . . . 0–0.5
5c,8c,11c,14c-20:4 (n-6) . . . . . . . . . 0–0.5
Other: 14:ISO, 0–0.3; 15:ISO + ANTISO, 0–1.5, 16:ISO, 0–0.5; 17:1, 0–1.0, 17: ISO + ANTISO, 0–0.5

**References** *Codex* Alinorm 97/17
*Riv. Ital. Sost. Grasse 52:* 79 (1975)
*Rev. Franc. Corps. Grac. 33:* 437 (1986)

# Rabbit Fat

Specific Gravity (SG)
  15.5/15.5°C
  25/25°C
  Other SG
Refractive Index (RI)
  25°C
  40°C
  Other RI
Iodine Value . . . . . . . . . . . . . . . . . . . . . . 72
Saponification Value
Titer °C
% Unsaponifiable
Melting Point °C
% hydrocarbons
% sterols
% Squalene
% Pristane

**Fatty Acid Composition (%)**
  14:0 . . . . . . . . . . . . . . . . . . . . . . . . . 4
  16:0 . . . . . . . . . . . . . . . . . . . . . . . . . 32
  16:1 . . . . . . . . . . . . . . . . . . . . . . . . . 6
  18:0 . . . . . . . . . . . . . . . . . . . . . . . . . 7
  Unassigned 18:1 . . . . . . . . . . . . . . . 23
  18:2 . . . . . . . . . . . . . . . . . . . . . . . . . 19
  18:3 . . . . . . . . . . . . . . . . . . . . . . . . . 2

**References** *J. Food Comp. Anal. 7:* 291 (1994)

# Ray, Starry Muscle Oil
*Raja radiata*

Specific Gravity (SG)
  15.5/15.5°C
  25/25°C
  Other SG
Refractive Index (RI)
  25°C
  40°C
  Other RI
Iodine Value
Saponification Value
Titer °C
% Unsaponifiable
Melting Point °C
% hydrocarbons
% sterols
% Squalene
% Pristane

**Fatty Acid Composition (%)**
  14:0 . . . . . . . . . . . . . . . . . . . . . . . . . 1
  16:0 . . . . . . . . . . . . . . . . . . . . . . . . . 20
  16:1 . . . . . . . . . . . . . . . . . . . . . . . . . 3
  16:2 (n-6) . . . . . . . . . . . . . . . . . . . . . 0.3
  16:2 (n-4) . . . . . . . . . . . . . . . . . . . . . 0.2
  16:3 (n-4) . . . . . . . . . . . . . . . . . . . . . 0.4
  18:0 . . . . . . . . . . . . . . . . . . . . . . . . . 5
  Unassigned 18:1 . . . . . . . . . . . . . . . 9
  11c-18:1 . . . . . . . . . . . . . . . . . . . . . . 6
  13c-18:1 . . . . . . . . . . . . . . . . . . . . . . 0.5
  18:2 . . . . . . . . . . . . . . . . . . . . . . . . . 1.6
  18:3 . . . . . . . . . . . . . . . . . . . . . . . . . 0.4
  6c,9c,12c,15c-18:4 (n-3) . . . . . . . . . . 0.4
  20:1 (n-9) . . . . . . . . . . . . . . . . . . . . . 2.5
  20:2 . . . . . . . . . . . . . . . . . . . . . . . . . 0.4
  8c,11c,14c,17c-20:4 (n-3) . . . . . . . . . 0.4
  5c,8c,11c,14c-20:4 (n-6) . . . . . . . . . . 2.7
  6c,9c,12c,15c,17c-20:5 . . . . . . . . . . . 6.6
  22:1 (n-9) . . . . . . . . . . . . . . . . . . . . . 0.6
  22:2 . . . . . . . . . . . . . . . . . . . . . . . . . 0.2
  7c,10c,13c,16c,19c-22:5 (n-3) . . . . . . . 2
  4c,7c,10c,13c,16c,19c-22:6 . . . . . . . . 26

**References** *JAOCS 70:* 1081 (1993)

## Redfish Oil
*Sebastes marinus*
Specific Gravity (SG)
  15.5/15.5°C
  25/25°C
  Other SG
Refractive Index (RI)
  25°C
  40°C
  Other RI
Iodine Value
Saponification Value
Titer °C
% Unsaponifiable
Melting Point °C
% hydrocarbons
% sterols
% Squalene
% Pristane

**Fatty Acid Composition (%)**
  14:0 . . . . . . . . . . . . . . . . . . . . . . . . . . . 4–6
  16:0 . . . . . . . . . . . . . . . . . . . . . . . . . . 10–14
  16:1 . . . . . . . . . . . . . . . . . . . . . . . . . . . 7–14
  18:0 . . . . . . . . . . . . . . . . . . . . . . . . . . . . 1–3
  Unassigned 18:1 . . . . . . . . . . . . . . . 17–22
  18:2 . . . . . . . . . . . . . . . . . . . . . . . . . . . 0.6–2
  18:3 . . . . . . . . . . . . . . . . . . . . . . . . . . . 0.2–1
  18:4 . . . . . . . . . . . . . . . . . . . . . . . . . . . . 1–3
  20:1 . . . . . . . . . . . . . . . . . . . . . . . . . . . 11–20
  5c,8c,11c,14c-20:4 (n-3) . . . . . . . . 0.1–0.5
  6c,9c,12c,15c,17c-20:5 . . . . . . . . . . 5–10
  7c,10c,13c,16c,19c-22:5 (n-3) . . . . 0.1–1
  4c,7c,10c,13c,16c,19c-22:6 . . . . . . . . 2–6

**References** Ackman, R.G. in *Objective Methods for Food Analysis,* National Academy of Sciences, Washington, DC, 1976

## Sablefish Lipids
Specific Gravity (SG)
  15.5/15.5°C
  25/25°C
  Other SG
Refractive Index (RI)
  25°C
  40°C
  Other RI
Iodine Value
Saponification Value
Titer °C
% Unsaponifiable
Melting Point °C
% hydrocarbons
% sterols
% Squalene
% Pristane

**Fatty Acid Composition (%)**
  14:0 . . . . . . . . . . . . . . . . . . . . . . . . . 5.2–5.7
  16:0 . . . . . . . . . . . . . . . . . . . . . . . 21.4–22.2
  16:1 . . . . . . . . . . . . . . . . . . . . . . . . . 8.7–10.1
  18:0 . . . . . . . . . . . . . . . . . . . . . . . . . . 3.7–3.9
  9c-18:1 . . . . . . . . . . . . . . . . . . . . . 22.4–23.4
  11c-18:1 . . . . . . . . . . . . . . . . . . . . . . 8.3–9.1
  18:2 . . . . . . . . . . . . . . . . . . . . . . . . . . . . 0.6
  18:3 . . . . . . . . . . . . . . . . . . . . . . . . . . . . 0–1
  20:0 . . . . . . . . . . . . . . . . . . . . . . . . . . 0.2–0.3
  20:1 . . . . . . . . . . . . . . . . . . . . . . . . . . 1.4–1.5
  8c,11c,14c-20:3 . . . . . . . . . . . . . . . . . . 0.2
  5c,8c,11c,14c-20:4 (n-6) . . . . . . . . 5.3–5.7
  6c,9c,12c,15c,17c-20:5 . . . . . . . 16.5–17.3
  7c,10c,13c,16c,19c-22:5 (n-3) . . . . . . 5–6
  4c,7c,10c,13c,16c,19c-22:6 . . . . . . 17–18
  24:1 . . . . . . . . . . . . . . . . . . . . . . . . . . . . 1–4

**References** *J. Food Comp. Anal. 4:* 128 (1991)

## Salmon, Oil
Specific Gravity (SG)
  15.5/15.5°C
  25/25°C
  Other SG . . . . . . . . . . (15/15) 0.924–0.926
Refractive Index (RI)
  25°C . . . . . . . . . . . . . . . . . . . 1.472–1.477
  40°C
  Other RI
Iodine Value . . . . . . . . . . . . . . . . . . 130–160
Saponification Value . . . . . . . . . . . . 183–186
Titer °C
% Unsaponifiable . . . . . . . . . . . . . . . . . 8–12
Melting Point °C
% hydrocarbons

% sterols
% Squalene
% Pristane

**Fatty Acid Composition (%)**
14:0 . . . . . . . . . . . . . . . . . . . . . . . . . . . 3.3
16:0 . . . . . . . . . . . . . . . . . . . . . . . . . . . 9.8
16:1 . . . . . . . . . . . . . . . . . . . . . . . . . . . 4.8
18:0 . . . . . . . . . . . . . . . . . . . . . . . . . . . 4.2
Unassigned 18:1 . . . . . . . . . . . . . . . . . . 17
18:2 . . . . . . . . . . . . . . . . . . . . . . . . . . . 1.5
18:3 . . . . . . . . . . . . . . . . . . . . . . . . . . . 1.1
18:4 . . . . . . . . . . . . . . . . . . . . . . . . . . . 2.8
20:1 . . . . . . . . . . . . . . . . . . . . . . . . . . . 3.9
5c,8c,11c,14c-20:4 (n-6) . . . . . . . . . . . 0.7
6c,9c,12c,15c,17c-20:5 . . . . . . . . . . . . 13
4c,7c,10c,13c,16c,19c-22:6 . . . . . . . 18.2

**References**

# Salmon, Atlantic Oil (whole body caught in wild, Canada)

*Salmo salar*

Specific Gravity (SG)
  15.5/15.5°C
  25/25°C
  Other SG
Refractive Index (RI)
  25°C
  40°C
  Other RI
Iodine Value
Saponification Value
Titer °C
% Unsaponifiable
Melting Point °C
% hydrocarbons
% sterols
% Squalene
% Pristane

**Fatty Acid Composition (%)**
14:0 . . . . . . . . . . . . . . . . . . . . . . . . . . . . 2
16:0 . . . . . . . . . . . . . . . . . . . . . . . . . . . 14
16:1 . . . . . . . . . . . . . . . . . . . . . . . . . . . . 6
18:0 . . . . . . . . . . . . . . . . . . . . . . . . . . . . 5
Unassigned 18:1 . . . . . . . . . . . . . . . . . . 13
18:2 . . . . . . . . . . . . . . . . . . . . . . . . . . . . 3

18:3 . . . . . . . . . . . . . . . . . . . . . . . . . . . . 2
20:1 . . . . . . . . . . . . . . . . . . . . . . . . . . . . 1
5c,8c,11c,14c-20:4 (n-6) . . . . . . . . . . . . 8
6c,9c,12c,15c,17c-20:5 . . . . . . . . . . . . . 5
22:1 . . . . . . . . . . . . . . . . . . . . . . . 0.2–11
7c, 10c, 13c, 16c-22:4 . . . . . . . . . . . . . . 1
4c, 7c,10c,13c,16c-22:5 (n-6) . . . . . . . . 2
7c,10c,13c,16c,19c-22:5 (n-3) . . . . . . . 3
4c,7c,10c,13c,16c,19c-22:6 . . . . . . . . 15

**References** *Lipids 21:* 117 (1986)
*Prog. Lipid Res. 26:* 281 (1987)

# Salmon, Atlantic Oil (muscle, Iceland)

*Salmo salar*

Specific Gravity (SG)
  15.5/15.5°C
  25/25°C
  Other SG
Refractive Index (RI)
  25°C
  40°C
  Other RI
Iodine Value
Saponification Value
Titer °C
% Unsaponifiable
Melting Point °C
% hydrocarbons
% sterols
% Squalene
% Pristane

**Fatty Acid Composition (%)**
14:0 . . . . . . . . . . . . . . . . . . . . . . . . . . . . 5
16:0 . . . . . . . . . . . . . . . . . . . . . . . . . . . 14
16:1 . . . . . . . . . . . . . . . . . . . . . . . . . . . . 8
18:0 . . . . . . . . . . . . . . . . . . . . . . . . . . . . 2
Unassigned 18:1 . . . . . . . . . . . . . . . . . . 19
11c-18:1 . . . . . . . . . . . . . . . . . . . . . . 3–4
13c-18:1 . . . . . . . . . . . . . . . . . . . . . . . . 1
18:2 . . . . . . . . . . . . . . . . . . . . . . . . . . . . 3
18:2 (n-4) . . . . . . . . . . . . . . . . . . . . . . 0.4
18:3 . . . . . . . . . . . . . . . . . . . . . . . . . . . . 1
18:4 . . . . . . . . . . . . . . . . . . . . . . . . . . . . 1
20:1 (n-9) . . . . . . . . . . . . . . . . . . . . . . . 11

20:2................................ 0.5
8c,11c,14c,17c-20:4 (n-3).......... 0.9
5c,8c,11c,14c-20:4 (n-6)........... 0.3
6c,9c,12c,15c,17c-20:5............. 4
22:1 (n-9) ........................ 7
7c,10c,13c,16c,19c-22:5 (n-3) ....... 1
4c,7c,10c,13c,16c,19c-22:6 .......... 5

**References** *JAOCS 70:* 1081 (1993)

## Sandeel Oil

Specific Gravity (SG)
  15.5/15.5°C
  25/25°C
  Other SG
Refractive Index (RI)
  25°C
  40°C
  Other RI
Iodine Value .................. 150–190
Saponification Value ............ 180–190
Titer °C
% Unsaponifiable .................. 1–6
Melting Point °C
% hydrocarbons
% sterols
% Squalene
% Pristane

**Fatty Acid Composition (%)**
  14:0............................. 6–7
  15:0............................. 1–2
  16:0............................ 10–19
  16:1............................. 5–10
  18:0............................. 1–3
  Unassigned 18:1 ................. 6–12
  18:2............................. 1.5–3
  18:3............................. 1–2
  18:4............................. 5
  20:1............................ 12–21
  20:2............................. 0.3
  5c,8c,11c,14c-20:4 (n-6).......... 0.5
  6c,9c,12c,15c,17c-20:5........... 7–11
  22:1............................ 11–22
  7c,10c,13c,16c,19c-22:5 (n-3) .... 0.6–1
  4c,7c,10c,13c,16c,19c-22:6 ....... 6–14
  Other........................... 1–17

**References**

## Sardine, Pilchard Oil

Specific Gravity (SG)
  15.5/15.5°C
  25/25°C................ 0.914–0.921
  Other SG
Refractive Index (RI)
  25°C
  40°C
  Other RI ............(65) 1.4634–1.4648
Iodine Value .................. 159–192
Saponification Value ............ 188–199
Titer °C
% Unsaponifiable ................ 0.1–1.3
Melting Point °C
% hydrocarbons
% sterols
% Squalene
% Pristane

**Fatty Acid Composition (%)**
  14:0............................ 4–12
  15:0............................ 0–0.6
  16:0............................ 9–22
  16:1............................ 6–13
  17:0............................ 0–1
  18:0............................ 2–7
  Unassigned 18:1 ................ 7–17
  18:2............................ 1–3
  18:3............................ 0.4–1
  18:4............................ 2–3
  20:1............................ 1–8
  5c,8c,11c,14c-20:4 (n-6).......... 1–3
  6c,9c,12c,15c,17c-20:5........... 9–35
  22:1............................ 1–8
  7c,10c,13c,16c,19c-22:5 (n-3) ...... 1–4
  4c,7c,10c,13c,16c,19c-22:6 ....... 4–13
  Other........................... 1–14

**References**

## Seal Blubber Oil, Harp

Specific Gravity (SG)
  15.5/15.5°C
  25/25°C
  Other SG
Refractive Index (RI)
  25°C
  40°C

Other RI
Iodine Value
Saponification Value
Titer °C
% Unsaponifiable
Melting Point °C
% hydrocarbons
% sterols
% Squalene
% Pristane

**Fatty Acid Composition (%)**
14:0 .......................... 3.7
14:1 .......................... 1.1
15:0 .......................... 0.2
16:0 ............................ 6
16:1 ........................... 18
17:0 .......................... 0.9
18:0 .......................... 0.9
9c-18:1 ...................... 20.8
11c-18:1 ...................... 5.2
18:2 .......................... 1.5
18:3 .......................... 0.6
18:4 ............................ 1
20:0 .......................... 0.1
20:1 ......................... 12.2
20:2 .......................... 0.2
8c,11c,14c-20:3 .............. 0.1
5c,8c,11c,14c-20:4 (n-6) ..... 0.5
6c,9c,12c,15c,17c-20:5 ....... 6.4
22:1 ............................ 2
7c,10c,13c,16c,19c-22:5 (n-3) ...... 4.7
4c,7c,10c,13c,16c,19c-22:6 ........ 7.6
Other: 17:1, 0.6; 22:4, 0.1

**References** *Lipids 30:* 1111 (1995)

## Seal, Antarctic Fur Seal
*Arctocephalus gazella*
Specific Gravity (SG)
   15.5/15.5°C
   25/25°C
   Other SG
Refractive Index (RI)
   25°C
   40°C
   Other RI
Iodine Value

Saponification Value
Titer °C
% Unsaponifiable
Melting Point °C
% hydrocarbons
% sterols
% Squalene
% Pristane

**Fatty Acid Composition (%)**
14:0 .......................... 3–6
16:0 ......................... 18–19
16:1 ......................... 9–11
17:1 ............................ 1
18:0 ......................... 1.7–2
Unassigned 18:1 ............. 32–37
18:2 ........................ 1.5–1.7
18:3 ........................ 0.4–0.7
20:1 ......................... 2.7–5
20:4 ........................ 0.4–0.7
20:5 ......................... 7–12
22:1 ........................ 0.6–1.3
22:5 ......................... 2–2.4
22:6 .......................... 5–8

**References** *Lipids 27:* 637 (1992)

## Seal Oil, Harp
Specific Gravity (SG)
   15.5/15.5°C
   25/25°C
   Other SG
Refractive Index (RI)
   25°C
   40°C
   Other RI
Iodine Value
Saponification Value
Titer °C
% Unsaponifiable
Melting Point °C
% hydrocarbons
% sterols
% Squalene
% Pristane

**Fatty Acid Composition (%)**
14:0 ......................... 3.5–5
15:0 .......................... 0.2

16:0 . . . . . . . . . . . . . . . . . . . . . . . . . . . 2–6
16:1 . . . . . . . . . . . . . . . . . . . . . . . . . 12–18
17:0 . . . . . . . . . . . . . . . . . . . . . . . . . . . 0.9
18:0 . . . . . . . . . . . . . . . . . . . . . . . . . 0.9–1
Unassigned 18:1 . . . . . . . . . . . . . . . 20–26
18:2 . . . . . . . . . . . . . . . . . . . . . . . . 1.4–1.5
18:3 . . . . . . . . . . . . . . . . . . . . . . . . 0.6–1.3
18:4 . . . . . . . . . . . . . . . . . . . . . . . . . 1–3.1
20:0 . . . . . . . . . . . . . . . . . . . . . . . . 0.1–0.2
20:1 . . . . . . . . . . . . . . . . . . . . . . . . . . 0–12
20:2 . . . . . . . . . . . . . . . . . . . . . . . . . . . 0.2
5c,8c,11c,14c-20:4 (n-6) . . . . . . . . 0.3–0.6
6c,9c,12c,15c,17c-20:5 . . . . . . . . . 6.4–6.8
7c,10c,13c,16c,19c-22:5 (n-3) . . . 3.7–4.7
4c,7c,10c,13c,16c,19c-22:6 . . . . 7.6–11.1
Other . . . . . . . . . . . . . . . . . . . . . . . . . 1.4

**References**  *J. Am. Oil Chem. Soc. 75:* 945 (1998)

## Seal Skin Oil

Specific Gravity (SG)
 15.5/15.5°C
 25/25°C
 Other SG . . . . . . . . . . . . . . . (20/20) 0.938
Refractive Index (RI)
 25°C
 40°C
 Other RI
Iodine Value . . . . . . . . . . . . . . . . . . . . 58–59
Saponification Value . . . . . . . . . . . . . . . . 180
Titer °C
% Unsaponifiable . . . . . . . . . . . . . . . . . . . 0.4
Melting Point °C
% hydrocarbons
% sterols
% Squalene . . . . . . . . . . . . . . . . . . . . . 36–80
% Pristane

**Fatty Acid Composition (%)**
 14:0 . . . . . . . . . . . . . . . . . . . . . . . . . . . . . 5
 16:0 . . . . . . . . . . . . . . . . . . . . . . . . . . . 7.4
 16:1 . . . . . . . . . . . . . . . . . . . . . . . . . . 19.3
 Unassigned 18:1 . . . . . . . . . . . . . . . . 27.3
 18:2 . . . . . . . . . . . . . . . . . . . . . . . . . . . 2.9
 18:3 . . . . . . . . . . . . . . . . . . . . . . . . . . . 1.3
 20:1 . . . . . . . . . . . . . . . . . . . . . . . . . . 13.6
 20:2 . . . . . . . . . . . . . . . . . . . . . . . . . . . 4.4

 6c,9c,12c,15c,17c-20:5 . . . . . . . . . . . . 6.3
 22:1 . . . . . . . . . . . . . . . . . . . . . . . . . . . . . 2
 7c,10c,13c,16c,19c-22:5 (n-3) . . . . . . 3.3
 4c,7c,10c,13c,16c,19c-22:6 . . . . . . . . 7.1

**References**  *J. Am. Oil Chem. Soc. 75:* 1015 (1998)

## Shark Liver Oil

Specific Gravity (SG)
 15.5/15.5°C
 25/25°C . . . . . . . . . . . . . . . . . 0.917–0.923
 Other SG
Refractive Index (RI)
 25°C . . . . . . . . . . . . . . . . . . . . 1.473–1.478
 40°C
 Other RI
Iodine Value . . . . . . . . . . . . . . . . . . . 150–300
Saponification Value . . . . . . . . . . . . 170–190
Titer °C
% Unsaponifiable
Melting Point °C
% hydrocarbons
% sterols
% Squalene
% Pristane

**Fatty Acid Composition (%)**
 14:0 . . . . . . . . . . . . . . . . . . . . . . . . . . . . . 2
 16:0 . . . . . . . . . . . . . . . . . . . . . . . . . . . . 21
 16:1 . . . . . . . . . . . . . . . . . . . . . . . . . . . . . 8
 18:0 . . . . . . . . . . . . . . . . . . . . . . . . . . . . . 2
 Unassigned 18:1 . . . . . . . . . . . . . . . . . . 45
 20:1 . . . . . . . . . . . . . . . . . . . . . . . . . . . . 12
 22:1 . . . . . . . . . . . . . . . . . . . . . . . . . . . . . 9
 Other . . . . . . . . . . . . . . . . . . . . . . . . . . . . 2

Cholesterol . . . . . . . . . . (400–1200 mg/kg)

**References**  *J. Am. Oil Chem. Soc. 74:* 497 (1997)
 *Chromatographia 39:* 329 (1994)

## Sheep Fat (subcutaneous)

Specific Gravity (SG)
 15.5/15.5°C
 25/25°C

Other SG
Refractive Index (RI)
  25°C
  40°C
  Other RI
Iodine Value
Saponification Value
Titer °C
% Unsaponifiable
Melting Point °C
% hydrocarbons
% sterols
% Squalene
% Pristane

**Fatty Acid Composition (%)**
10:0 . . . . . . . . . . . . . . . . . . . . . . . . . 0.4
12:0 . . . . . . . . . . . . . . . . . . . . . . . . . 0.4
14:0 . . . . . . . . . . . . . . . . . . . . . . . . . . 5
15:0 . . . . . . . . . . . . . . . . . . . . . . . . . 0.6
16:0 . . . . . . . . . . . . . . . . . . . . . . . . . 25
16:1 . . . . . . . . . . . . . . . . . . . . . . . . . 1.6
17:0 . . . . . . . . . . . . . . . . . . . . . . . . . 1.0
18:0 . . . . . . . . . . . . . . . . . . . . . . . . . 23
Unassigned 18:1 . . . . . . . . . . . . . . . . . . 38
18:2 . . . . . . . . . . . . . . . . . . . . . . . . . . 1
20:1 . . . . . . . . . . . . . . . . . . . . . . . . . 0.2
Other: 16:1t, 0.8

**References** Pearson, A.M. in *Advances in Food Research*, Vol. 23 (Chichester, C.O., ed.) Academic Press, N.Y., 1977, p. 28

# Shrimp, Alaska

Specific Gravity (SG)
  15.5/15.5°C
  25/25°C
  Other SG
Refractive Index (RI)
  25°C
  40°C
  Other RI
Iodine Value
Saponification Value
Titer °C
% Unsaponifiable
Melting Point °C
% hydrocarbons
% sterols
% Squalene
% Pristane

**Fatty Acid Composition (%)**
14:0 . . . . . . . . . . . . . . . . . . . . . . . . . 2.5
16:0 . . . . . . . . . . . . . . . . . . . . . . . . . 16
16:1 . . . . . . . . . . . . . . . . . . . . . . . . . . 6
18:0 . . . . . . . . . . . . . . . . . . . . . . . . . 2.6
Unassigned 18:1 . . . . . . . . . . . . . . . . . . 19
18:2 . . . . . . . . . . . . . . . . . . . . . . . . . 1.5
18:3 . . . . . . . . . . . . . . . . . . . . . . . . . 1.4
18:4 . . . . . . . . . . . . . . . . . . . . . . . . . . 1
20:1 . . . . . . . . . . . . . . . . . . . . . . . . . 2.4
5c,8c,11c,14c-20:4 (n-6) . . . . . . . . . . . 0.4
6c,9c,12c,15c,17c-20:5 . . . . . . . . . . . . 22
22:1 . . . . . . . . . . . . . . . . . . . . . . . . . 1.6
7c,10c,13c,16c,19c-22:5 (n-3) . . . . . . . . 1
4c,7c,10c,13c,16c,19c-22:6 . . . . . . . . . 16

**References** Ackman, R.G. in *Objective Methods for Food Analysis,* National Academy of Sciences, Washington, DC, 1976

# Shrimp

*Pennaeus spp.*

Specific Gravity (SG)
  15.5/15.5°C
  25/25°C
  Other SG
Refractive Index (RI)
  25°C
  40°C
  Other RI
Iodine Value
Saponification Value
Titer °C
% Unsaponifiable
Melting Point °C
% hydrocarbons
% sterols
% Squalene
% Pristane

**Fatty Acid Composition (%)**
14:0 . . . . . . . . . . . . . . . . . . . . . . . . . 1–2
16:0 . . . . . . . . . . . . . . . . . . . . . . . . . 13–16

16:1 .......................... 5–7
18:0 .......................... 7–8
9c-18:1 ....................... 6–8
11c-18:1 ...................... 3
18:2 .......................... 1–3
5c,8c,11c,14c-20:4 (n-6) ....... 6–7
6c,9c,12c,15c,17c-20:5 ........ 17–22
7c,10c,13c,16c,19c-22:5 (n-3) .. 2
4c,7c,10c,13c,16c,19c-22:6 .... 13–15

Cholesterol .......... (1500–1600 mg/kg)

References *J. Food Sci. 54:* 237 (1989)

## Shrimp Ecuador White
*Pennaeus vannanei*
Specific Gravity (SG)
 15.5/15.5°C
 25/25°C
 Other SG
Refractive Index (RI)
 25°C
 40°C
 Other RI
Iodine Value
Saponification Value
Titer °C
% Unsaponifiable
Melting Point °C
% hydrocarbons
% sterols
% Squalene
% Pristane

**Fatty Acid Composition (%)**
 14:0 .......................... 5
 16:0 .......................... 18
 16:1 .......................... 2
 18:0 .......................... 7
 9c-18:1 ....................... 12
 11c-18:1 ...................... 3
 18:2 .......................... 9
 5c,8c,11c,14c-20:4 (n-6) ....... 6
 6c,9c,12c,15c,17c-20:5 ........ 16
 7c,10c,13c,16c,19c-22:5 (n-3) .. 1
 4c,7c,10c,13c,16c,19c-22:6 .... 12

Cholesterol .......... (1370–1690 mg/kg)

References *J. Food Sci. 54:* 237 (1989)

## Shrimp Louisiana Brown
*Pennaeus aztecus aztecus*
Specific Gravity (SG)
 15.5/15.5°C
 25/25°C
 Other SG
Refractive Index (RI)
 25°C
 40°C
 Other RI
Iodine Value
Saponification Value
Titer °C
% Unsaponifiable
Melting Point °C
% hydrocarbons
% sterols
% Squalene
% Pristane

**Fatty Acid Composition (%)**
 14:0 .......................... 1
 16:0 .......................... 16
 16:1 .......................... 6
 18:0 .......................... 8
 9c-18:1 ....................... 8
 11c-18:1 ...................... 3
 18:2 .......................... 3
 5c,8c,11c,14c-20:4 (n-6) ....... 7
 6c,9c,12c,15c,17c-20:5 ........ 17
 7c,10c,13c,16c,19c-22:5 (n-3) .. 2
 4c,7c,10c,13c,16c,19c-22:6 .... 15

Cholesterol .......... (1560–1620 mg/kg)

References *J. Food Sci. 54:* 237 (1989)

## Smelt, American Oil
## (Fillets, Cayuga Lake, NY)
*Osmerus mordax*
Specific Gravity (SG)
 15.5/15.5°C
 25/25°C

Other SG
Refractive Index (RI)
  25°C
  40°C
  Other RI
Iodine Value
Saponification Value
Titer °C
% Unsaponifiable
Melting Point °C
% hydrocarbons
% sterols
% Squalene
% Pristane

**Fatty Acid Composition (%)**
  14:0. . . . . . . . . . . . . . . . . . . . . . . . . . . . 5
  16:0. . . . . . . . . . . . . . . . . . . . . . . . . . . 14
  16:1. . . . . . . . . . . . . . . . . . . . . . . . . . . . 9
  18:0. . . . . . . . . . . . . . . . . . . . . . . . . . . . 1
  Unassigned 18:1 . . . . . . . . . . . . . . . . . 18
  18:2. . . . . . . . . . . . . . . . . . . . . . . . . . . . 4
  18:3. . . . . . . . . . . . . . . . . . . . . . . . . . . . 5
  6c,9c,12c,15c-18:4 (n-3). . . . . . . . . . 1.7
  5c,8c,11c,14c-20:4 (n-6). . . . . . . . . . 3.5
  6c,9c,12c,15c,17c-20:5. . . . . . . . . . . 13
  4c, 7c,10c,13c,16c-22:5 (n-6). . . . . . . . 1
  4c,7c,10c,13c,16c,19c-22:6 . . . . . . . . 23

**References** *JAOCS 54:* 424 (1977)

## Smelt, Greater Silver Oil
*Argentina silus*

Specific Gravity (SG)
  15.5/15.5°C
  25/25°C
  Other SG
Refractive Index (RI)
  25°C
  40°C
  Other RI
Iodine Value
Saponification Value
Titer °C
% Unsaponifiable
Melting Point °C
% hydrocarbons
% sterols
% Squalene
% Pristane

**Fatty Acid Composition (%)**
  14:0. . . . . . . . . . . . . . . . . . . . . . . . . . . . 5
  16:0. . . . . . . . . . . . . . . . . . . . . . . . . . . 15
  16:1. . . . . . . . . . . . . . . . . . . . . . . . . . . . 6
  18:0. . . . . . . . . . . . . . . . . . . . . . . . . . . . 2
  Unassigned 18:1 . . . . . . . . . . . . . . 16–17
  18:2. . . . . . . . . . . . . . . . . . . . . . . . . . . . 1
  18:3. . . . . . . . . . . . . . . . . . . . . . . . . . . . 1
  6c,9c,12c,15c-18:4 (n-3). . . . . . . . . . . . 2
  20:1 (n-9) . . . . . . . . . . . . . . . . . . . . . . 10
  20:2. . . . . . . . . . . . . . . . . . . . . . . . . . 0.5
  20:3. . . . . . . . . . . . . . . . . . . . . . . . 0–0.2
  6c,9c,12c,15c,17c-20:5. . . . . . . . . . . . 5
  22:1 (n-9) . . . . . . . . . . . . . . . . . . . . . . 14
  7c,10c,13c,16c,19c-22:5 (n-3) . . . . . . . 1
  4c,7c,10c,13c,16c,19c-22:6 . . . . . . 9–10

**References** *JAOCS 70:* 1081 (1993)

## Snapper, Red Oil (fillet)

Specific Gravity (SG)
  15.5/15.5°C
  25/25°C
  Other SG
Refractive Index (RI)
  25°C
  40°C
  Other RI
Iodine Value
Saponification Value
Titer °C
% Unsaponifiable
Melting Point °C
% hydrocarbons
% sterols
% Squalene
% Pristane

**Fatty Acid Composition (%)**
  14:0. . . . . . . . . . . . . . . . . . . . . . . . . . 2.5
  15:0. . . . . . . . . . . . . . . . . . . . . . . . . . 1.5
  16:0. . . . . . . . . . . . . . . . . . . . . . . . . . . 14
  16:1. . . . . . . . . . . . . . . . . . . . . . . . . . . . 4
  18:0. . . . . . . . . . . . . . . . . . . . . . . . . . . 11

Unassigned 18:1 .................. 18
18:2 ........................... 1–2
20:1 ............................. 1
5c,8c,11c,14c-20:4 (n-6). . . . . . . . . . . 3.6
6c,9c,12c,15c,17c-20:5. . . . . . . . . . . . . 6
7c,10c,13c,16c,19c-22:5 (n-3) . . . . . . . . 4
4c,7c,10c,13c,16c,19c-22:6 . . . . . . . . . 24

**References** *J. Food Sci 52:* 1209 (1987)

## Sole, Lemmon Oil
*Microstomus kitt*
Specific Gravity (SG)
  15.5/15.5°C
  25/25°C
  Other SG
Refractive Index (RI)
  25°C
  40°C
  Other RI
Iodine Value
Saponification Value
Titer °C
% Unsaponifiable
Melting Point °C
% hydrocarbons
% sterols
% Squalene
% Pristane

**Fatty Acid Composition (%)**
  14:0 ........................... 2
  16:0 .......................... 14
  16:1 ........................... 4
  16:2 (n-6) ..................... 0.6
  16:2 (n-4) ..................... 0.4
  16:3 (n-4) ..................... 0.5
  16:4 (n-1) ..................... 0.2
  18:0 ........................... 3
  9c-18:1 ........................ 4
  11c-18:1 ....................... 5
  18:2 ........................... 0.5
  18:2 (n-4) ..................... 0.3
  18:3 ........................... 0.5
  6c,9c,12c,15c-18:4 (n-3). . . . . . . . . . . 0.8
  20:1 (n-9) ..................... 4

20:2 ........................... 0.7
20:3 ........................... 0.6
8c,11c,14c,17c-20:4 (n-3). . . . . . . . . . 0.4
5c,8c,11c,14c-20:4 (n-6). . . . . . . . . . . 3
6c,9c,12c,15c,17c-20:5. . . . . . . . . . . . 12
22:1 (n-9) ..................... 2
22:2 ........................... 1
7c,10c,13c,16c,19c-22:5 (n-3) . . . . . . . . 3
4c,7c,10c,13c,16c,19c-22:6 . . . . . . . . . 15

**References**

## Sprat Oil
Specific Gravity (SG)
  15.5/15.5°C
  25/25°C
  Other SG
Refractive Index (RI)
  25°C
  40°C
  Other RI
Iodine Value .................. 125–147
Saponification Value
Titer °C
% Unsaponifiable .................. 1–2
Melting Point °C
% hydrocarbons
% sterols
% Squalene
% Pristane

**Fatty Acid Composition (%)**
  16:0 ........................ 16–17
  16:1 .......................... 6–8
  18:0 .......................... 2–3
  Unassigned 18:1 ............... 15–17
  18:2 ........................... 2
  18:3 ........................... 2
  20:1 ......................... 10–11
  6c,9c,12c,15c,17c-20:5. . . . . . . . . . . 6–7
  22:1 ......................... 13–16
  7c,10c,13c,16c,19c-22:5 (n-3) . . . . . . 0.8
  4c,7c,10c,13c,16c,19c-22:6 . . . . . . . 7–11
  Other: 22:4, 0.5–1; other 14–16

**References**

## Squid

Specific Gravity (SG)
    15.5/15.5°C
    25/25°C
    Other SG
Refractive Index (RI)
    25°C
    40°C
    Other RI
Iodine Value
Saponification Value
Titer °C
% Unsaponifiable
Melting Point °C
% hydrocarbons
% sterols
% Squalene
% Pristane

**Fatty Acid Composition (%)**
    14:0 . . . . . . . . . . . . . . . . . . . . . . . . . . . 1
    16:0 . . . . . . . . . . . . . . . . . . . . . . . . . . 19
    16:1 . . . . . . . . . . . . . . . . . . . . . . . . . . 0.5
    16:2 (n-6) . . . . . . . . . . . . . . . . . . . . . 0.3
    16:3 (n-4) . . . . . . . . . . . . . . . . . . . . . 0.3
    16:4 (n-1) . . . . . . . . . . . . . . . . . . . . . 0.3
    18:0 . . . . . . . . . . . . . . . . . . . . . . . . . . . 5
    Unassigned 18:1 . . . . . . . . . . . . . . . . . . 2
    11c-18:1 . . . . . . . . . . . . . . . . . . . . . . . 1
    13c-18:1 . . . . . . . . . . . . . . . . . . . . . . 0.3
    18:2 . . . . . . . . . . . . . . . . . . . . . . . . . . 0.6
    18:2 (n-4) . . . . . . . . . . . . . . . . . . . . . 0.3
    18:3 . . . . . . . . . . . . . . . . . . . . . . . . . . 0.2
    6c,9c,12c,15c-18:4 (n-3). . . . . . . . . . 0.6
    20:1 (n-9) . . . . . . . . . . . . . . . . . . . . . . 6
    20:2 . . . . . . . . . . . . . . . . . . . . . . . . . . 0.4
    20:3 . . . . . . . . . . . . . . . . . . . . . . . . . . 0.9
    8c,11c,14c,17c-20:4 (n-3). . . . . . . . . 0.4
    5c,8c,11c,14c-20:4 (n-6). . . . . . . . . . . 1
    6c,9c,12c,15c,17c-20:5. . . . . . . . . . . . 2
    22:1 (n-9) . . . . . . . . . . . . . . . . . . . . . 13
    22:2 (n-6) . . . . . . . . . . . . . . . . . . . . . 0.3
    7c,10c,13c,16c,19c-22:5 (n-3) . . . . . 0.4
    4c,7c,10c,13c,16c,19c-22:6 . . . . . . . 31
    24:1 . . . . . . . . . . . . . . . . . . . . . . . . . . 0.2

**References** *JAOCS 70:* 1081 (1993)

## Sucker, White
*Catostromus commersonni*

Specific Gravity (SG)
    15.5/15.5°C
    25/25°C
    Other SG
Refractive Index (RI)
    25°C
    40°C
    Other RI
Iodine Value
Saponification Value
Titer °C
% Unsaponifiable
Melting Point °C
% hydrocarbons
% sterols
% Squalene
% Pristane

**Fatty Acid Composition (%)**
    14:0 . . . . . . . . . . . . . . . . . . . . . . . . . 2–3
    16:0 . . . . . . . . . . . . . . . . . . . . . . . . . . 15
    16:1 . . . . . . . . . . . . . . . . . . . . . . . . . . 19
    18:0 . . . . . . . . . . . . . . . . . . . . . . . . . . . 2
    Unassigned 18:1 . . . . . . . . . . . . . . . . 14
    18:2 . . . . . . . . . . . . . . . . . . . . . . . . . . . 3
    18:3 . . . . . . . . . . . . . . . . . . . . . . . . . . . 2
    6c,9c,12c,15c-18:4 (n-3). . . . . . . . . . . 2
    20:1 . . . . . . . . . . . . . . . . . . . . . . . . . . . 1
    5c,8c,11c,14c-20:4 (n-6). . . . . . . . . . . 4
    6c,9c,12c,15c,17c-20:5. . . . . . . . . . . 10
    7c,10c,13c,16c,19c-22:5 (n-3) . . . . . . 3
    4c,7c,10c,13c,16c,19c-22:6 . . . . . . . 15

**References** *JAOCS 54:* 424 (1977)

## Swordfish

Specific Gravity (SG)
    15.5/15.5°C
    25/25°C
    Other SG
Refractive Index (RI)
    25°C
    40°C

Other RI
Iodine Value
Saponification Value
Titer °C
% Unsaponifiable
Melting Point °C
% hydrocarbons
% sterols
% Squalene
% Pristane

**Fatty Acid Composition (%)**
  14:0 .............................. 4
  15:0 .............................. 2
  16:0 .............................. 8
  16:1 .............................. 9
  18:0 .............................. 8
  Unassigned 18:1 ................. 20
  18:2 .............................. 2
  6c,9c,12c,15c-18:4 (n-3) ......... 1
  20:1 .............................. 6
  5c,8c,11c,14c-20:4 (n-6) ......... 5
  6c,9c,12c,15c,17c-20:5 ........... 4
  22:1 .............................. 1
  7c, 10c, 13c, 16c-22:4 ........... 1
  4c, 7c,10c,13c,16c-22:5 (n-6) .... 2
  7c,10c,13c,16c,19c-22:5 (n-3) .... 6
  4c,7c,10c,13c,16c,19c-22:6 ...... 19
  24:1 .............................. 2

**References** *J. Food Sci* 52: 1209 (1987)

## Tallow (beef)

Specific Gravity (SG)
  15.5/15.5°C .............. 0.938–0.952
  25/25°C .................. 0.903–0.907
  Other SG
Refractive Index (RI)
  25°C
  40°C ..................... 1.450–1.458
  Other RI
Iodine Value ................... 33–47
Saponification Value ........... 190–200
Titer °C ....................... 40–47
% Unsaponifiable ............... 0–0.5
Melting Point °C ............... 45–48
% hydrocarbons

% sterols
% Squalene
% Pristane

**Fatty Acid Composition (%)**
  14:0 ............................ 1–6
  16:0 ........................... 20–37
  16:1 ............................ 1–9
  17:0 ............................ 1–3
  18:0 ........................... 25–40
  Unassigned 18:1 ............... 31–50
  18:2 ............................ 1–5

Cholesterol ............... (1090 mg/kg)

**References** *J. Am. Oil Chem. Soc.* 67: 980 (1990)
  *USDA Agriculture Handbook 8-4* (1979)


## Tallow (mutton)

Specific Gravity (SG)
  15.5/15.5°C .............. 0.938–0.955
  25/25°C
  Other SG
Refractive Index (RI)
  25°C
  40°C ..................... 1.452–1.458
  Other RI
Iodine Value ................... 35–46
Saponification Value
Titer °C ....................... 43–58
% Unsaponifiable
Melting Point °C ............... 44–51
% hydrocarbons
% sterols
% Squalene
% Pristane

**Fatty Acid Composition (%)**
  14:0 ............................ 2–4
  16:0 ........................... 20–27
  16:1 ........................... 1.4–4.5
  18:0 ........................... 22–34
  Unassigned 18:1 ............... 30–42
  18:2 ........................... 1.9–2.4
  Other 15:1, 0.5–1

Cholesterol ............... (1020 mg/kg)

**References** *J. Am. Oil Chem. Soc. 67:* 980 (1990)
*USDA Agriculture Handbook 8-4* (1979)

## Trout Lipids

Specific Gravity (SG)
  15.5/15.5°C
  25/25°C
  Other SG
Refractive Index (RI)
  25°C
  40°C
  Other RI
Iodine Value
Saponification Value
Titer °C
% Unsaponifiable
Melting Point °C
% hydrocarbons
% sterols
% Squalene
% Pristane

**Fatty Acid Composition (%)**
  14:0 . . . . . . . . . . . . . . . . . . . . . . . . 3–4
  16:0 . . . . . . . . . . . . . . . . . . . . . . . 21–24
  16:1 . . . . . . . . . . . . . . . . . . . . . . . . 4–10
  18:0 . . . . . . . . . . . . . . . . . . . . . . . . 3–8
  Unassigned 18:1 . . . . . . . . . . . . . . 18–31
  18:2 . . . . . . . . . . . . . . . . . . . . . . . . 7–16
  18:3 . . . . . . . . . . . . . . . . . . . . . . . . 1–2
  20:1 . . . . . . . . . . . . . . . . . . . . . . . . 0–3
  5c,8c,11c,14c-20:4 (n-6) . . . . . . . . . . 0–2
  6c,9c,12c,15c,17c-20:5 . . . . . . . . . . . 0–6
  7c,10c,13c,16c,19c-22:5 (n-3) . . . . 0–0.4
  4c,7c,10c,13c,16c,19c-22:6 . . . . . . . . 1–7

**References** Ackman, R.G. in *Objective Methods for Food Analysis,* National Academy of Sciences, Washington, DC, 1976

## Trout, Lake
*Salvelinus namaycush namaycush*

Specific Gravity (SG)
  15.5/15.5°C
  25/25°C
  Other SG
Refractive Index (RI)
  25°C
  40°C
  Other RI
Iodine Value
Saponification Value
Titer °C
% Unsaponifiable
Melting Point °C
% hydrocarbons
% sterols
% Squalene
% Pristane

**Fatty Acid Composition (%)**
  14:0 . . . . . . . . . . . . . . . . . . . . . . . . 2–3
  15:0 . . . . . . . . . . . . . . . . . . . . . . . . 0.2
  16:0 . . . . . . . . . . . . . . . . . . . . . . . . . 13
  16:1 . . . . . . . . . . . . . . . . . . . . . . . . . . 8
  18:0 . . . . . . . . . . . . . . . . . . . . . . . . 2–3
  9c-18:1 . . . . . . . . . . . . . . . . . . . . . . 26
  11c-18:1 . . . . . . . . . . . . . . . . . . . . 5–6
  18:2 . . . . . . . . . . . . . . . . . . . . . . . . . . 4
  18:3 . . . . . . . . . . . . . . . . . . . . . . . . . . 3
  6c,9c,12c,15c-18:4 (n-3) . . . . . . . . . . 1
  20:1 (n-7) . . . . . . . . . . . . . . . . . . . . 0.3
  20:1 (n-9) . . . . . . . . . . . . . . . . . . . . . 1
  20:2 (n-6) . . . . . . . . . . . . . . . . . . . . . 1
  8c,11c,14c-20:3 . . . . . . . . . . . . . . . 0.3
  11c,14c,17c-20:3 (n-3) . . . . . . . . . . . 1
  8c,11c,14c,17c-20:4 (n-3) . . . . . . . . . 2
  5c,8c,11c,14c-20:4 (n-6) . . . . . . . . . . 2
  6c,9c,12c,15c,17c-20:5 . . . . . . . . . . . 4
  21:5 (n-3) . . . . . . . . . . . . . . . . . . . . 0.4
  22:1 (n-11) . . . . . . . . . . . . . . . . . . . 0.2
  4c, 7c,10c,13c,16c-22:5 (n-6) . . . . . . 0.7
  7c,10c,13c,16c,19c-22:5 (n-3) . . . . . . 2
  4c,7c,10c,13c,16c,19c-22:6 . . . . . . . 10

**References** *J. Food Comp. Anal. 2:* 13 (1989)

## Trout, Ocean

Specific Gravity (SG)
  15.5/15.5°C
  25/25°C

Other SG
Refractive Index (RI)
 25°C
 40°C
 Other RI
Iodine Value
Saponification Value
Titer °C
% Unsaponifiable
Melting Point °C
% hydrocarbons
% sterols
% Squalene
% Pristane

**Fatty Acid Composition (%)**
 14:0............................2
 16:0...........................19
 16:1............................3
 18:0............................8
 Unassigned 18:1..................10
 18:2............................1
 20:1............................2
 5c,8c,11c,14c-20:4 (n-6)............4
 6c,9c,12c,15c,17c-20:5..............7
 4c, 7c,10c,13c,16c-22:5 (n-6).........2
 7c,10c,13c,16c,19c-22:5 (n-3).......39
 24:1............................1

**References** *J. Food Sci. 52:* 1209 (1987)


## Trout, Siscowet
*Salvelinus namaycush siscowet*
Specific Gravity (SG)
 15.5/15.5°C
 25/25°C
 Other SG
Refractive Index (RI)
 25°C
 40°C
 Other RI
Iodine Value
Saponification Value
Titer °C
% Unsaponifiable
Melting Point °C
% hydrocarbons

% sterols
% Squalene
% Pristane

**Fatty Acid Composition (%)**
 14:0............................2
 16:0...........................13
 16:1............................9
 18:0............................3
 9c-18:1.........................31
 18:2............................3
 18:3............................2
 20:1 (n-7)......................0.3
 20:2 (n-6)......................0.8
 8c,11c,14c-20:3.................0.3
 11c,14c,17c-20:3 (n-3)...........0.6
 8c,11c,14c,17c-20:4 (n-3)........1.6
 5c,8c,11c,14c-20:4 (n-6).........2.4
 6c,9c,12c,15c,17c-20:5............4
 21:5 (n-3)......................0.1
 22:1 (n-11).....................0.1
 4c, 7c,10c,13c,16c-22:5 (n-6)....0.4
 7c,10c,13c,16c,19c-22:5 (n-3)....2.3
 4c,7c,10c,13c,16c,19c-22:6........7

**References** *J. Food Comp. Anal. 2:* 13 (1989)


## Tuna (white meat)
*Thunnus alalunga*
Specific Gravity (SG)
 15.5/15.5°C
 25/25°C
 Other SG
Refractive Index (RI)
 25°C
 40°C
 Other RI
Iodine Value
Saponification Value
Titer °C
% Unsaponifiable
Melting Point °C
% hydrocarbons
% sterols
% Squalene
% Pristane

**Fatty Acid Composition (%)**
    14:0 . . . . . . . . . . . . . . . . . . . . . . . . . . . 3
    15:0 . . . . . . . . . . . . . . . . . . . . . . . . . . . 1
    16:0 . . . . . . . . . . . . . . . . . . . . . . . . . . 21
    16:1 . . . . . . . . . . . . . . . . . . . . . . . . . . . 2
    17:0 . . . . . . . . . . . . . . . . . . . . . . . . . . . 2
    18:0 . . . . . . . . . . . . . . . . . . . . . . . . . . . 6
    Unassigned 18:1 . . . . . . . . . . . . . . . . . . 15
    18:2 . . . . . . . . . . . . . . . . . . . . . . . . . . 1–2
    18:3 . . . . . . . . . . . . . . . . . . . . . . . . . . . 2
    18:4 . . . . . . . . . . . . . . . . . . . . . . . . . . 0.1
    20:1 . . . . . . . . . . . . . . . . . . . . . . . . . . 0.7
    20:4 . . . . . . . . . . . . . . . . . . . . . . . . . . . 4
    5c,8c,11c,14c-20:4 (n-6). . . . . . . . . . . . 4
    20:5 . . . . . . . . . . . . . . . . . . . . . . . . . . . 8
    6c,9c,12c,15c,17c-20:5. . . . . . . . . . . . . 8
    22:2 . . . . . . . . . . . . . . . . . . . . . . . . . . 0.5
    7c,10c,13c,16c,19c-22:5 (n-3) . . . . . . . . 2
    22:6 . . . . . . . . . . . . . . . . . . . . . . . . . . 29
    4c,7c,10c,13c,16c,19c-22:6 . . . . . . . . . 29

**References**  *J. Food Comp. Anal. 7:* 119 (1994)

## Turtle (Green) Oil

Specific Gravity (SG)
    15.5/15.5°C
    25/25°C . . . . . . . . . . . . . . . . 0.914–0.916
    Other SG
Refractive Index (RI)
    25°C . . . . . . . . . . . . . . . . . . . . . . . . 1.467
    40°C . . . . . . . . . . . . . . . . . . 1.461–1.465
    Other RI
Iodine Value . . . . . . . . . . . . . . . . . . . . 58–88
Saponification Value . . . . . . . . . . . 210–214
Titer °C
% Unsaponifiable . . . . . . . . . . . . . . . . . 5–15
Melting Point °C
% hydrocarbons
% sterols
% Squalene
% Pristane

**Fatty Acid Composition (%)**
    Other . . . . . . . . . . . . . . . . . . . Saturate, 45

**References**

## Whale Oil
### *Balsenidae spp.*

Specific Gravity (SG)
    15.5/15.5°C
    25/25°C
    Other SG . . . . . . . . . (15/15) 0.917–0.926
Refractive Index (RI)
    25°C . . . . . . . . . . . . . . . . . . . . 1.468–1.472
    40°C
    Other RI
Iodine Value . . . . . . . . . . . . . . . . . . 97–115
Saponification Value . . . . . . . . . . . 188–202
Titer °C
% Unsaponifiable . . . . . . . . . . . . . . . . 10–40
Melting Point °C
% hydrocarbons
% sterols
% Squalene
% Pristane

**Fatty Acid Composition (%)**
    Other . . . . . . . . . . . . . . . . Saturate, 18–28

**References**

## Whale Oil, Minke

Specific Gravity (SG)
    15.5/15.5°C
    25/25°C
    Other SG
Refractive Index (RI)
    25°C
    40°C
    Other RI
Iodine Value
Saponification Value
Titer °C
% Unsaponifiable
Melting Point °C
% hydrocarbons
% sterols
% Squalene
% Pristane

**Fatty Acid Composition (%)**
    14:0 . . . . . . . . . . . . . . . . . . . . . . . . . . . 5

16:0 .............................. 8
16:1 .............................. 9
18:0 .............................. 2
Unassigned 18:1 ................. 18
18:2 ............................ 1.4
18:3 ............................ 1.3
18:4 ............................ 1.6
20:0 ............................ 0.3
20:1 ............................. 17
5c,8c,11c,14c-20:4 (n-6) ......... 0.3
6c,9c,12c,15c,17c-20:5 ........... 4.3
22:1 ............................. 11
7c,10c,13c,16c,19c-22:5 (n-3) .... 2.3
4c,7c,10c,13c,16c,19c-22:6 ....... 7.9
Other ........................... 1.6

**References** *Lipids 30:* 1111 (1995)

## Whale Oil, Pacific Beaked
*Beradius bairdii*
Specific Gravity (SG)
   15.5/15.5°C
   25/25°C
   Other SG
Refractive Index (RI)
   25°C
   40°C
   Other RI
Iodine Value
Saponification Value
Titer °C
% Unsaponifiable
Melting Point °C
% hydrocarbons
% sterols
% Squalene
% Pristane

**Fatty Acid Composition (%)**
12:0 ............................ 0.5
14:0 .............................. 6
14:1 .............................. 1
16:0 .............................. 7
16:1 ............................. 22
18:0 .............................. 1
9c-18:1 .......................... 26
11c-18:1 .......................... 4

18:2 ............................ 0.5
22:1 ............................ 5.6
Other: 7c-18:1, 3

**References** *Lipids 13:* 860 (1978)

## Whitefish Oil
*Coregonus clupeaformis*
Specific Gravity (SG)
   15.5/15.5°C
   25/25°C
   Other SG
Refractive Index (RI)
   25°C
   40°C
   Other RI
Iodine Value
Saponification Value
Titer °C
% Unsaponifiable
Melting Point °C
% hydrocarbons
% sterols
% Squalene
% Pristane

**Fatty Acid Composition (%)**
14:0 .............................. 3
16:0 ............................. 14
16:1 ............................. 24
18:0 .............................. 3
Unassigned 18:1 ................. 24
18:2 .............................. 2
18:3 .............................. 2
5c,8c,11c,14c-20:4 (n-6) ......... 1.6
6c,9c,12c,15c,17c-20:5 ............ 9
7c,10c,13c,16c,19c-22:5 (n-3) ..... 2
4c,7c,10c,13c,16c,19c-22:6 ........ 7

## Whiting
*Merlangius merlangus*
Specific Gravity (SG)
   15.5/15.5°C
   25/25°C
   Other SG

Refractive Index (RI)
   25°C
   40°C
   Other RI
Iodine Value
Saponification Value
Titer °C
% Unsaponifiable
Melting Point °C
% hydrocarbons
% sterols
% Squalene
% Pristane

**Fatty Acid Composition (%)**
   14:0 . . . . . . . . . . . . . . . . . . . . . . . . . 2–15
   15:0 . . . . . . . . . . . . . . . . . . . . . . . . . . 0.5
   16:0 . . . . . . . . . . . . . . . . . . . . . . . . . 9–17
   16:1 . . . . . . . . . . . . . . . . . . . . . . . . . . . 3
   16:2 (n-6) . . . . . . . . . . . . . . . . . . . . . 0.3
   16:2 (n-4) . . . . . . . . . . . . . . . . . . . . . 0.3
   16:3 (n-4) . . . . . . . . . . . . . . . . . . . . . 0.3
   16:4 (n-1) . . . . . . . . . . . . . . . . . . . . . 0.4
   18:0 . . . . . . . . . . . . . . . . . . . . . . . . . . 3–8
   Unassigned 18:1 . . . . . . . . . . . . . . 11–15
   18:2 . . . . . . . . . . . . . . . . . . . . . . . . . . . 1
   18:3 . . . . . . . . . . . . . . . . . . . . . . . . . . . 1
   6c,9c,12c,15c-18:4 (n-3) . . . . . . . . . . 1–3
   20:1 (n-9) . . . . . . . . . . . . . . . . . . . . . . 6
   20:2 . . . . . . . . . . . . . . . . . . . . . . . . . . 0.2
   8c,11c,14c,17c-20:4 (n-3) . . . . . . . . . 0.3
   5c,8c,11c,14c-20:4 (n-6) . . . . . . . . . . 1–4
   6c,9c,12c,15c,17c-20:5 . . . . . . . . . . 9–13
   22:1 (n-9) . . . . . . . . . . . . . . . . . . . . . . 6
   7c, 10c, 13c, 16c-22:4 . . . . . . . . . . . . 0–4
   4c, 7c,10c,13c,16c-22:5 (n-6) . . . . . . 0–2
   7c,10c,13c,16c,19c-22:5 (n-3) . . . . . 0–4
   4c,7c,10c,13c,16c,19c-22:6 . . . . . 24–30

**References**

## Shark Liver Oil (Deep Sea)
*Centrophorus squamosus*

Specific Gravity (SG)
   15.5/15.5°C
   25/25°C
   Other SG

Refractive Index (RI)
   25°C
   40°C
   Other RI
Iodine Value
Saponification Value
Titer °C
% Unsaponifiable
Melting Point °C
% hydrocarbons
% sterols
% Squalene
% Pristane

**Fatty Acid Composition (%)**
   14:0 . . . . . . . . . . . . . . . . . . . . . . . . . . . 2
   16:0 . . . . . . . . . . . . . . . . . . . . . . . . . . 21
   16:1 . . . . . . . . . . . . . . . . . . . . . . . . . . . 8
   17:1 . . . . . . . . . . . . . . . . . . . . . . . . . . . 1
   18:0 . . . . . . . . . . . . . . . . . . . . . . . . . . . 2
   Unassigned 18:1 . . . . . . . . . . . . . . . 45
   20:1 . . . . . . . . . . . . . . . . . . . . . . . . . . 12
   22:1 . . . . . . . . . . . . . . . . . . . . . . . . . . . 9

**References**

## Shark Liver Oil (Basking)
*Centorhinus maximus*

Specific Gravity (SG)
   15.5/15.5°C
   Other SG . . . . . . . . . . . . . . . (15/4) 0.8922
Refractive Index (RI)
   25°C
   40°C
   Other RI . . . . . . . . . . . . . . . . . . (20) 1.4819
Iodine Value
Saponification Value
Titer °C
% Unsaponifiable . . . . . . . . . . . . . . . . . . . 36
Melting Point °C
% hydrocarbons . . . . . . . . . . . . . . . . . . . 96
% sterols . . . . . . . . . . . . . . . . . . . . . . . . 2.5
% Squalene . . . . . . . . . . . . . . . . . . . . . . 98
% Pristane

**References**

# Triglyceride Molecular Species
of Selected Oils and Fats

## Almond Kernel Oil
*Prunus dulcis*

Triglyceride Composition

| | |
|---|---|
| POO | 10 |
| SOO | 2 |
| PPL | 0.3 |
| OOO | 36 |
| POL | 8 |
| OOL | 25 |
| PLL | 2 |
| OLL | 13 |
| LLL | 2 |

## Alpine Current Seed Oil
*Ribes nigrum*

Triglyceride Composition

| | |
|---|---|
| OOO | 0.2 |
| POL + SOL | 2 |
| OOL | 5 |
| OOLng | 1 |
| PLLng + LLL | 15 |
| PLn + Lng + LLLng | 5 |
| PLnLn + PLngLng | 14 |
| PLL + SLL | 5 |
| PLLn + SLLn | 3 |
| OLL | 14 |
| OLLn | 10 |
| OLLng | 4 |
| OLugLug | 0.4 |
| OLnLng + OLLt | 2 |
| LLnLn + LLLt | 4 |
| LLnLt | 2 |

## Amaranth Seed Oil (Various)
*Amaranthus hypochondriacus, cruentus, edulis*

Triglyceride Composition

| | |
|---|---|
| PSO | 0.49–1.49 |
| PPP | 0–0.95 |
| PSS | 0.15 |
| SOS | 0–0.55 |
| PPO | 0.56–1.07 |
| SOO | 0.73–1.7 |
| PPL | 7.01–9 |
| POL | 16.47–18.67 |
| PLLn | 0.85–1.34 |
| LLL | 5.44–8.66 |
| LLLn | 1.06–2.4 |
| LLnLn | 0.31–1.11 |

## Apricot Kernel Oil
*Prunus armeniaca*

Triglyceride Composition

| | |
|---|---|
| POO | 4 |
| SOO | 1 |
| OOO | 26 |
| POL | 5 |
| OOL | 33 |
| PLL | 2 |
| OLL | 23 |
| LLL | 5 |

## Argan Seed Oil
*Argania spinosa*

Triglyceride Composition

| | |
|---|---|
| PSO | 1.9–2 |
| SOS | 0.3 |
| PPO | 1.8–3 |
| POO | 7.1–14 |
| SOO | 3–5 |
| PPL | 1.5–1.6 |
| PSL | 1.6 |
| SSL | 0.5–3.9 |
| OOO | 7.6–16 |
| POL | 12.4–14 |
| SOL | 3–6.2 |
| OOL | 16–20 |
| PLL | 5–6 |
| SLL | 2–2.6 |
| OLL | 13–17.3 |
| LLL | 5–7 |

## Baillonella Toxisperma Kernel Oil
*Baillonella toxisperma*

Triglyceride Composition

| | |
|---|---|
| PPS | 13 |
| PSS | 12 |
| SOS | 1 |
| PPO | 5 |
| POO | 19 |
| PPL | 0.5 |
| OOO | 17 |
| POL | 3 |
| OOL | 3 |
| OLL | 1 |

## Blackcurrant Oil
*Ribes nigrum*

Triglyceride Composition

| | |
|---|---|
| POL | 1 |
| OOL + OEL | 3 |
| PLL + SLL | 5 |
| PLLn + SLLn | 2 |
| OLL + ELL | 12 |
| LLLn | 12 |
| LLL + POLng + OOLn | 20 |
| LLnLng + LLLt | 6 |
| LLnLn | 3 |
| LLnLt + LngLnLn | 2 |

## Borage Oil
*Borrago officinalis*

Triglyceride Composition

| | |
|---|---|
| PPO | 0.3 |
| POO | 1 |
| PPL | 1 |
| OOO | 1 |
| POL | 6 |
| POLng | 14 |
| OOL | 5 |
| OOLng | 2 |
| PLL | 8 |
| OLL | 20 |
| OLLng + PLngLng | 16 |
| OLngLng | 1 |
| LLngLng | 6 |

## Buchanania Lanzan
*Buchanania lanzan, B. latifolia*

Triglyceride Composition

| | |
|---|---|
| PSO | 9 |
| PPP | 2 |
| PPS | 1 |
| SOS | 1 |
| PPO | 23 |
| POO | 31 |
| SOO | 6 |
| PPL | 2 |
| PSL | 1 |
| OOO | 11 |
| POL | 7 |
| OOL | 3 |

## Camellia Oleifera Seed Oil
*Camellia oleifera*

Triglyceride Composition

| | |
|---|---|
| PSO | 1 |
| POO | 18 |
| SOO | 9 |
| PPL | 0.1 |
| OOO | 54 |
| POL | 3 |
| OOL | 11 |
| PLL | 0.4 |
| OLL | 2 |
| ALO | 1 |

## Camellia Sinensis Seed Oil
*Camellia sinensis*

Triglyceride Composition

| | |
|---|---|
| PSO | 1.5 |
| PPO | 3 |
| POO | 19 |
| SOO | 5 |

| PPL | 2 |
|---|---|
| OOO | 25 |
| POL | 13 |
| OOL | 17 |
| PLL | 4 |
| OLL | 10 |
| ALO | 0.6 |

## Cape Marigold Seed Oil
*Dimorphotheca pluvialis*

| Triglyceride Composition | |
|---|---|
| DOO | 4 |
| DLL | 3 |
| DDP | 6 |
| DDO | 28 |
| DDL | 21 |
| DDD | 3 |

## Cashew Nut Oil
*Anarcadium occidentale*

| Triglyceride Composition | |
|---|---|
| PSO | 3–6 |
| SOS | 2–3 |
| PPO | 2–5 |
| POO | 15–19 |
| SOO | 11–12 |
| PPL | 1–2 |
| OOO | 19–29 |
| POL | 8–11 |
| SOL | 3.5–5 |
| OOL | 11.8–17 |
| PLL | 1.7–2.6 |
| OLL | 3–5 |
| LLL | tr–0.5 |

## Castor Oil
*Ricinus communis*

| Triglyceride Composition | |
|---|---|
| ROO | 3 |
| RLL | 4 |
| RRO | 9 |
| RRL | 12 |
| RRR | 69 |

## Celastrus Orbiculatus
*Celastrus orbiculatus*

| Triglyceride Composition | |
|---|---|
| OTHER | Includes C36 and C38 Acetyloglycerides |

## Cherry Kernel Oil
*Prunus avium*

| Triglyceride Composition | |
|---|---|
| POO | 4 |
| PPL | 1 |
| OOO | 16–16.1 |
| POL | 7.7–8 |
| OOL | 15–15.4 |
| PLL + EOO | 9.8–10 |
| OLL | 18.5–19 |
| LLL | 3–3.3 |
| EEL | 0.5 |
| ELL | 12.9–13 |
| ELO | 6.9–7 |
| ELP | 3.9–4 |
| EOP | tr |

## Cloudberry Seed Oil
*Rubus chamaemorus*

| Triglyceride Composition | |
|---|---|
| PPO | 0.2 |
| POO | 0.3 |
| OOO | 3 |
| POL | 1 |
| POLn | 1 |
| OOL | 4 |
| OOLn + OELn | 1 |
| PLL | 2 |
| PLLn | 12 |
| OLL + ELL | 11 |
| OLLn + ELLn | 15 |
| LLL | 14 |
| LLLn + LELn | 24 |

## Cocoa Butter
*Theolbroma cocoa*

### Triglyceride Composition

| | |
|---|---|
| PSO | 36–40 |
| PPS | 1 |
| PSS | 1 |
| SOS | 23–26 |
| PPO | 14–18 |
| POO | 3–4 |
| SOO | 3–6 |
| PPL | 1–2 |
| PSL | 3 |
| SSL | 2 |
| SOL | 0.3–1 |
| AOS | 1–2 |
| AOP | 0–1 |
| OTHER | 1–8 |

## Coconut Oil
*Cocos nucifera*

### Triglyceride Composition

| | |
|---|---|
| PPP | 0.38 |
| PPO | 0.21 |
| LaMM | 11.25 |
| MMM | 7.15 |
| MMP | 2.36 |
| CCC | 1.89 |
| CCL | 0.06 |
| CCO | 1.65 |
| CCP | 23.16 |
| CCS | 18.22 |
| PLC | 0.29 |
| LaCC | 8.75 |
| MCC | 15 |
| LLaC | 0.03 |
| OLaC | 1.11 |
| LMC | 0.22 |
| LnPC | 0.33 |
| OMC | 2.34 |
| LOC | 0.04 |
| OPC | 1.84 |
| OOC | 0.19 |
| LLaO | 0.08 |
| SOC | 1.54 |
| OOLa | 0.73 |
| SSC | 0.3 |
| SOLa | 0.54 |
| SMP | 0.38 |

## Cottonseed Oil
*Gossypium spp.*

### Triglyceride Composition

| | |
|---|---|
| PPP | 1.2 |
| PPS | 0.3 |
| PSS | tr |
| POO | 0.3 |
| SOO | 0.3 |
| PPL | 3.3 |
| OOO | 0.7 |
| POL | 7.3 |
| SOL | 0.8 |
| OOL | 3.8 |
| PLL | 19.5 |
| SLL | 2.3 |
| OLL | 10.3 |
| LLL | 13.5 |

## Crepis Alpina Seed Oil
*Crepis alpina*

### Triglyceride Composition

| | |
|---|---|
| LLL | 2 |
| CCC | 37 |
| CCL | 33 |
| CCO | 4 |
| CLL | 7 |
| CCP | 12 |
| CCS | 4 |
| PLC | 2 |
| SLC | 1 |

## Evening Primrose Oil
*Oenothera biennis*

### Triglyceride Composition

| | |
|---|---|
| OOO | 0.3 |

POL......1
OOL......1
PLL + SLL + ALL......12
OLL......15
Ohhug......4
LLL + PLLng......50
LLLng......15
LLngLng......1

## Fungal Oil
*Mortierella alpina*

Triglyceride Composition

| | |
|---|---|
| ALO | 1–3 |
| ALS | 4–14 |
| AOS | 1–5 |
| AOO | 0.4–2 |
| AAA | 6–24 |
| GAA | 1–2 |
| LAA | 7–14 |
| OAA | 3–9 |
| PAA | 11–25 |
| SAA | 10–23 |
| OGA | 0.3–1 |
| SGA | 0.3–1 |
| LLA | 0.2–1 |
| PLA | 2–12 |
| PPA | 0.3–1.5 |
| PSA | 0.7–2 |
| SSA | 0.2–0.5 |

## Hannoa Undulata Seed Oil
*Hannoa undulata (simarubacea)*

Triglyceride Composition

| | |
|---|---|
| PSO | 8 |
| SOS | 7 |
| POO | 11 |
| SOO | 25 |
| OOO | 23 |
| POL | 3 |
| SOL | 11 |
| OOL | 7 |
| OLL | 3 |

AOS......0.1
AOO......3

## Hazelnut Oil (Filbert)
*Corylus avellana*

Triglyceride Composition

| | |
|---|---|
| PPP | 0.1–2.7 |
| PPO | 1–2 |
| POO | 10–18 |
| SOO | 2–7 |
| PPL | 0–1 |
| OOO | 36–57 |
| POL | 3–6 |
| OOL | 10–24 |
| PLL | 0.5–2 |
| OLL | 2–11 |
| LLL | 0.5–4 |

## Hibiscus Cannabinus
*Hibiscus cannabinus*

Triglyceride Composition

| | |
|---|---|
| PPP | 1.7 |
| PPS | 0.2 |
| PSS | tr |
| PPO | 3.2 |
| POO | 1.5 |
| SOO | 0.3 |
| PPL | 2.2 |
| OOO | 3.1 |
| POL | 4.5 |
| SOL | 0.2 |
| OOL | 4.3 |
| PLL | 11.9 |
| SLL | 0.6 |
| OLL | 11.3 |
| LLL | 3.9 |

## Hibiscus sabdariffa
*Hibiscus sabdariffa*

Triglyceride Composition

PPP......1.1

PPS . . . . . . . . . . . . . . . . . . . . . . . . . 0.3
PSS . . . . . . . . . . . . . . . . . . . . . . . . . . tr
PPO . . . . . . . . . . . . . . . . . . . . . . . . . 2.2
POO . . . . . . . . . . . . . . . . . . . . . . . . . 1.1
SOO . . . . . . . . . . . . . . . . . . . . . . . . . 1.3
PPL . . . . . . . . . . . . . . . . . . . . . . . . . 2.4
OOO . . . . . . . . . . . . . . . . . . . . . . . . . 4.9
POL . . . . . . . . . . . . . . . . . . . . . . . . 10.6
SOL . . . . . . . . . . . . . . . . . . . . . . . . . 1.4
OOL . . . . . . . . . . . . . . . . . . . . . . . . 10.5
PLL . . . . . . . . . . . . . . . . . . . . . . . . . . 8
SLL . . . . . . . . . . . . . . . . . . . . . . . . . . 1
OLL . . . . . . . . . . . . . . . . . . . . . . . . . 7.8
LLL . . . . . . . . . . . . . . . . . . . . . . . . . 4.2

## Hollyhock
*Althea rosea*

Triglyceride Composition

| | |
|---|---|
| PPS | 0.1 |
| PSS | tr |
| PPO | 2.0 |
| POO | 0.7 |
| SOO | 0.3 |
| PPL | 3.6 |
| OOO | 2.3 |
| POL | 12.8 |
| SOL | 0.5 |
| OOL | 8.2 |
| PLL | 12 |
| SLL | 0.5 |
| OLL | 8 |
| LLL | 7 |

## Illipe (mowrah) Butter
*Madhuca latiflora/longiflora*
*(Bassia latiflora)*

Triglyceride Composition

| | |
|---|---|
| PSO | 22.2 |
| SOS | 10.6 |
| PPO | 18.9 |
| POO | 12.6 |
| SOO | 6.7 |
| OOO | 2.2 |

## Irvingia Gabonensis Kernel Fat (Dika Fat)
*Irvingia gabonensis*

Triglyceride Composition

| | |
|---|---|
| LaLaLa | 8 |
| LaLaM | 31 |
| LaMM | 44 |
| MMM | 15 |
| MMP | 2 |

## Kokum Butter
*Garcinia indica*

Triglyceride Composition

| | |
|---|---|
| PSO | 7.4–14 |
| SOS | 59–72.3 |
| PPO | 0.5–2 |
| POO | 0.5–2 |
| SOO | 15.1–21 |
| OOO | 2.1 |

## Korean Pine Seed Oil
*Pinus koraiensis*

Triglyceride Composition

| | |
|---|---|
| PSO | 0.1 |
| SOS | tr |
| PPO | 0.1 |
| POO | 2.6 |
| SOO | 1.3 |
| PPL | 0.6 |
| PSL | 0.5 |
| OOO | 7.6 |
| POL | 8.3 |
| SOL | 3.5 |
| OOL | 7.4 |
| PLL | 3.8 |
| PLLn | 5.5 |
| OLL | 18.1 |
| OLLn | 6.5 |
| LLL | 8.1 |
| LLLn | 10.7 |

## Lime Seed Oil
*Citrus aurantifolia*

Triglyceride Composition

| | |
|---|---:|
| PPO | 3 |
| POO | 4 |
| PPL | 14 |
| PSL | 4 |
| OOO | 2 |
| POL | 14 |
| SOL | 3 |
| OOL | 4 |
| PLL | 11 |
| OLL | 7 |
| LLL | 4 |
| LnLO | 5 |
| LnOP | 6 |
| LnLL | 4 |
| LnLnL | 1 |

## Linseed Oil (Flax)
*Linum usitatissimum*

Triglyceride Composition

| | |
|---|---:|
| POO | 1 |
| SOO | 1 |
| OOO | 3 |
| POL | 2 |
| SOL | 1 |
| OOL | 3 |
| OLL | 1 |
| LLL | 1 |
| LnLO | 5 |
| LnOP | 4 |
| LnOO | 7 |
| LnLL | 4 |
| LnLP | 7 |
| LnLS | 1 |
| LnLnL | 14 |
| LnLnP | 7 |
| LnLnS | 3 |
| LnLnO | 8 |
| LnLnLn | 21 |
| OTHER | 4.8 |

## Mahua Fat
*Madhuca latifolia*

Triglyceride Composition

| | |
|---|---:|
| PSO | 22 |
| SOS | 11 |
| PPO | 19 |
| POO | 13 |
| SOO | 7 |
| OOO | 2 |

## Moringa Peregrina Seed Oil
*Moringa peregrina*

Triglyceride Composition

| | |
|---|---:|
| PSO | 0.2 |
| PPP | 1 |
| PPO | 0.4 |
| POO | 29 |
| SOO | 9 |
| PPL | 0.2 |
| PSL | 5 |
| OOO | 45 |
| POL | 1.3 |
| OOL | 5 |
| PLL | 0.6 |
| LLL | 0.3 |
| LnLO | 0.4 |
| LnLnL | 0.2 |

## Ochoco Butter (kernel fat)
*Scyphocephalium ochocoa*

Triglyceride Composition

| | |
|---|---:|
| LaLaLa | 0.5 |
| LaLaM | 5 |
| LaMM | 38 |

MMM . . . . . . . . . . . . . . . . . . . . . . . . . . . 54
MMP . . . . . . . . . . . . . . . . . . . . . . . . . . . . 2
MMO . . . . . . . . . . . . . . . . . . . . . . . . . . . . 2

## Okra Seed Oil
*Hibiscus esculentus*

Triglyceride Composition

| | |
|---|---|
| PPP | 0.8–2 |
| PPS | 0.1–0.2 |
| PSS | tr |
| PPO | 1.8–6 |
| POO | 0.6–5 |
| SOO | 0.3–1 |
| PPL | 2.2 |
| PSL | 4 |
| OOO | 1–5.2 |
| POL | 11–21 |
| SOL | 0.6–2 |
| OOL | 7–13 |
| PLL | 7.4–16 |
| SLL | 0.4 |
| OLL | 5.5–10 |
| LLL | 4.8–5.5 |

## Olive Oil (for quality grade reference values see IOOC documentation)
*Olea europaea*

Triglyceride Composition

| | |
|---|---|
| PSO | 2 |
| SOS incl AOP | 0.6 |
| PPO incl PLS | 3 |
| POO incl SOL | 23 |
| SOO incl AOL | 8 |
| OOO | 45 |
| POL incl PPOO | 4 |
| OOL incl PoOO | 10 |
| PLL incl LnOO | 2 |
| OLL incl PoLO | 1 |
| AOO | 0.7 |
| LnLO | 0.2 |
| LnOP | 0.6 |
| GOO | 0.5 |

## Peach Kernel Oil
*Prunus persica*

Triglyceride Composition

| | |
|---|---|
| POO | 7 |
| SOO | 2 |
| PPL | 0.3 |
| OOO | 31 |
| POL | 9 |
| OOL | 28 |
| PLL | 2 |
| OLL | 17 |
| LLL | 4 |

## Peanut/Groundnut Oil
*Arachis hypogaea*

Triglyceride Composition

| | |
|---|---|
| PSO | 0.6 |
| PPO | 1 |
| POO | 6 |
| SOO | 4 |
| PPL | 2 |
| OOO | 5 |
| POL | 13 |
| OOL | 22 |
| PLL | 8 |
| OLL | 26 |
| LLL | 6 |
| AOO | 4 |
| BOL | 1 |
| BOO | 1 |

## Pecan Nut Oil
*Carya illinoensis*

Triglyceride Composition

| | |
|---|---|
| POO | 3–5 |
| SOO | 0.1–0.7 |
| OOO | 4–10 |
| POL incl PoOO | 8–10 |
| SOL | 0.1–1 |
| OOL | 24–29 |
| SLL | 0.3–1 |
| OLL | 24–29 |

| | |
|---|---|
| LLL | 12–17 |
| LnLO | 0.5–1 |
| LnOO | 6–9 |
| LnLL | 1–3 |
| LnLnL | 0.1–1 |
| LnLnLn | 0.3–1 |

## Phulwara Butter
*Madhuca butyraceae*

Triglyceride Composition

| | |
|---|---|
| PPP | 8 |
| PPS | 1 |
| SOS | 0.4 |
| PPO | 53 |
| POO | 14 |
| SOO | 1 |
| PPL | 5 |
| PSL | 1 |
| OOO | 1 |
| POL | 2 |

## Pili Nut Oil
*Canarium ovatum*

Triglyceride Composition

| | |
|---|---|
| PSO | 11.9 |
| PPP | 16.57 |
| PPS | 7.55 |
| PSS | 0.55 |
| SOS | 0.89–2.24 |
| PPO | 0.72 |
| POP + MSO | 21.2 |
| POO | 34.99 |
| POO + PSL | 25.5 |
| SOO | 7.53–12.47 |
| PPL | 0.08–5.58 |
| OOO | 7.79 |
| POL | 7.79 |
| OOL + PoOO | 2.38 |
| PLL + PLnO | 2.43 |
| PLLn | 0.14 |
| OLL + OOLn | 0.81 |
| LLL | 0.52 |
| LLLn | 0.13 |
| LLnLn | 0.4 |

| | |
|---|---|
| LLP | 0.47 |
| OLL | 0.8 |
| OLP | 4.1 |
| SLL | 11.14 |
| PPP | 16.57 |
| OOO | 14.63 |
| SPO | 8.02 |

## Poga Oleosa Kernel Oil
*Poga oleosa*

Triglyceride Composition

| | |
|---|---|
| PSO | 1 |
| PPO | 1 |
| SOO | 11 |
| OOO | 37 |
| POL | 6 |
| OOL | 18 |
| PLL | 2 |
| OLL | 8 |
| LLL | 3 |

## Prune Kernel Oil
*Prunus cerasifera*

Triglyceride Composition

| | |
|---|---|
| POO | 5 |
| SOO | 4 |
| OOO | 55 |
| POL | 3 |
| OOL | 22 |
| PLL | 0.6 |
| OLL | 9 |
| LLL | 2 |

## Ricinodendron Heudelotii Kernel Oil
*Ricinodendron heudelotii*

Triglyceride Composition

| | |
|---|---|
| PLL | 4 |
| OLL | 3 |
| LLL | 6 |
| EEE | 15 |

EEO................................ 6
EEP................................ 6
EEL............................... 42
ELL................................ 9
ELO................................ 6
ELP................................ 3

## Safou Oil
*Dacryodes edulis*

### Triglyceride Composition

PSO................................ 3
PPO............................... 27
POO............................... 20
PPL............................... 20
POL............................... 15
PLL............................... 10

## Sesame Seed Oil
*Sesamum inidicum*

### Triglyceride Composition

PSO............................ 0–0.6
SOS............................ 0.3–4
PPO............................ 0–0.6
POO.............................. 0–3
SOO............................. 2–10
PPL.............................. 0–2
PSL.............................. 0–1
SSL............................ 0.5–5
OOO.............................. 4–7
POL.............................. 0–8
SOL............................. 4–21
OOL............................ 15–20
PLL............................. 0–11
SLL............................. 3–10
OLL............................ 18–25
LLL............................. 5–20
ALO............................ 0–0.3
LnLL........................... 0–0.5

## Sesame Seed Oil
*Sesamum radiatum*

### Triglyceride Composition

PSO................................ 1

SOS.............................. 0.3
PPO.............................. 0.3
POO................................ 2
SOO................................ 2
PPL................................ 1
PSL................................ 1
SSL................................ 1
OOO................................ 3
POL................................ 6
SOL................................ 5
OOL............................... 16
PLL................................ 9
SLL................................ 5
OLL............................... 24
LLL............................... 18
ALO.............................. 0.3
ALS.............................. 0.3
LnLO............................. 0.6
LnLL.............................. 2
LnLP............................. 0.8

## Soybean Oil
*Glycine max*

### Triglyceride Composition

PSO........................... 0.5–0.7
SOS.............................. 0.2
PPO........................... 0.5–0.8
POO........................... 2.1–3.4
SOO........................... 1.0–1.2
PPL........................... 0.9–3.1
PSL........................... 2.3–3.1
SSL........................... 0.7–1.1
OOO........................... 1.4–3.3
POL........................... 6.4–9.4
SOL........................... 1.8–4.2
OOL........................... 6.3–11.8
PLL.......................... 0.8–10.23
SLL........................... 2.6–6.4
OLL.......................... 16–25.9
LLL.......................... 17.6–20.6
AOO.............................. 0.5
PBL........................... 0.3–0.5
LnLO.......................... 3.7–4.8
LnOP............................. 0.3
LnOO............................. 0.6
LnLL.......................... 7.9–8.1
LnLP.......................... 2.4–3.7

| | |
|---|---|
| LnLS | 2.3 |
| LnLnL | 1.3–3.1 |
| LnLnP | 0.1 |
| LnLnS | 0.1 |
| LnLnO | 0.4 |

## Soybean Oil (high palmitic; HP)
### GMO

| Triglyceride Composition | |
|---|---|
| PSO | 1.6 |
| PPO | 2.2 |
| POO | 2 |
| PPL | 9.9 |
| PSL | 3.3 |
| SSL | 1.6 |
| POL | 7.7 |
| SOL | 2.1 |
| OOL | 2.1 |
| PLL | 14 |
| SLL | 4.5 |
| OLL | 6.1 |
| LLL | 6.9 |
| PBL | 1.5 |
| LnLO | 3.4 |
| LnOP | 2.7 |
| PPLn | 2.2 |
| LnLL | 4.9 |
| LnLP | 8.1 |
| LnLnL | 2 |
| LnLnP | 1.9 |

## Soybean Oil (high saturate; Hsat)
### GMO

| Triglyceride Composition | |
|---|---|
| PSO | 3.7 |
| SOS | 3.2 |
| PPO | 3.1 |
| PPL | 9.6 |
| PSL | 13.8 |
| SSL | 8.1 |
| POL | 5.1 |
| SOL | 4.5 |

| | |
|---|---|
| PLL | 9.6 |
| SLL | 6.8 |
| OLL | 3.3 |
| LLL | 3.9 |
| PBL | 3.3 |
| LnOP | 4.5 |
| PPLn | 3.2 |
| LnLL | 3.5 |
| LnLP | 6.7 |
| LnLnP | 3 |

## Soybean Oil (high stearic; HS)
### GMO

| Triglyceride Composition | |
|---|---|
| PSO | 2.5 |
| SOS | 4.1 |
| POO | 1.5 |
| SOO | 3.1 |
| PPL | 1.7 |
| PSL | 7.4 |
| SSL | 10.1 |
| OOO | 1.4 |
| POL | 3.7 |
| SOL | 9.8 |
| OOL | 2.5 |
| PLL | 5 |
| SLL | 9.7 |
| OLL | 6 |
| LLL | 5.9 |
| PBL | 2 |
| LnLO | 2.3 |
| LnOP | 4.4 |
| LnOS | 1.9 |
| LnLL | 3.6 |
| LnLP | 2.3 |
| LnLnL | 1.5 |

## Soybean Oil (HP/LLn)
### GMO

| Triglyceride Composition | |
|---|---|
| PSO | 2.9 |
| PPO | 3.2 |
| POO | 4 |
| SOO | 2.8 |
| PPL | 6.8 |

| | |
|---|---|
| PSL | 3.7 |
| OOO | 3 |
| POL | 9.8 |
| SOL | 3.6 |
| OOL | 4.9 |
| PLL | 14.2 |
| SLL | 5 |
| OLL | 9.7 |
| LLL | 9.4 |
| LnLO | 2.9 |
| LnLL | 3.2 |
| LnLP | 3.2 |

## Soybean Oil (low linolenic; LLn)
*GMO*

Triglyceride Composition

| | |
|---|---|
| PSO | 2.9 |
| POO | 4 |
| SOO | 3.4 |
| PPL | 3 |
| PSL | 3.1 |
| OOO | 4.6 |
| POL | 7.3 |
| SOL | 4.5 |
| OOL | 8.5 |
| PLL | 9.9 |
| SLL | 4.5 |
| OLL | 16.1 |
| LLL | 17.7 |
| LnLO | 3 |
| LnLL | 3.8 |

## Soybean Oil (low saturate; Lsat)
*GMO*

Triglyceride Composition

| | |
|---|---|
| SOO | 3 |
| OOO | 4 |
| POL | 3.5 |
| SOL | 3.4 |
| OOL | 8.4 |
| PLL | 4.5 |
| SLL | 3.7 |

| | |
|---|---|
| OLL | 17.8 |
| LLL | 23.2 |
| LnLO | 6.3 |
| LnOO | 3 |
| LnLL | 11.3 |
| LnLP | 3 |
| LnLnL | 3.7 |

## Soybean Oil (Lsat/LLn)
*GMO*

Triglyceride Composition

| | |
|---|---|
| POO | 3.2 |
| SOO | 3.5 |
| OOO | 5.6 |
| POL | 4.2 |
| SOL | 4.6 |
| OOL | 11.7 |
| PLL | 5 |
| SLL | 4.3 |
| OLL | 21.6 |
| LLL | 24.5 |
| LnLO | 3.2 |
| LnLL | 4.1 |

## Sunflower Seed Oil
*Helianthus annuus*

Triglyceride Composition

| | |
|---|---|
| PSO | 0.6 |
| PPO | 0.5 |
| POO | 1.6 |
| SOO | 1.2 |
| PPL | 0.8 |
| PSL | 1.2 |
| SSL | 0.7 |
| OOO | 2.5 |
| POL | 6.4 |
| SOL | 4.2 |
| OOL | 11.8 |
| PLL | 8.9 |
| SLL | 6.4 |
| OLL | 25.9 |
| LLL | 20.6 |
| AOO | 0.5 |
| PBL | 0.5 |

## Sunflower Seed Oil (high linoleic; HL)
*GMO*

Triglyceride Composition

| | |
|---|---|
| PSO | 0.6 |
| PPO | 0.7 |
| POO | 0.7 |
| PPL | 1.1 |
| PSL | 0.8 |
| SSL | 0.7 |
| OOO | 0.7 |
| POL | 3.2 |
| SOL | 1 |
| OOL | 2.5 |
| PLL | 12.9 |
| SLL | 2.9 |
| OLL | 19.1 |
| LLL | 37.5 |
| AOO | 0.7 |

## Sunflower Seed Oil (high oleic; HO)

Triglyceride Composition

| | |
|---|---|
| PSO | 1.1 |
| PPO | 0.2 |
| POO | see SOL |
| SOO | 4.3–11 |
| OOO | 73.9 |
| POL | 1.6 |
| SOL | 10 |
| OOL | 6.7 |
| PLL | 1 |
| OLL | 2.3 |
| LLL | 1.5 |
| AOO | 1.2 |

## Sunflower Seed Oil (high palmitic/high linoleic; HP/HL)
*GMO*

Triglyceride Composition

| | |
|---|---|
| PSO | 1.4 |
| PPO | 3.5 |
| POO | 3 |
| PPL | 11.9 |
| PSL | 2.7 |
| SSL | 1.2 |
| OOO | 1.3 |
| POL | 6.6 |
| SOL | 1.9 |
| OOL | 1.8 |
| PLL | 17.4 |
| SLL | 5.1 |
| OLL | 4.9 |
| LLL | 7.6 |
| PBL | 1.4 |
| PoLL | 2.6 |
| PoPL | 3.5 |
| PoPO | 1.4 |

## Sunflower Seed Oil (high palmitic/high oleic; HP/HO)
*GMO*

Triglyceride Composition

| | |
|---|---|
| PSO | 3.7 |
| SOS | 2.3 |
| PPO | 13.8 |
| POO | 31.2 |
| SOO | 5 |
| OOO | 12.9 |
| POL | 6.6 |
| OOL | 4.4 |
| SLL | 3.5 |
| AOO | 2.6 |
| BOO | 2.9 |
| PoOO | 2.7 |

## Sunflower Seed Oil (high stearic/high oleic; HS/HO)
*GMO*

Triglyceride Composition

| | |
|---|---|
| PSO | 3.7 |
| SOS | 3.6 |
| POO | 10.5 |
| SOO | 25.1 |
| OOO | 41.9 |

| | |
|---|---|
| OOL | 3.2 |
| AOO | 3.5 |
| BOO | 3.1 |

## Teaseed Oil
*Thea sinensis*

Triglyceride Composition

| | |
|---|---|
| PPP | 1 |
| PPO | 9 |
| POO | 25 |
| PPL | 3 |
| OOO | 21 |
| POL | 15 |
| OOL | 19 |
| PLL | 2 |
| OLL | 6 |
| LLL | 0.5 |

## Vernonia Seed Oil
*Vernonia galamensis*

Triglyceride Composition

| | |
|---|---|
| OOL + LLS | 0.5 |
| VVV | 43 |
| VVL | 21 |
| VVO | 8 |
| VVP | 8 |
| LLV | 4 |
| VVS | 6 |
| OLV | 2 |
| PLV | 2 |
| SLV | 1 |
| POV | 1 |
| SOV | 1 |
| LAV | 0.5 |

## Walnut Oil
*Juglans regia*

Triglyceride Composition

| | |
|---|---|
| POL | 2 |
| OOL | 5 |
| PLL | 4 |
| SLL | 2 |

| | |
|---|---|
| OLL | 10 |
| OLLn | 0.1 |
| LLL | 53 |
| LnOP | 0.1 |

## Winged Bean Oil
*phocarpus tetragonolobus*

PSO Triglyceride Composition

| | |
|---|---|
| OOE + LLB | 9 |
| OOS + LLB | 5 |
| PSL | 3 |
| SSLg | 1 |
| OOO | 11 |
| POB | 3 |
| POL | 6 |
| OOL | 6 |
| OOLn + LLL | 1 |
| PLL | 2 |
| OLL | 3 |
| OLLg | 5 |
| LLL + OLLn | 1 |
| BOL | 21 |
| BOO | 16 |
| PBL | 5 |
| PSB | 2 |

## Herring Oil

Triglyceride Composition

| | |
|---|---|
| PSO | 21.1 |
| PPS | 2.4 |
| PSS | 3.1 |
| SSS | 0.5 |
| SOS | 1.9 |
| PPO | 7.6 |
| POO | 26.4 |
| SOO incl. PPP | 5.3 |
| PSL incl. PoSO | 3.8 |
| OOO | 5.7 |
| POL | 6.5 |
| SOL | 0.8 |
| OOL | 1.7 |
| PLL | 0.5 |
| SLL | 1.6 |
| OLL | 0.5 |

| | |
|---|---|
| LnOP | 0.9 |
| LnOO | 0.7 |
| MOP | 1.2 |
| MOO incl. PPoO | 3.2 |
| OTHER | 4.6 |

## Norway Pout Oil
Triglyceride Composition

| | |
|---|---|
| PSO | 14.5 |
| PPS | 3.3 |
| PSS | 3.1 |
| SSS | 1.6 |
| SOS | 4 |
| PPO | 10.6 |
| POO | 23 |
| SOO incl. PPP | 11.5 |
| PSL incl. PoSO | 4.8 |
| OOO | 4.4 |
| POL | 2.7 |
| SOL | 1.4 |
| OOL | 0.9 |
| LnOP | 0.4 |
| MOP | 3 |
| MOO incl. PPoO | 3.8 |
| PoOO | 1.1 |
| OTHER | 6 |

Acacia arabica, 3
*Acacia auriculiformis*, 3
Acacia Auriculiformis Seed Oil, 3
*Acacia coriacea*, 3
Acacia Coriacea Seed Oil, 3
*Acacia lenticularis*, 3
Acacia Lenticularis Seed Oil, 3
*Acacia mellifera*, 4
Acacia Mellifera Seed Oil, 4
*Acacia minhassai*, 4
Acacia Minhassai Seed Oil, 4
*Acacia mollissima*, 4
Acacia Mollissima Seed Oil, 4
*Acacia tortilis*, 5
Acacia Tortilis Seed Oil, 5
*Acer spp.*, 6
Achras sapota, 5
Acioa edulis, 5
Acorn Oil, 6
*Actinodaphne hookeri*, 125
*Adansonia spp.*, 20
*Adonsonia digitata*, 6
Adonsonia Digitata Seed Oil, 6
*Aesculus hippocastanum*, 77
Aesculus sinensis, 6
African Mango, 7
*Afzelia bella*, 7
Afzelia Bella Seed Oil, 7
*Aiphanes acanthophylla*, 111
*Albizia lebbeck*, 7
Albizia Lebbeck Seed Oil, 7
*Albizia zygia*, 8
Albizia Zygia Seed Oil, 8
*Aleurites fordii (aleurites montana; Vernicia montana)*, 164
*Aleurites moluccana*, 33
*Aleurites montana*, 8
Aleurites Montana Seed Oil, 8
Alewife Oil, 173
Alfalfa Oil (Utah), 9
*Allium cepa*, 108
*Allium sativum*, 69
Almond Kernel Oil, 10, 213
*Alosa pseudoharengus*, 173
Alpine Current Seed Oil, 10, 213
*Althea rosea*, 77, 218
Alyogine Hakeifolia, 9
Alyogine Huegelii, 9
Amaranth Seed Oil, 11

Amaranth Seed Oil (Various), 10, 213
*Amaranthus caudatus*, 11
*Amaranthus hypochondriacus, cruentus,edulis*, 10, 213
Amaranthus mangostanus, 12
*Ambloplites rupestris*, 173
Ambrette Seed Oil: raw seed oil, 12
*Amoora rohituka*, 12
Amoora Rohituka Seed Oil, 12
*Anarcadium occidentale*, 35, 215
Anchovy Oil, 173
Andenopus Breviflorus Seed Oil, 13
*Andeopus breviflorus*, 13
*Anecastrum romanozoffianum*, 121
*Anethum graveolens*, 13
Anise Seed Oil, 14
*Apium graveolens*, 39
Apple Seed Oil, 14
Apricot Kernel Oil, 14, 213
Arabidopsis Thaliana Seed Oil, 15
*Arachis hypogaea*, 118, 220
*Arctocephalus gazella*, 197
*Argania spinosa*, 15, 213
Argan Seed Oil, 15, 213
*Argemone mexicana L*, 16
Argemone Oil, 16
*Argentina silus*, 201
*Astrocarpum spp*, 163
*Astrocarpum vulgare*, 164
*Attalea speciosa Martius syn Orbignya spp.*, 17
*Avena sativa*, 106
Avocado (pulp) Oil, 16
*Azima Tetracantha*, 17
*Azima tetracantha*, 17

Babassu Palm Oil (Brazil), 17
Bacury Seed Fat, 18
Baguacu Pulp Oil, 18
Baguacu Seed Oil, 19
Bahera Seed Oil , 19
*Baillonella toxisperma*, 20, 214
Baillonella Toxisperma Kernel Oil, 20, 214
*Balsenidae spp.*, 207
Baobab Seed Oil, 20
Barley Oil, 20
*Basella rubra*, 21
Basella Rubra Seed Oil, 21
Basil Seed Oil, 21

Bass, Rock Oil, 173
Bass, Sea, 174
Beechnut Kernel Oil, 22
Beef Flank Fat, 174
Bengal Gram (Chickpea) Oil, 22
*Beradius bairdii*, 208
*Bertholletia excelsia, Nobilis, B. myrtaceae*, 29
Bitter Almond Kernel Oil, 23
Bittersweet Oil, 23
Bitter Vetch Seed Oil, 23
*Black - Brassica juncea, white/yellow - Sinapsis alba*, 101
Blackcurrant Oil, 24, 214
Black Gram Oil, 24
*Bliphia sapida*, 25
Bliphia Sapida Seed Oil, 25
*Bombax constantum*, 25
Bombax Constantum Seed Oil, 25
*Bombax munguba*, 25
Bombax Munguba Seed Oil, 25
Borage Oil, 26, 214
Borage Oil (Dwarf), 26
*Borago officinalis*, 26
*Borago pygmea*, 26
Borneo Tallow, 27
*Borrago officinalis*, 214
*Brachyandra calophylla*, 27
Brachyandra Calophylla Seed Oil, 27
Brachystegia Nigerica, 27
*Brassica campestris*, 135
*Brassicaceae columbia*, 15
*Brassica chinensis*, 28
Brassica Chinensis Seed Oil, 28
*Brassica napus*, 133
*Brassica oleracea*, 28
Brassica Oleracea Seed Oil, 28
Brazil Nut Oil, 29
*Brunfelsia americana*, 29
Brunfelsia Americana Seed Oil, 29
Buchanania Lanzan, 214
*Buchanania lanzan, B. latifolia*, 30, 214
Buchanania Lanzan Seed Oil, 30
*Buchanania latifolia*, 43
Buffalo Gourd Seed Oil, 30
*Butea frondosa/monosperma*, 110
Butterfat, 174
Butternut Oil, 30
*Buttia capitata*, 111

*Butyrospermum parkii/Vitellaria paradoxa*, 145

*Calendula officianales*, 30
*Calendula officinalis*, 96
Calendula Seed Oil, 30
California Laurel Seed Oil, 31
*Calocarpum mammosum*, 95
*Calophyllum inophyllum*, 62
*Camelina sativa*, 31
Cameline Oil (False Flax), 31
*Camellia japonica*, 161
*Camellia oleifera*, 31, 214
Camellia Oleifera Seed Oil, 31, 214
*Camellia sinensis*, 32, 214
Camellia Sinensis Seed Oil, 32, 214
*Camilla sasanqua*, 160
Camphor Kernel Fat (Camphor Tree), 32
*Canarium ovatum*, 121, 221
Canarium tramdenum, 33
*Canavalia ensiformis*, 81
Candlenut (Lumbang) Oil, 33
*Cannabis sativa*, 75
Cantaloupe Seed Oil, 33
Capelin Oil, 175
Cape Marigold Seed Oil, 34, 215
*Capsicum annuum*, 136
Caraway Seed Oil, 34
*Carica papaya*, 115
Carob Bean Oil, 34
Carp Lipids, 175
*Carthamus oxycanthus*, 127
*Carthamus tinctorius*, 138
*Carum carvi*, 34
*Carya illinoensis*, 119, 220
*Caryaovata ovata*, 77
Casca-de tatu Seed Oil, 34
Cashew Nut Oil, 35, 215
*Cassia alata*, 35
Cassia Alata (Ringworm Shrub), 35
*Cassia occidentalis*, 36
Cassia Occidentalis (Wild Coffee), 36
*Cassia siamea*, 37
Cassia Siamea Seed Oil, 37
*Cassia siberiana*, 37
Cassia Siberiana Seed Oil, 37
*Castanea mollisima*, 40
Castor Oil, 38, 215
Catfish Lipids, 176

*Catostromus commersonni*, 203
Cay-Cay Fat, 38
*Ceiba pentandra (Bombax spp)*, 84
Celastrus Orbiculatus, 215
*Celastrus orbiculatus*, 38, 215
Celastrus Orbiculatus Seed Oil, 38
*Celastrus parniculatus*, 63
*Celastrus scandens*, 23
Celery Seed Oil, 39
*Centorhinus maximus*, 209
*Centrophorus squamosus*, 209
*Centroscyllium ritteri*, 181
*Ceratonia siliqua*, 34
Cherry Kernel Oil, 39, 40, 215
Chestnut Oil, 40
Chia Oil, 41
Chicken Egg Lipids, whole egg, 176
Chicken Egg Lipids, yolk, 176
Chicken Fat, 177
Chickling Vetch Seed Oil, 41
Chinese Melon Seed Oil (Bitter Gourd), 41
Chinese Soapberry Seed Oil, 42
Chinese Vegetable Tallow (Mesocap fat; Chinese Tallow Tree), 42
Chinook Salmon Lipids, 177
Chirongi Oil, 43
Chrysanthemum coronarium, 43
*Chrysanthemum corymbosum*, 158
*Cicer arietinum*, 22
Cimicifuga racemosa, 43
*Cinnamomum camphora*, 32
Citrullus Colocynthis, 44
*Citrullus lanatus/vulgaris*, 166
*Citrulus colocynthis*, 161
*Citrus aurantifolia*, 89, 219
*Citrus grandis (paradisi)*, 72
*Citrus sinensis*, 109
*Citrus spp*, 88
Cloudberry Seed Oil, 44, 215
Cocoa Butter, 44, 216
Coconut Oil, 45, 216
*Cocos nucifera*, 45, 216
Cod, Atlantic, 178
Cod, Pacific, 178
Cod Liver Oil, 178
*Coffea arabica*, 46
Coffee Bean Oil (Raw, Brazil), 46
Coffee Bean Oil (Roasted), 46
Cohune Nut Oil (Palm Oil), 47

Coincya Longirostra, 47
*Coincya longirostra*, 47
Coincya Monensis, 48
Coincya Rupestris, 48
Coincya Transtagana, 47
*Colocynthis citrullus*, 63
Comphrey Seed Oil, 48
Connarus paniculatus, 49
Corchorus olitorius, 49
Cordia Rothii Seed Oil, 50
*Coregonus clupeaformis*, 208
Coriander Seed Oil, 50
*Coriandrum sativum*, 50
Corn Oil (High Oleic), 50
Corn Oil (Low Saturate), 51
Corn Oil (Maize), 51
*Corylus avellana*, 74, 217
*Corynocapus laevigatus*, 84
Cottonseed Oil, 52, 216
Couepia longipendula, 52
Cowpea Oil, 53
Crab Lipids, King, 179
Crab Lipids, Queen, 179
*Crambe abyssinica*, 53
Crambe Oil, 53
*Crepis alpina*, 53, 216
Crepis Alpina Seed Oil, 53, 216
*Crotalaria juncea*, 54
Crotalaria Juncea Seed Oil, 54
Croton Seed Oil, 54
*Croton tiglium*, 54
Cryptolepis Buchnani, 54
*Cryptostegia grandiflora*, 55
Cryptostegia Grandiflora Seed Oil, 55
*Cucumeropsis edulis*, 55
Cucumeropsis Edulis Seed Oil, 55
*Cucumeropsis manni*, 55
Cucumeropsis Manni Seed Oil, 55
*Cucumis melo*, 33
*Cucumis sativus*, 56
Cucumis Sativus Seed Oil, 56
*Cucurbita foetidissima*, 30
*Cucurbita palmata*, 98
*Cucurbita pepo*, 56, 130
Cucurbita Pepo Seed Oil, 56
*Cupania anacardioides*, 147
Cupania Anacardioides Seed Oil, 56
*Cuphea fruiticosa*, 58
*Cuphea lythracea*, 62

Cuphea Seed Oil (Capric acid rich), 57
Cuphea Seed Oil (Caprylic acid rich), 57
Cuphea Seed Oil (Lauric acid rich), 57, 58
Cuphea Seed Oil (Linoleic acid rich), 58
*Cuphea viscosissima*, 58
Cuphea Viscosissima Seed Oil, 58
*Cuphea wrightii*, 57
Cupu Assu Kernel Oil, 59
*Cydonia Mill*, 132
*Cyprinus carpio*, 175

*Dacryodes edulis*, 140, 222
*Dacryodes rostrata*, 82
*Daniella ogea*, 59
Daniellia Ogea Seed Oil, 59
*Deania calceus*, 180
*Delavaya toxocarpa*, 60
Delavaya Toxocarpa Seed Oil, 60
Delphinium ajacis, 60
Dhupa Fat (Malabar Tallow), 61
*Dimocarpus longan*, 61
Dimocarpus Longan Seed Oil, 61
*Dimorphotheca pluvialis*, 34, 101
*Dimorphotheca pluvialis*, 215
*Diospyros mespiliformis*, 62
Diospyros Mespiliformis Seed Oil, 62
Diploptychea Painteri Seed Oil, 62
*Diploptychia painteri*, 57
*Dipteryx odorata (Erythrina spp)*, 163
Dogfish, Birdbeak Muscle Oil, 180
Dogfish, Sping Liver Oil, 181
Dogfish, Spur Liver Oil, 181
*Dolichos biflorus*, 78
Dolphin Dorsal Blubber , 180
Domba Fat, 62
Dover Sole Lipids, 181
Dukudu Seed Oil, 63

Eel Lipids, 182
Egusi Seed Oil, 63
*Elaeis guineensis*, 110, 114
*Elaeis guineensis Dura*, 113
*Elaeis oleifera*, 112
*Elaeis oleifera Elaeis melanococca*, 112
*Eleusine coracana*, 67
Elm Seed Oil, 63
Emu Oil, 182
*Entandrophragma angolense*, 64
Entandrophragma angolense Seed Oil, 64

*Enterolobium cyclocarpium*, 64
Enterolobium Cyclocarpium Seed Oil, 64
Ephedra Gerardiana, 64
*Erisma calcaratum*, 81
*Eruca sativa*, 158
Erythrophleum fordii, 65
*Esox lucius*, 191
*Euphorbia lagascae*, 65
Euphorbia Lagascae Seed Oil, 65
*Euphoria longana*, 91
Evening Primrose Oil, 66, 216

*Fagus orientalis Lipsky*, 22
Fennel Seed Oil, 66
Fenugreek Seed Oil, 67
*Fevillea cordifloria*, 143
*Ficus carica*, 67
Fig Seed Oil, 67
Finger Millet, 67
*Foeniculum officinale*, 66
Fokienia hodginsii, 68
Foxtail Millet, 68
Fucus Serratus , 68
Fungal Oil, 69, 217

Gamboge Butter (Kernel Fat), 69
*Garcinia indica*, 85, 218
*Garcinia morella*, 69
Garlic Oil, 69
*Gevuina avellana*, 74
Ghee (Buffalo Milk) Butter, 183
*Glycine max*, 149, 222
*Glyricidia sepium*, 70
Glyricidia Sepium Seed Oil, 70
*GMO*, 149, 150, 151, 152, 154, 155, 156, 223, 224, 225
Gnetum sp., 70
*Gomphera globosa*, 70
Gomphera Globosa Seed Oil, 70
Gooseberry Seed Oil, 71
Goose Fat, 183
*Gossypium spp.*, 52, 216
Grapefruit Seed Oil, 72
Grape Seed Oil, 71
Green Gram, 72
Guava Seed Oil, 73
Guinea Fowl Egg Fat , 183
*Guizotia abyssinica*, 104
Haddock, 184

Halibut, Greenland Oil, 184
*Hannoa undulata (simarubacea)*, 73, 217
Hannoa Undulata Seed Oil, 73, 217
Hazelnut Oil (Chilean), 74
Hazelnut Oil (Filbert), 74, 217
*Heisteria silvanii*, 34
*Helianthus annuus*, 154, 224
Hempseed Oil, 75
Herring Oil, 184, 226
*Hesperis matronalis*, 156
*Heteranthus epilobiifolia*, 58, 76
Heteranthus Epilobiifolia Seed Oil, 76
*Heterodon koehneana*, 57
*Hevea brasiliensis*, 137
*Hibiscus abelmoschus*, 12
Hibiscus Cannabinus, 76, 217
*Hibiscus cannabinus*, 76, 217
Hibiscus Coatesii, 76
*Hibiscus esculentus*, 107, 220
Hibiscus sabdariffa, 76
*Hibiscus sabdariffa*, 217
Hickory Nut Oil, 77
Hollyhock, 77, 218
*Hoplostethus atlanticus*, 189
*Hordeum vulgare*, 20
Horse Chestnut Oil, 77
Horse Fat, 185
Horsegram, 78

Illipe (mowrah) Butter, 78, 218
*Inia geoffrensis*, 180
Ipomoea aquatica, 79
Ironwood/Nahar Fat (Indian Rose Chestnut), 79
*Irvingia gabonensis*, 7, 80, 218
Irvingia Gabonensis Kernel Fat (Dika Fat), 80, 218
*Irvingia oliveri*, 38
Isano (boleko) Seed Oil, 80
*Isochrysis galbana*, 97
*Isotoma longiflora*, 80
Isotoma Longiflora Seed Oil, 80

Jaboty Tallow (Fat, Butter), 81
Jack Bean Oil, 81
Japan Tallow (Wax), 81
*Jatropha curcas*, 82, 121
Jatropha Oil (See also Physic Nut Oil), 82
Java Almond Fat, 82

Java Olive Oil, 82
*Jessenia bataua M*, 116
Jojoba Oil, 83
*Juglans cinerea*, 30
*Juglans regia*, 165, 226

Kaiphal Oil, 83
Kanya Tallow (Fat), 83
Kapok Seed Oil, 84
Karaka Seed Oil, 84
Karanja Oil, Pongam Oil, 84
Katio Fat, 85
Khakan Fat, Pelu Fat, 85
Kokum Butter, 85, 218
Kombo Butter, 86
Korean Pine Seed Oil, 86, 218
Kusum Oil (Macassar Oil), 86

*Lallemantia iberica*, 87
Lallemantia Oil, 87
Lamb Shoulder Fat, 185
Lard (Rendered Pork Fat), 186
*Larix sibirica*, 87
Larix Sibirica Seed Oil, 87
*Lathyrus cicera*, 23
*Lathyrus sativus*, 41
Laurel Berry (Bay Berry) Oil, 87
*Laurus nobilis*, 87
Lawrencia Viridigrisea, 88
Lemon Seed Oil, 88
*Lesquerella fendleri*, 88
Lesquerella Fendleri Seed Oil, 88
*Lesquerella perforata*, 88
Lesquerella Perforata Seed Oil, 88
*Lesquerella recurvata*, 89
Lesquerella Recurvata Seed Oil, 89
*Licania rigida*, 107
Lime Seed Oil, 89, 219
*Limnanthes alba*, 97
*Limnanthes douglasii*, 97
*Lindera benzoin*, 152
*Lindera umbellata*, 89
Lindera Umbellata Seed Oil, 89
Linseed Oil (Flax), 90, 219
Linseed Oil (Low linolenic flax), 90
*Linum usitatissimum*, 90, 219
*Litchi chinensis*, 91
Litchi Chinensis Seed Oil, 91
Longan Seed Oil, 91

*Louchocarpus sericens*, 92
Louchocarpus Sericens Seed Oil, 92
*Luffa cylindrica*, 92
Luffa Cylindrica Seed Oil, 92
Lupin (Lupine) Seed Oil, 92, 93
Lupine Seed Oil, 93
*Lupinus albus*, 92
*Lupinus angustifolius*, 93
*Lupinus luteus*, 93
Lupu Fat, 94
*Lycopersicon lycopersicum*, 162

Maasbanker Oil, 186
Macadamia Nut Oil, 94
*Macadamia tetraphylla/ternifolia*, 94
*Macassar schleicheratrijuga*, 86
Mackerel, Atlantic Oil, 187
Mackerel Oil, 187
*Madhuca butyraceae*, 120, 221
*Madhuca indica*, 95
*Madhuca latiflora/longiflora (Bassia latiflora)*, 78, 218
*Madhuca latifolia*, 95, 219
*Madhuca latifolia and M. longifolia*, 101
*Madhuca mottleyana*, 85
Mahua Fat, 95, 219
Mahua Oil, 95
*Mallotus villosus*, 175
*Malus deomestica L.*, 14
Mammy Apple Seed Oil, 95
*Mangifera indica*, 95, 96
Mango Pulp Oil, 95
Marigold Seed Oil, 96
Marine Microalga Fatty Acid Extract, 97
Meadowfoam Seed Oil (Alba), 97
Meadowfoam Seed Oil (Douglas), 97
*Medicago sativa*, 9
Mediterranean Seagrass, 128
*Melia/Azadirachta indica*, 103
Menhaden Oil, 188
*Merlangius merlangus*, 208
*Mesua ferrea*, 79
*Microstomus kitt*, 202
*Milletia thonningii*, 98
Milletia Thonningii Seed Oil, 98
Mock Orange, 98
*Momordica charantia*, 41
*Momordica cochinchinensis*, 98
Momordica Cochinchinensis Seed Oil, 98

Monkey Pod Seed Oil, 99
*Moringa oleifera*, 99
Moringa Oleifera Seed Oil, 99
*Moringa peregrina*, 100, 219
Moringa Peregrina Seed Oil, 100, 219
*Morone americanus*, 190
*Mortierella alpina*, 69, 217
Mowrah Butter, 101
Mullet Oil, 188
Munch Seed Oil, 101
Mustard Seed Oil, 101
Mustard Seed Oil Oriental, 102
*Myristica fragrans*, 105
*Myristica malabarica*, 83

Nectarine Seed Oil, 102
Neem (Margosa) Oil, 103
Neetsfoot Oil, 103
Neou Seed Oil, 104
Nephelium lappaceum, 103
*Nephelium lappaceum*, 103, 133
*Nigella sativa*, 104
Nigella Seed Oil, 104
Niger Seed Oil, 104
Norway Pout Oil, 189, 227
*Numida meleagris*, 183
Nutmeg Butter, 105

Oat Bean Oil, 105
Oat Oil, 106
Ochoco Butter (kernel fat), 106, 219
*Ocimum basilicum/canum/gratissimum/ sanctum*, 21
*Oenothera biennis*, 66, 216
Oil Bean Oil, 106
Oiticica Oil, 107
Okra Seed Oil, 107, 220
*Olea europaea*, 108, 220
Olive (Wild) Oil, Kandarakkara Oil, 107
Olive Oil (for quality grade reference values see IOOC documentation), 108, 220
*Oneguekoa gore E.*, 80
Onion Seed Oil, 108
Onosmodium Hispidissimum, 109
Orange Roughy, 189
Orange Seed Oil, 109
*Oryza sativa*, 136
*Osmerus mordax*, 200
Otoba Butter, American Nutmeg Butter, 110

Ouricuri Tallow, 110
Oyster Lipids (American), 190
Oyster Lipids (European), 190

*Pachyrhizus spp.*, 168
Palas Oil, 110
Palm Kernel Oil, 110, 111, 112
Palm Oil, 112, 113, 114
Palm Olein, 114
Palm Stearin, 115
*Papaver somniferum*, 127
Papaya Seed Oil, 115
*Parinarium mocrophyllum*, 104
*Parkia biglandulosa*, 116
Parkia Biglandulosa Seed Fat, 116
Parsley Seed Oil, 116
Pataua Palm Oil (Pulp), 116
*Paullinia elegans*, 147
Paullinia Elegans Seed Oil, 116
Peach Kernel Oil, 117, 220
Peanut/Groundnut Oil, 118, 220
Peanut Oil (high oleic), 118
Pear Seed Oil, 119
Pecan Nut Oil, 119, 220
*Pennaeus aztecus aztecus*, 200
*Pennaeus spp.*, 199
*Pennaeus vannanei*, 200
*Pentaclethera macrophylla*, 105
*Pentaclethra macrophylla*, 106
*Pentadesma butyracea*, 83
Perch, White Oil, 190
*Perilla frutescens*, 120
Perilla Oil, 120
*Persea americana*, 16
*Petroselinum sativum*, 116
*phocarpus tetragonolobus*, 226
Phulwara Butter, 120, 221
Physic Nut Oil, Ratanjyor Oil, 121
Pike, Northern Oil, 191
Pili Nut Oil, 121, 221
*Pimpinella acuminata*, 121
Pimpinella Acuminata Seed Oil, 121
*Pimpinella anisum*, 14
*Pindarea fastuosa*, 18, 19
Pindo Palm Kernel Oil, 121
Pine Nut Oil, 122, 123, 124
*Pinus Banksiana spp*, 123
*Pinus cembroides edulis*, 124
*Pinus halepensis spp*, 122

*Pinus koraiensis*, 86, 218
*Pinus monophylla*, 122
*Pinus pinaster*, 124
*Pinus pinea*, 122
*Pinus ponderosa spp*, 123
*Piper nigrum*, 124
Piper Nigrum Seed Oil, 124
Pisa Oil, 125
Pistachio Nut Oil, 125
*Pistacia atlantica*, 126
Pistacia Atlantica Fruit Oil, 126
*Pistacia vera*, 125
*Platonia insignis*, 18
Plum Kernel Oil, 126
*Poga oleosa*, 127, 221
Poga Oleosa Kernel Oil, 127, 221
Poli Oil, Wild Safflower Seed Oil, 127
Pompano, 191
*Pongamia glabra*, 84
*Ponicum miliaceum*, 128
Poppyseed Oil, 127
*Posidonia oceanica*, 128
Pout, Norway Oil, 192
Premier Jus (Beef/sheep tallow), 192
Proso Millet, 128
*Prosopis africana*, 128
Prosopis Africana Seed Oil, 128
Prune Kernel Oil, 129, 221
*Prunus armeniaca*, 14, 213
*Prunus avium*, 39, 215
*Prunus cerasifera*, 129, 221
*Prunus cerasus*, 40
*Prunus domestica*, 126
*Prunus dulcis*, 10, 23
*Prunus persica*, 117, 220
*Prunus persica var nectarina*, 102
*Pseudotsuga menziesii*, 129
Pseudotsuga Menziesii Seed Oil, 129
*Psidium guajava*, 73
*Psophocarpus tetragonolobus*, 168
*Pterocarpus osun*, 130
Pterocarpus Osun Seed Oil, 130
*Pterocarpus santalinoides*, 130
Pterocarpus Santalinoides Seed Oil, 130
Pumpkin Seed Oil, 130
*Pycnanthus kombo*, 86
*Pyrus domestica*, 119

*Quamoclit coccinea*, 132

# Index

*Quamoclit phoenicea*, 131
Quamoclit Seed Oil, 131, 132
Quince Seed Oil, 132

Rabbit Fat, 193
Radyera farragei, 132
*Raja radiata*, 193
Rambutan Tallow, 133
Rapeseed Oil, 133
Rapeseed Oil (low erucic, Canola), 134
Rapeseed Oil (low linolenic, Canola), 134
*Raphanus sativus*, 135
Raphanus Sativus Seed Oil, 135
Ravison Oil, 135
Ray, Starry Muscle Oil, 193
Redfish Oil, 194
Red Pepper Seed Oil, 136
*Reinhardtius hippoglossoides*, 184
*Rhus succedanea*, 81
*Ribes alpinum*, 136
Ribes Alpinum Seed Oil, 136
*Ribes grossularia*, 71
*Ribes nigrum*, 10, 24, 213, 214
Rice Bran Oil, 136
*Ricinodendron heudelotii*, 137, 221
Ricinodendron Heudelotii Kernel Oil, 137, 221
*Ricinus communis*, 38, 215
Rubber Seed Oil, 137
*Rubus chamaemorus*, 44, 215
Rye Germ Oil, 138

Sablefish Lipids, 194
Safflower Oil (high linoleic), 138
Safflower Seed Oil, 138
Safflower Seed Oil (high oleic), 139
Safou Oil, 140, 222
Sal Fat, 140
*Salicornia bigelovii; glasswort; maroh samphire*, 140
Salicornia Seed Oil, 140
Salmon, Atlantic Oil (muscle, Iceland), 195
Salmon, Atlantic Oil (whole body caught in wild, Canada), 195
Salmon, Oil, 194
*Salmo salar*, 195
*Salvadora oleoides and S. persica*, 85
*Salvelinus namaycush namaycush*, 205
*Salvelinus namaycush siscowet*, 206
*Salvia hispanica*, 41

*Samanea saman*, 99, 141
Samanea Saman Seed Oil (Monkey pod), 141
Sandeel Oil, 196
*Sapindaceae (Soapberry) Family*, 56, 116
*Sapindus drummondii*, 167
*Sapindus mukorossi*, 42, 141, 146
Sapindus mukorossi, 141
*Sapindus trifoliatus*, 146
*Sapium sebiferum*, 42, 153
Sardine, Pilchard Oil, 196
*Schizochytrium aggregatum*, 142
Schizochytrium aggregatum (ATCC 28209) Fungal Lipids, 142
*Schizonepeta tenuifolia*, 142
*Sciadopytis verticillata*, 143
Sciadopytis Verticillata Seed Oil, 143
*Scomber scombrus*, 187
*Scyphocephalium ochocoa*, 106, 219
Seal, Antarctic Fur Seal, 197
Seal Blubber Oil, Harp, 196
Seal Oil, Harp, 197
Seal Skin Oil, 198
*Sebastes marinus*, 194
*Secale cereale*, 138
Sequa Oil, 143
Sesame Seed Oil, 143, 144, 222
*Sesamum indicum*, 143
*Sesamum inidicum*, 222
*Sesamum radiatum*, 144, 222
*Sesbania pachycarpa*, 145
Sesbania Pachycarpa Seed Oil, 145
*Setaria italica*, 68
Shark Liver Oil, 198
Shark Liver Oil (Basking), 209
Shark Liver Oil (Deep Sea), 209
Sheanut Butter, 145
Sheep Fat (subcutaneous), 198
*Shorea robusta*, 140
*Shorea stenoptera*, 27
Shrimp, 199
Shrimp, Alaska, 199
Shrimp Ecuador White, 200
Shrimp Louisiana Brown , 200
*Sida humilis*, 145
Sida Humilis Seed Oil, 145
*Simarouba glauca*, 146
Simarouba Oil (Paradise Tree), 146
*Simmondsia chinensis*, 83

Smelt, American Oil (Fillets, Cayuga Lake, NY), 200
Smelt, Greater Silver Oil, 201
Snapper, Red Oil (fillet), 201
Soapberry (Chinese) Seed Oil, 146
Soapberry Seed Oil, 147
Soap Tree Seed Oil, 146
*Solanum melongena*, 147
Solanum Melongena Seed Oil, 147
Sole, Lemmon Oil, 202
*Sorghum bicolor*, 148
Sorghum Seed Oil, 148
*Sorghum vulgare*, 148
Soybean Oil, 149, 222
Soybean Oil (high palmitic; HP), 149, 223
Soybean Oil (high saturate; Hsat), 150, 223
Soybean Oil (high stearic; HS), 150, 223
Soybean Oil (HP/LLn), 150, 223
Soybean Oil (low linolenic; LLn), 151, 224
Soybean Oil (low saturate; Lsat), 151, 224
Soybean Oil (Lsat/LLn), 152, 224
Soybean Oil (Tropical area), 152
*Specific Gravity (SG)*, 47, 50
Spicebush Kernel Fat, 152
Sprat Oil, 202
*Squalus acanthias*, 181
Squid, 203
*Sterculia foetida*, 82
Sterculia Tomentosa, 153
Sterculia tragacantha, 153
Stillingia Seed Kernel Oil (Chinese Tallow Tree), 153
Sucker, White, 203
Sunflower Seed Oil, 154, 224
Sunflower Seed Oil (high linoleic; HL), 154, 225
Sunflower Seed Oil (high oleic; HO), 155, 225
Sunflower Seed Oil (high palmitic/high linoleic; HP/HL), 155, 225
Sunflower Seed Oil (high palmitic/high oleic; HP/HO), 156, 225
Sunflower Seed Oil (high stearic/high oleic; HS/HO), 156, 225
Sweet Rocket Oil (Dame's Violet), 156
Swordfish, 203
*Syagrus coronata/Orbignya cohune*, 110
*Symphytum officinale*, 48
Tabebuia argentia, 157

Tall Oil (Crude from pine wood pulping), 157
Tallow (beef), 204
Tallow (mutton), 204
Tamarind Kernel Oil, 157
*Tamarindus indica*, 157
Tanacetum Seed Oil, 158
Taramira Seed Oil (Rocket Salad), 158
*Taxus baccata*, 158
Taxus Baccata Seed Oil, 158
*Taxus chinensis*, 159
Taxus Chinensis Seed Oil, 159
*Taxus cuspidata*, 159
Taxus Cuspidata Seed Oil, 159
*Taxus grandis*, 160
Taxus Grandis Seed Oil, 160
Teaseed Oil, 160, 226
Teaseed Oil (Tsubaki Oil), 161
Teaseed Oil (Turkish), 161
*Terminalia bellirica*, 19
*Thea sinensis*, 160, 226
*Theobroma bicolor+B260*, 94
*Theobroma cocoa*, 44
*Theobroma grandiflora*, 59
*Theolbroma cocoa*, 216
Thumba Oil, 161
*Thunnus alalunga*, 206
Tobacco Seed Oil, 162
Tomato Seed Oil, 162
Tonka Bean Oil, 163
*Trachurus trachurus*, 186
*Trichosanthes kirilowii*, 163
Trichosanthes Kirilowii Seed Oil, 163
*Trigonella foenum-graecum*, 67
*Trisopterus esmarki*, 192
*Triticum aestinum/durum*, 167
Trout, Lake, 205
Trout, Ocean, 205
Trout, Siscowet, 206
Trout Lipids, 205
Tucum (Aoiara) Kernel Oil, 163
Tucum Pulp Oil, 164
Tuna (white meat), 206
Tung Oil, 164
Turtle (Green) Oil, 207

Ucuhuba Butter Oil, 164
*Ulmus americana*, 63
*Umbellularia californica*, 31

# Index

*Valeria indica*, 61
*Vernonia anthelmintica*, 165
*Vernonia galamensis*, 165, 226
Vernonia Seed Oil, 165, 226
*Vigna mungo*, 24
*Vigna radiata*, 72
*Vigna unguiculata*, 53
*Virola otoba*, 110
*Virola surinamensis*, 164
*Vitis vinifera*, 71

Walnut Oil, 165, 226
Watermelon Seed Oil, 166
*Welwitschia mirabilis*, 166
Welwitschia Mirabilis Seed Oil, 166

Western Soapberry Seed Fat (Wild Chinaberry), 167
Whale Oil, 207
Whale Oil, Minke, 207
Whale Oil, Pacific Beaked, 208
Wheat Germ Oil, 167
Whitefish Oil, 208
Whiting, 208
Winged Bean Oil, 168, 226

*Ximenia americana*, 107

Yam Bean, 168

*Zea mays*, 51

INTERNATIONAL PRODUCT LICENSE AGREEMENT AND WARRANTY
**INDIVIDUAL USER**

This product is protected by United States Copyright Law. Please read the following agreement carefully. It is assumed that by opening the disk package you have read, understood, and agreed to the product license agreement stated below.

1. Product License. AOCS grants the original purchaser the nonexclusive right to use this CD-ROM software and the information contained in it for personal use. The license for the single user grants the right to use this product on one computer, but if you want to use this product on other computers, you must purchase an appropriate number of products.

This license does not represent a sale and this software may not be rented or leased.

Use of this product at more than one location by more than one user without an appropriate license is specifically prohibited as is downloading or transmitting this software electronically from one computer to another.

The copyrighted contents of this CD-ROM may not be reproduced without the accompanying copyright notices. Republication or resale of these contents is specifically prohibited. You may not disassemble or in any manner decode the software.

2. Limited warranty. AOCS warranties to the original purchaser that this software disk is free of defects in material and of faulty workmanship under normal use for 60 days from the date of purchase. If this physical media is believed to be defective during that period, please notify AOCS as the address and/or number below. If AOCS determines that there has been a defect in the physical CD or faulty workmanship, the CD will be replaced without charge. Return of the CD without authorization by AOCS may delay or nullify agreement. Otherwise AOCS disclaims all other warranties, expressed or implied, including but not limited to any implied warranty of merchantability or fitness or a particular purpose. AOCS does not warrant that the functions contained in the software will meet your requirements or that the operation of the software will be error free.

AOCS and its licensors disclaim any warranty, guarantee or representation as to the correctness, accuracy, reliability, timeliness of the contents of the CD or to its use. You as the user assume the entire risk and no information or help provided by AOCS, its employees, its licensors, agents, distributors or others associated with the development, production, manufacturing and distribution shall be considered to create any warranty or liability on behalf of AOCS.

The only liability and remedy available under this limited warranty are limited solely to replacement of this CD-ROM if defective and shall not include or extend to any claim for or right to recover any other damages, such as loss of profit, or any direct or indirect consequential or incidental damages. If any competent jurisdiction determines that the above warranty disclaimer is invalid in any way, then the purchaser and AOCS agree that the maximum amount of damage of any kind recoverable shall not exceed the purchase price of the product, as determined by a bill of sale or other proof of purchase that must be provided by the purchaser. The remedies available to you against AOCS under this agreement are exclusive.

If you have any questions about this agreement or about this product, you may contact AOCS Technical Services, P.O. Box 3489, Champaign, IL 61826-3489, USA, or e-mail us at: technical@aocs.org.

To view the CD-ROM contents, you will be prompted for a password. This password is: AOCSsecI2006. Sharing this password constitutes a violation of this product license agreement.

To navigate the CD files use the Adobe Acrobat Reader Search function in basic or advanced mode. Contents and index pages contain active links to individual entries.